In Zero Trust We Trust

Avinash Naduvath, CCIE® Security
No. 59092

Cisco Press

Hoboken, New Jersey

In Zero Trust We Trust

A Practical Guide to Adopting Zero Trust Architectures

Avinash Naduvath

Published by:
Cisco Press
Hoboken, New Jersey

1 2024

Library of Congress Control Number: 2023923554

ISBN-13: 978-0-13-823740-0
ISBN-10: 0-13-823740-9

Cover credit: DadBusiness/Shutterstock

Warning and Disclaimer

Trademark Acknowledgments

All terms mentioned in this book that are known to be trademarks or service marks have been appropriately capitalized. Cisco Press or Cisco Systems, Inc., cannot attest to the accuracy of this information. Use of a term in this book should not be regarded as affecting the validity of any trademark or service mark.

Special Sales

For information about buying this title in bulk quantities, or for special sales opportunities (which may include electronic versions; custom cover designs; and content particular to your business, training goals, marketing focus; or branding interests), please contact our corporate sales department at corpsales@pearsoned.com or (800) 382-3419.

For government sales inquiries, please contact governmentsales@pearsoned.com.

For questions about sales outside the U.S., please contact intlcs@pearson.com.

Feedback Information

At Cisco Press, our goal is to create in-depth technical books of the highest quality and value. Each book is crafted with care and precision, undergoing rigorous development that involves the unique expertise of members from the professional technical community.

Readers' feedback is a natural continuation of this process. If you have any comments regarding how we could improve the quality of this book, or otherwise alter it to better suit your needs, you can contact us through email at feedback@ciscopress.com. Please make sure to include the book title and ISBN in your message.

We greatly appreciate your assistance.

Vice President, IT Professional: Mark Taub

Alliances Manager, Cisco Press: Caroline Antonio

Director, ITP Product Management: Brett Bartow

Executive Editor: James Manly

Managing Editor: Sandra Schroeder

Development Editor: Chris Cleveland

Senior Project Editor: Mandie Frank

Copy Editor: Bart Reed

Technical Editor: Cindy Green-Ortiz

Editorial Assistant: Cindy Teeters

Designer: Chuti Prasertsith

Composition: codeMantra

Indexer: Timothy Wright

Proofreader: Barbara Mack

CISCO.

Americas Headquarters
Cisco Systems, Inc.
San Jose, CA

Asia Pacific Headquarters
Cisco Systems (USA) Pte. Ltd.
Singapore

Europe Headquarters
Cisco Systems International BV Amsterdam,
The Netherlands

Cisco has more than 200 offices worldwide. Addresses, phone numbers, and fax numbers are listed on the Cisco Website at **www.cisco.com/go/offices**.

Cisco and the Cisco logo are trademarks or registered trademarks of Cisco and/or its affiliates in the U.S. and other countries. To view a list of Cisco trademarks, go to this URL: www.cisco.com/go/trademarks. Third party trademarks mentioned are the property of their respective owners. The use of the word partner does not imply a partnership relationship between Cisco and any other company. (1110R)

About the Author

Avinash Naduvath is a renowned security architect in the Customer Experience (CX) Security Services division at Cisco Systems. As part of CX-Security, he has delivered multiple solutions to help secure customer networks. The range of services included incepting secure architectures, designs, technology advisories, best practice recommendations, and security assessments.

Prior to his current role in Cisco, Avinash was part of the technical services for security in Cisco-Bangalore and has helped troubleshoot and secure networks for multiple customers. He is a subject matter expert in next-generation firepower technology. Previous to this, Avinash was part of the professional services team in Cisco-Bangalore as a network consulting engineer.

Avinash has over 10 years of experience in the information security domain, having worked on multiple aspects of security such as secure engineering and secure architecture. He has a passion for offensive security and has spoken on various topics at conferences such as Cisco SECCON and the Offensive Summit held at Cisco. Avinash has also contributed to and created multiple automation projects that have helped accelerate the security business. He is currently based in Singapore and enjoys presenting topics relevant to Zero Trust and its adoption.

He holds a master's degree in software systems from BITS Pilani, and is a Certified Information Systems Security Professional (CISSP), Cisco Certified Internetwork Expert—Security (CCIE), CompTIA Advanced Security (CASP+) practitioner, SABSA Charted Architect–Foundations and has acquired Cloud Security Alliance's Certified Competence in Zero Trust (CCZT) among many security-based certifications he has accumulated during the course of his career.

Avinash is a Certified Forrester's Zero Trust Adoption practitioner and is also the author of the award-winning fictional novel *Mindbender* (Literary Titan Silver Book Awardee and a Feathered Quill finalist).

About the Technical Reviewer

Cindy Green-Ortiz is a Cisco senior security architect, cybersecurity strategist, architect, and entrepreneur. She works in the Customer Experience, Global Enterprise segment for Cisco. She holds the CISSP, CISM, CSSLP, CRISC, PMP, and CSM certifications, along with two degrees—a BS CIS magna cum laude and an AS CIS with honors.

She has been with Cisco for 6+ years. Cindy had been in the cybersecurity field for 40 years, where she has held D-CIO, D-CISO, and corporate security architecture leadership roles, founding two technology businesses as the CEO. Cindy is a Cisco Chairman's Club winner (Club Cisco). She is an active blogger for Cisco and has published whitepapers for Cisco and the U.S. Department of Homeland Security. She has spoken to many groups, including PMI International Information Systems & Technology Symposium-Cybersecurity Keynote; Cisco SECCON, and Cisco Live. Cindy is president emeritus and serves now as treasurer of Charlotte InfraGard and is a cofounder of InfraGard CyberCamp.

Dedications

I would like to dedicate this book to my wonderful wife and my bubbly son. Both of them have patiently endured the writing process with me and have been a constant source of encouragement.

Acknowledgments

To begin, I would like to thank Nadhem Al-Fardan, Distinguished Architect, at Cisco who essentially planted the seed of writing a book in my mind. His guidance on narrating a story rather than listing down steps and recommendations was the core of how I formulated the chapters and the rhetoric of the book. I would like to thank Viktor Pozgay, head of Threat Scenario-Lead Risk Assessment at Standard Chartered Bank, who was the first technical reviewer of the book. Thank you Viktor for taking the time out of your busy schedule to review all the chapters. I would like to extend my heartfelt thanks to my manager, Xu Guang Chen, Customer Delivery Leader at Cisco, and Siew Fay Ho, Customer Delivery Director at Cisco, for their constant support and encouragement. "So, what about your book?" is a good enough motivation to continue creating content and completing the book.

I would like to thank James Manly for giving me the opportunity to pursue this project. Your trust in me has helped me feel confident enough to continue with my vision for the book. A big thank-you to Chris Cleveland, development editor, for providing the best support developing the book, and to Cindy Green-Ortiz, Principal Architect at Cisco, for her unbiased and honest technical review of the book and its contents.

Finally, thanks to all the unknown YouTubers, content creators, trainers, and authors who have influenced my narration style and contributed to my research. Additionally, thanks to all my colleagues at Cisco as well as the Cisco Press employees who were directly or indirectly working behind the scenes for the success of the book.

Contents at a Glance

Reader Services

Register your copy at www.ciscopress.com/title/ISBN for convenient access to downloads, updates, and corrections as they become available. To start the registration process, go to www.ciscopress.com/register and log in or create an account*. Enter the product ISBN 9780138237400 and click Submit.

*Be sure to check the box that you would like to hear from us to receive exclusive discounts on future editions of this product.

Contents

Command Syntax Conventions

The conventions used to present command syntax in this book are the same conventions used in the IOS Command Reference. The Command Reference describes these conventions as follows:

- **Boldface** indicates commands and keywords that are entered literally as shown. In actual configuration examples and output (not general command syntax), boldface indicates commands that are manually input by the user (such as a **show** command).

- *Italic* indicates arguments for which you supply actual values.

- Vertical bars (|) separate alternative, mutually exclusive elements.

- Square brackets ([ ]) indicate an optional element.

- Braces ({ }) indicate a required choice.

- Braces within brackets ([{ }]) indicate a required choice within an optional element.

Introduction

There have been so many discoveries in the cybersecurity space that the expression "jumping on the bandwagon" is now a commonplace expression when it comes to adopting technology. Rarely do we witness a concept or a technology being discussed a long time ago and then dying out, only to gain traction years after its inception. Zero Trust is one such concept. It would not be false to say that there were some early thinkers, like me, who do not implicitly trust anything or anyone and would have gladly implemented a super-secure network. Over time, as I have learned the hard way, one realizes that there is a delicate balance to maintain between convenience and security and, at times, almost everyone flips to the convenience side. In the early 2000s, enterprises would not have taken a Zero Trust mentality seriously. As time passed by, data got exfiltrated "mission impossible" style. As attackers became more sophisticated and as their cash flow increased exponentially, we began to think maybe, just maybe, we should not have been so complacent. I strongly believe that was the time when the concept of Zero Trust finally began picking up.

I recall my first engagement for a customer who was just getting started with a cybersecurity augmentation program and was interested in pursuing the Zero Trust journey. I had been mildly exposed to the Zero Trust concept at the time and watched some videos about it. Looking back, I can see that knowing the concept was one thing, but implementing designing and operationalizing an entire Zero Trust Network Architecture (ZTNA as we call it) was something else. Imagine learning to swim in a five-foot-deep pool and then attempting to swim in the open ocean with sharks—that is how it felt at the time.

When we engaged the customer to understand their current state, our strategy was simple—try and get as much information as we could from the customer relating to their current network and security and then build an architectural road map for them. We believed this was an important factor to consider when it comes to Zero Trust engagement. When we actually engaged with the customer, we realized that nothing had really changed, and we still needed to understand, design, and implement the right security controls, same as any other architecture assessment engagement. However, what I recall clearly was us asking the question "Why do you want to adopt Zero Trust?" We will discuss the importance of this question later in the book, but that question changed the direction of the conversation.

As architects, we are used to identifying the scope of engagement and trying to maintain a balance between the best security controls and alignment with the customer's business. As Zero Trust consultants, we realize that the "why" factor is equally important to give us a baseline of what the enterprise actually requires. Unlike a standard security augmentation project where a set of security controls is identified and implemented, Zero Trust has wider implications and has many moving parts that need to align. When we went into the discussion with the customer about how and why they chose to go on this journey, some of the reasons were the usual suspects, such as "compliance" and "secure by design," but others, such as "end-to-end encryption," really got us thinking about why this concept was never considered earlier. We heard about their overall strategy, how they discovered Zero Trust, their initial thoughts, and how Zero Trust aligned with their business use cases, and so on, and the discussions were far from technical. It was a glimpse into an entire enterprise's vision, and it was intriguing to see how the big picture was driving their mindset change.

The reality is that "necessity is the mother of invention." Historically, cyber-attacks were less, data was not considered as important, and enterprises just did not care as much about securing their assets. Data has evolved and information is the new currency. With the evolution of attack complexity and attacker capabilities, the stakes are higher and enterprises cannot just build their networks within a huge castle. On the flip side, enterprises also cannot make the lives of their employees and customers more difficult under the disguise of advanced security. There needed to be a change in the way security was perceived, and that is how this Zero Trust movement came to be (and I call it "movement" for a reason). Zero Trust is not some new technology; it is an actual mindset and philosophy.

An important takeaway from my first customer encounter is that irrespective of the access model being implemented, as long as it is aligned with the overall business vision and strategy, it will always have the desired result. In more technical terms, it is all about identifying the right use case for the customer to pivot to this new (or somewhat old) paradigm shift. After understanding some of the gaps in the customer's network, we provided them a clear road map and formulated their Zero Trust strategy. The customer was fairly satisfied, and as consultants so were we. We felt we really made a difference by helping an enterprise start off on their journey toward Zero Trust.

That's where my personal Zero Trust exploration began, and over the course of multiple customer engagements, I began considering some common themes. Senior leadership from select enterprises had already done their research on Zero Trust and were looking for a trusted partner to walk with them on this journey. These enterprises had engineers and mid-level managers who were trying their best to understand and build an architecture to fit their leadership's vision. On the flip side, we spoke to leadership who were sitting on the fence and considering their options. Their teams were trying to show them the value-add of the concept and how it would help the enterprise on the whole, and we as advisors were enabling them to do so. The common theme was that Zero Trust is not something that can be measured with a data sheet. A common question we got was how well were other customers doing after adopting and implementing Zero Trust in their enterprises. The metrics, discussed in detail in the book, are uniquely different. Where would someone have to start to understand how to measure the efficacy of the strategy and implementation?

Another aspect that I felt generally lacking was an overall adoption lifecycle framework. As of today, there is no Zero Trust lifecycle framework or a reusable consulting process from the inception of the idea to the signing off on a successful implementation project. Who are the people to speak to, what metrics would satisfy stakeholders, who would support the project? These were questions multiple managers were trying to answer in enterprises we spoke with. When we do software design, we have the software design lifecycle, but Zero Trust has no such lifecycle in place.

I find this simple comparison useful because it helps push the idea about what we are trying to put forward to various stakeholders. Consider Zero Trust as an open field, and everyone (I mean everyone) gets their own tools to create a building at different parts of the field in an effort to build a city. Not everyone who can build can build well, so everyone builds their own building, and the overall city looks disjointed. Perhaps there are no parks, or there are some private parks that cannot otherwise be accessed by others. The city shouts out restrictions, and the atmosphere is nervous and borderline belligerent. This is how security is today. Not everyone has a big picture, and most personnel are Subject Matter Experts in their field just doing what they do best toward a mission-less destination. If you were the mayor of this city, you would be firefighting factions every day.

Consider the alternative. You get a specific set of people to decide what they want to achieve as society. High-level ideas like "live in harmony," "work and play together," and "welcome all guests" come to mind. You get like-minded folks and start planning where to place each building rather than just starting to build. Once you agree on where to build a specific building, you decide who is the best person to build it. Once the city is built, you identify ways in which you can get in more people—but after performing the right background check. You dedicate a common independent body to decide entry and exit to the city. Doesn't this seem to be common characteristics of how basic city politics should be? That is why there are working societies and cities in the world and that is exactly what Zero Trust is all about. It is almost never about the technology or the products you wish to implement. It begins much earlier than the first discussions with a CxO and lasts much later after the last Zero Trust project was implemented. It is a mindset and a movement, and it cannot just be approved after a single presentation to senior leadership.

That is what I want this book to be. Irrespective of whether you are a CxO or a mid-level manager or even an engineer trying to convince your manager to start talking about Zero Trust, I want this book to give you a starting point. There are a plethora of books out there that step right into Zero Trust architecture by explaining concepts and then listing different methodologies that Zero Trust can be achieved. I want to stress that all those books are awesome, some of which I have used myself to begin my journey, but they are worthwhile to read to increase our theoretical knowledge of the concept. What I see generally lacking is the practical aspect of where to begin and what to do next. You cannot just wake up one day and decide to start implementing Zero Trust in your AWS network; it takes time and a lot of coaxing.

If you are a CxO, I want you to truly look inside the enterprise and see how Zero Trust helps you augment the enterprise. If you are a mid-senior level manager, I want you to see how your team can help propagate the Zero Trust story to both leadership and end users. If you are an individual contributor to the Zero Trust initiative, I want you to see how Zero Trust is achieved and what the key mechanisms are to consider. If you are an architect creating a Zero Trust story for a customer, I want you to see end to end how many people and how much time must be invested before you even begin your first pitch. The main question is not "How can I achieve a Zero Trust architecture?" It is and should always be "Why am I looking at Zero Trust as my framework and how does it align with my vision?" Once these business drivers are established with the customer, the specific use cases and mechanisms for the entire Zero Trust architecture will become clearer. Here are some questions this book aims to answer:

- How should one approach or even begin with Zero Trust conversations, and with whom?

- How can we validate feasibility to move to Zero Trust?

- Is there a standard format, guide, or framework to follow when recommending an architecture? If not, what is the general approach?

- What are the common use cases for the customer to consider adoption of Zero Trust, and how do we design an architecture for those use cases?

- How do we make sure the architecture still caters to the customer's use cases over time?

Context is key in Zero Trust, so the goal is also to make sure security controls augmenting existing context to a flow can be derived and the right context-based control can be implemented by the policy engine. Context or attribute-based access control is the future of access control. Those who adopt it early can disrupt their business much faster than others.

Let's discuss the format of the book. Most books in the market follow a standard concept and implementation format. My vision is to help everyone at every stage to benefit from the content. Hence, the entire book will be formulated as a conversation with various people in a fictitious enterprise. We will start off with a conversation with the CIO and, over time, move to various other key stakeholders. At each stage, we will try to practically complete tasks that are required to provide a tangible outcome so that you as the reader can understand what the key topic of discussion is, why you are having it, and how it helps move forward. The conversations will be in a different font to help isolate the theory from the conversation so that the practical aspect of the engagement is clear as well. The focus, while moving toward Zero Trust Network Architecture, is to set up a framework that is tailor-made for each enterprise and to make sure certain use cases are met. This will, in turn, achieve specific business targets. An enterprise might not want to achieve Level 5 maturity and might be comfortable being at a Level 3, as long as the business supports it and the risk is acceptable.

The motive of the book is to guide you to ask the right questions, visualize the right path, and help you implement that path either for yourself or for your customers who have begun their Zero Trust journeys. Remember, this book is a conversation with various people at various times to showcase how much time the journey really takes. As the reader of this book, you should be consuming and re-creating information specific to your use case and customers.

I hope this book helps reduce the traffic and helps you speed up (within the speed limit) on the wonderful Zero Trust highway toward your secure enterprise vision.

Book Structure

The book is organized into six parts/phases:

Phase 1: Mindset

Chapter 1, When It All Begins: In this chapter, the reader is introduced to the Zero Trust topic and how the book is structured. Since the narrative has a background story to it, the main characters and their pertinent history are shared so that the overarching story makes sense.

Chapter 2, The Zero Trust Kaleidoscope: This chapter introduces the reader to common perceptions of Zero Trust and how various product enterprises pivot the basic concept of Zero Trust to suit their needs.

Another aspect the chapter covers is why Zero Trust took quite some time to get traction in the security community. At the end of the chapter, the clear similarities in all narratives will be revealed.

Chapter 3, Defining Zero Trust: In this chapter, the reader will dive deeper into the trenches of Zero Trust standards and finally reach the core idea of what Zero Trust is fundamentally. Frameworks like NIST will be explored in this section. This is an important step in all engagements with Zero Trust, to let the stakeholders understand what they are signing up for. The chapter will cover basic tenets of Zero Trust and some catalysts that speed up the adoption process.

Phase 2: Align to the Business Vision and Mission and Craft Metrics for Success

Chapter 4, Always Start with "Why": This chapter aims to direct the reader to common business and technical drivers for most leadership stakeholders. The core of the chapter showcases that the Zero Trust initiative cannot be successful without the support of leadership and that the initiative will always be top-down. In addition to the common business and technical drivers, the reader will also be introduced to common use cases specific to why one may choose to adopt Zero Trust architectures.

Chapter 5, Measuring Zero Trust Success: You cannot manage what you cannot measure. This chapter dives into explaining how the reader will build performance and risk metrics to effectively measure the success of the Zero Trust initiative based on feedback from various stakeholders. The chapter will also explore the various types of measurement methodologies that can be utilized to create customized Zero Trust metrics. Some of these include strategic, tactical, and operational measurements, along with qualitative and quantitative metrics.

Chapter 6, Understanding Zero Trust Maturity: Once metrics have been identified, it is time to look into the enterprise and understand where you stand from a people, process, and technology perspective. This chapter utilizes an established maturity framework to show how the maturity of an enterprise is measured to identify gaps and then start building architectures that encompass the remediations.

Phase 3: Identify Key Stakeholders and Enable a Zero Trust Team

Chapter 7, Zero Trust Avengers, Assemble!: No man is a silo, and no initiative can be complete with the help of just one team. The multifaceted nature of Zero Trust demands that we explore creating a highly motivated Agile team to support the overall initiative. This chapter lists all the key personnel, at both the leadership and subject matter expert levels, that are needed to build and manage a Zero Trust architecture.

Phase 4: Develop the Target Zero Trust Architecture

Chapter 8, Building a Zero Trust Architecture: You have spoken to your leaders and have built a team and a framework to measure the Zero Trust initiative. It's time to get your hands dirty and start designing your architecture. This chapter will talk about how Zero Trust overlays the existing network and security processes. Key terminologies for Zero Trust architectures will be introduced. Software-defined perimeters (SDPs) will be discussed. The chapter will also introduce the concept of a Zero Trust policy and show that none of the policy constructs are new, they are just structured differently. Finally, the basic flows of a Zero Trust architecture are introduced. This will be a key baseline to expand to multiple business flows as one starts flow discovery.

Chapter 9, Critical Security Mechanisms for Zero Trust Architectures: This chapter will touch the key technical requirements needed for the apt functioning of the Zero Trust architecture. Topics such as Identity and Access Management, segmentation, application development, and more will be detailed.

Phase 5: Present the Zero Trust Strategy and Metrics

Chapter 10, Presenting the Zero Trust Strategy: Once the architecture is in place and the metrics are designed, the Zero Trust team will present the strategy to the board. Their main motive here is to convince the board as to the return of investment and reduction of risk. Strategies are listed on how the presentation can be approached. The reader is expected to tailor their presentation based on the various tactics provided.

Phase 6: Implement, Monitor, Feedback, Repeat

Chapter 11, Implementation and Continuous Monitoring: Now that you are armed with support from your leadership, a strong Zero Trust team, and a viable architecture blueprint, this chapter will explain the next implementation steps into operationalizing the Zero Trust architecture. A typical implementation approach is highlighted based on various engagements with enterprises.

Chapter 12, The Road Ahead: This is it. Your Zero Trust implementation has been completed, but your journey is just beginning. With multiple innovations in technology and various aspects of Zero Trust pivoting to support the enterprise, this chapter introduces the Zero Trust lifecycle framework, which is extremely useful for any enterprise embarking on the Zero Trust journey. It is the culmination and combination of all the concepts elaborated on throughout the book. The chapter brings the concepts together and looks forward toward a secure future and how Zero Trust as a philosophy takes the security community toward that collective vision.

Mindset

This phase targets the mindset of all stakeholders. It is an important aspect of the overall Zero Trust lifecycle and should not be skipped. Directly jumping into strategy and implementation or even product selection usually leads to loose coupling with business drivers. This phase makes sure the stakeholders have their say in what they want, and you try and explain to them what Zero Trust has to offer to support their business. The following chapters are covered in this phase:

Chapter 1   When It All Begins

Chapter 2   The Zero Trust Kaleidoscope

Chapter 3   Defining Zero Trust

When It All Begins

The first phase of any Zero Trust journey is to incept the idea of "change" in the minds of your target audience. It involves changing the mindset of key stakeholders who have the power to influence strategy. Think of it like a sales pitch, not for a product or a service, but for an idea about a new paradigm in security that can change the way you protect your assets. If you are a mid-level manager, you should be advocating the Zero Trust strategy and its benefits to enterprise leadership. If you are a CxO,[1] you should be presenting the strategy to both the board of directors as well as to your enterprise for adoption. If you are an independent architect, you should be partnering with customers and helping them align Zero Trust and overall business needs. Before deep diving into more details around a strategy, there needs to be a common starting point that most enterprises or consultants can refer to when they are asked to design and architect Zero Trust. For a deeper insight into the importance of leadership and leadership's role in embarking on the Zero Trust journey, see Chapter 7, "Zero Trust Avengers, Assemble!"

Zero Trust, as a strategical approach to security, is a complex discussion point for most leaders and architects. As a construct, however, it is fairly simple. As we will explore in deeper detail in Chapter 3, "Defining Zero Trust," the overall construct of Zero Trust is essentially knowing *who* can access *what* and building policy for each *transaction* independently while keeping the principle of least privilege in mind. Unlike usual security projects, however, Zero Trust cannot be completed in one project cycle. It needs careful insertion into the minds of key stakeholders with a detailed roadmap spanning multiple technologies and teams. It starts months before an actual conversation takes place and matures to be a continuous cycle of self-improvement that does not end in a span of years. By the end of this book, you will have walked the path of an enterprise considering adoption of Zero Trust from scratch, which will give you the key stepping stones required to achieve a seamless migration to Zero Trust. This entire book will be a conversation between you, the reader, and various stakeholders in the enterprise. The interview that follows will serve as a starting point to identify some key aspects, such as the following:

- The protect surface, which includes critical infrastructure

- Business and technical drivers

- Current security posture and existing transformational projects

- Pain points and concerns

- Perceived threats to the business and technology

This is the common starting point to get most conversations rolling, but topics should not be limited to these themes. The conversation you will encounter throughout this book is a fictitious conversation with a CIO, which has been created by combining multiple conversational experiences that many security experts have had with actual executive leaders.

Interview Strategies

At the inception stage, you need to consider why interview strategies are critical. Having a conversation with an executive-level leader will be daunting for most. Irrespective of whether you are part of the same organization or an external entity, executive leaders value time, and the direction of your conversations will vary based on how open the interviewee is. Your agenda should not be to sell a product or service. You wouldn't be speaking to an executive if that was the case. Your discussion has the potential to change the entire path of the enterprise, not only from a security perspective but from an overall sustenance perspective as well. Hence, conversations must be crisp and to the point, and each aspect must circle back to alignment, as that is what leaders want—alignment to business. To begin with, here are some key considerations before interviewing a CxO:

- **Research the company and industry:** Perform research on the target company and its industry. Questions must be relevant to what the leaders are planning to achieve, and industry knowledge will help understand why certain decisions need to be made from their perspective. This will also help you understand the leader's role within the organization itself and how decision-making occurs.

- **Be prepared with questions:** Based on your research, prepare a list of questions you want to ask the interviewee. Do not wing it. Make sure the questions are relevant to the company and industry, and focus on their experience, leadership style, and vision for the future. Get your list reviewed by peers who have had discussions with leadership before.

- **Listen actively:** During the conversation, make sure you are actively listening to the interviewee's responses. This will help you follow up with relevant questions and show that you are engaged in the conversation. This also ensures that you do not repeat questions by asking for information that was shared previously. Repeating questions sheds a bad light on the interviewer.

- **Ask for examples:** When the CxO is discussing their experience or leadership style, ask for specific examples to illustrate their points. This will help you understand their approach to problem-solving and decision-making.

- **Show enthusiasm:** Show the leaders that you are excited about the opportunity to interview them and learn about their role. This will help establish a positive rapport and make the interview more productive.

- **Be respectful:** Remember that the CxO is a senior executive within the organization and deserves respect. Avoid asking overly personal or inappropriate questions and focus on the topics that are relevant to the interview.

- **Follow up:** After the interview, send a thank-you email or note to the interviewee. This will help you stand out and show that you are serious about the opportunity.

Why is preparation for the first talk so important? Simply put, the stakes are high, and the impact of what you say could change a lot of processes, people, and technology. Hence, you should be in a position to provide sound technical and analytical advice to leadership, to help them make the right decision.

The enterprise hierarchy is also a mirror into what drives strategy in the organization. For example, if you are having a conversation with the CIO of the company, it is likely that Zero Trust is being positioned as an information technology (IT) strategy and that it is being considered as a strategic enabler for the business. Alternatively, if you end up speaking to the CISO about Zero Trust and its use cases, it is likely that Zero Trust is being positioned as a security strategy in response to a compliance requirement or a security conversation. Chapter 4, "Always Start with 'Why,'" covers some common use cases that trigger a Zero Trust mind shift. Similarly, a hierarchy where the CISO reports to the CIO generally would end up being a detractor to Zero Trust due to generally conflicting ideas of a CISO and a CIO. This would require you to change your strategy when approaching both executives. On the other hand, an independent CISO directly reporting to the CEO depicts a more strategic approach and a commitment to cybersecurity as an important part of overall business strategy. Understand your target audience.

Your first of many assignments will be to conduct an interview with the CIO of a bank who has just heard about Zero Trust. The entire book will have conversations interspersed between technical themes to bring in a perspective of realism as well as a timeline for the conversations. Step into the shoes of the consultant and walk the Zero Trust journey along with the bank.

Key Zenith Trust Bank Stakeholders

Before you begin this interview with the CIO of Zenith Trust Bank, you need to conduct some research on what Zenith Trust Bank is and what it does. This section will constitute the notes similar to what you should have collated based on research on social media and the Internet pertaining to the relevant stakeholders' description. All names and descriptions are fictional. Any resemblance in name or profile is purely coincidental. Figure 1-1 illustrates the overall organizational hierarchy in Zenith Trust Bank.

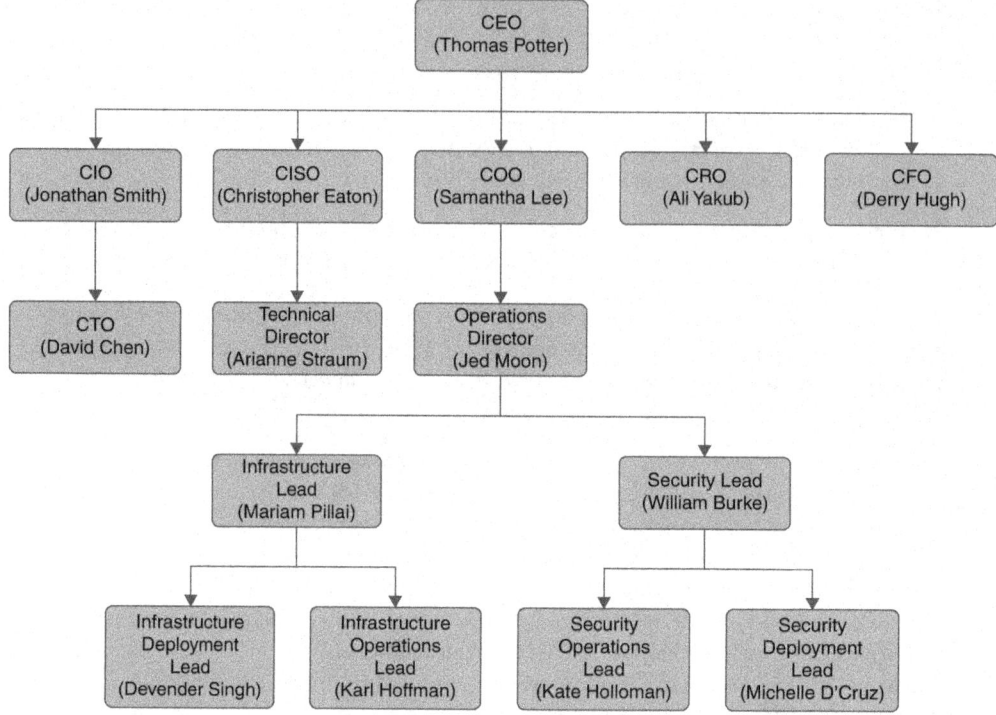

Figure 1-1   *Zenith Trust Bank Hierarchical Organization Chart*

Over the course of Zenith Trust Bank's Zero Trust journey, you will speak to Christopher, Jonathan, David, and Samantha along with all their respective teams.

Jonathan Smith, CIO

Jonathan Smith, the current chief information officer (CIO) of Zenith Trust Bank, had humble beginnings as an engineer when he joined the bank straight out of college. With his sharp technical skills and a natural aptitude for problem-solving, he quickly made his way up the ranks and became a vital member of the bank's IT team.

Jonathan's early years at the bank were marked by his ability to identify and fix technological issues before they became major problems. He earned a reputation for being a hard worker who was always willing to go above and beyond to ensure the bank's systems were running smoothly. As he gained more experience and expertise, Jonathan realized that the bank's technology infrastructure was holding it back from achieving its true potential. He recognized that the bank needed to modernize its IT systems if it was going to keep up with the rapidly changing banking landscape.

Jonathan took a proactive approach to this challenge, implementing key working models such as agile development and DevOps to speed up the bank's IT projects and increase efficiency. He also invested heavily in cloud-based infrastructure, which allowed the bank to scale up and down quickly to meet customer demand. Under Jonathan's leadership, the bank's IT department became more agile, adaptable, and customer-focused. He introduced innovative products and services, such as mobile banking and online account opening, that helped the bank attract new customers and retain existing ones.

One of Jonathan's most significant business actions was his role in leading the bank's digital transformation. He recognized early on that digital channels were becoming the primary way customers interacted with the bank, and he worked tirelessly to ensure the bank was ready for this shift. Jonathan's efforts paid off when the bank's digital services became a major revenue generator for the institution, with customers praising the ease and convenience of the bank's mobile and online platforms.

Today, Jonathan is a well-respected CIO, known for his ability to drive innovation and growth through technology. He continues to stay on the cutting edge of technological advancements, ensuring that the bank is always one step ahead of its competitors.

Christopher Eaton, CISO

Christopher Eaton is a seasoned cybersecurity professional who has spent over 20 years in the security industry. He started his career as a security analyst at a small software company. In the year 2005, Christopher joined a Fortune 500 financial services company as a security architect, where he was responsible for designing and implementing the company's security infrastructure. Over the next few years, he made a name for himself as a risk-taker who was unafraid to challenge conventional wisdom and try new approaches to security.

In the year 2010, Christopher was promoted to the role of chief information security officer (CISO) at the same company. In this role, he was responsible for overseeing all aspects of the company's cybersecurity strategy, including risk management, incident response, and compliance. Under Christopher's leadership, the company's cybersecurity program became known for its aggressive approach to risk management. Christopher was willing to take on high-risk projects and initiatives that other companies shied away from, such as implementing emerging technologies like blockchain.

In the year 2015, Christopher was recruited by a rival financial services company to serve as their CISO. Once again, he brought his risk-taking approach to the role and implemented innovative security solutions that helped the company stay ahead of the curve. In the year 2020, Christopher was offered a position at Zenith Trust Bank, where he currently serves as CISO. In this role, he continues to push the boundaries of cybersecurity and is known for his willingness to take on high-risk, high-reward initiatives that have the potential to transform the industry.

Throughout his career, Christopher has become known as a visionary leader who is unafraid to challenge conventional thinking and take bold risks in pursuit of innovation. While some of his decisions have been controversial, there is no denying that his approach has helped to elevate the cybersecurity profession and drive progress in the industry.

Samantha Lee, COO

Samantha Lee is a seasoned business executive with over 18 years of experience in the technology industry. After several years in software development, Samantha transitioned into project management, where she oversaw the successful delivery of complex software projects for multiple companies. Her ability to manage large, cross-functional teams and deliver projects on time and on budget quickly earned her a reputation as a rising star in the industry.

In the year 2010, Samantha was recruited by a leading technology consulting firm to serve as a senior engagement manager. In this role, she was responsible for leading consulting engagements with some of the firm's largest clients, developing strategic roadmaps, and managing project delivery across multiple workstreams. After several years of success in the consulting industry, Samantha was approached by Zenith Trust Bank to join their executive team as vice president of operations. In this role, she was responsible for overseeing the company's global operations, including supply chain management, manufacturing, and logistics. Under Samantha's leadership, the company's operations became known for their efficiency and agility, enabling the company to rapidly scale its production and distribution capabilities to meet growing demand. She also played a key role in driving innovation within the company, developing new processes and technologies that improved the company's ability to bring new products to market quickly.

In the year 2018, Samantha was promoted to the role of chief operating officer (COO), where she currently serves. In this role, she oversees all aspects of the company's operations, including product development, supply chain management, and customer service. She is known for her ability to manage complex projects and drive change within the organization, while maintaining a focus on delivering high-quality products and services to customers.

Throughout her career, Samantha has been recognized for her strategic thinking, operational expertise, and leadership skills. She is a sought-after speaker and thought leader in the technology industry and is widely respected for her ability to drive results in high-pressure, fast-paced environments.

David Chen, CTO

David Chen is a seasoned technology executive with over 25 years of experience in the industry. After 10 years in software development, David transitioned into technical leadership roles, where he led teams of engineers in the design and development of cutting-edge software products. He quickly became known for his ability to bridge the gap between technical and business stakeholders, and his talent for translating complex technical concepts into language that non-technical stakeholders could understand skyrocketed him to technical stardom.

In the year 2008, David was recruited by a leading technology company to serve as director of engineering. In this role, he was responsible for leading a team of engineers in the development of a new cloud-based software platform. He successfully led the project from inception to launch, driving innovation and ensuring that the platform met the needs of both technical and business stakeholders.

After several years of success as director of engineering, David was promoted to the role of chief technology officer (CTO) at the same company. In this role, he was responsible for overseeing all aspects of the company's technology strategy, including research and development, product development, and technology operations. Under David's leadership, the company's technology platform became known for its scalability, reliability, and innovation. David played a key role in driving the company's digital transformation, leading the development of new technologies and processes that enabled the company to deliver products and services to customers more quickly and efficiently.

In the year 2019, David was recruited by Zenith Trust Bank to serve as their CTO. In this role, he oversees all aspects of the company's technology strategy, including product development, data science, and infrastructure. He is known for his ability to drive innovation and leverage technology to drive business results, and he is widely respected for his technical expertise and strategic thinking.

Throughout his career, David has been recognized for his ability to bridge the gap between technical and business stakeholders as well as for his talent for driving innovation and delivering results. He is widely respected for his contributions to the field of technology.

The Interview

The following is an interview between Glenn Taylor, who is a security architect working at a vendor company (Prolink Solutions) that specializes in IT and security products, and the CIO of Zenith Trust Bank (Jonathan Smith), the CISO of Zenith Trust Bank (Christopher Eaton), the COO of Zenith Trust Bank (Samantha Lee), and the CTO of Zenith Trust Bank (David Chen). Prolink Solutions was summoned by Zenith Trust Bank to speak in detail about an overall IT strategy with respect to Zero Trust adoption.

[Glenn, John, Christopher, and David are sitting in a conference room in the Zenith Trust Bank head office. Samantha has dialed into the meeting via a remote telepresence session. Glenn begins the conversation once everyone is seated.]

Glenn: Hi Mr. Smith, Mr. Chen, Ms. Lee, and Mr. Eaton. It's an honor to be able to have a chat with you. My name is Glenn Taylor, and I am a security architect at Prolink Solutions. I would like to understand how Prolink Solutions can help you with your digital transformation journey.

Mr. Smith: Please call me John, and I am keen to work with your firm. It is a small world. I almost had got an offer to work at Prolink Solutions a long time ago, but I took this role instead. Looks like I made the right decision, right?

[Everyone in the room smiles.]

Mr. Chen: Hello Glenn.

Mr. Eaton: Hi Glenn.

Ms. Lee*[remotely]:* Hello Glenn.

[Glenn gives a preliminary brief on overall security issues in the field today and how Prolink Solutions is helping many other customers achieve their security vision.]

Glenn: Thanks for providing Prolink Solutions the opportunity to give you a preliminary brief on how we support our customers. From what we understand, Zenith Trust Bank's vision is to be the customer's partner in their journey to financial independence and, in turn, to be the customer's first choice when choosing a trustworthy bank. Zenith Trust Bank also wishes to be the leading bank, which makes banking seamless and available to everyone, everywhere, and you have spent quite some years in order to achieve this goal. Would you like to start off with a few words about some key topics that pertain to the enterprise and its vision?

Mr. Smith: Thanks for the presentation, and we are excited to have Prolink Solutions on board with us. As you have rightfully mentioned, we have a vision to be the number-one banking partner for all our customers. Overall, I believe there are a lot of initiatives running in the network itself, but one key aspect that has been stuck at the back of my mind is to increase the overall accessibility and security of our IT infrastructure.

Glenn: Yes, we understand that was one of the key top-of-mind topics for you for today's discussion. If you do not mind, could you brief us in some of the key business use cases that exist today in Zenith Trust Bank?

Mr. Smith: Alright, as a bank, some of our core functionality is money management for our customers. We focus on all kinds of customers, ranging from individuals and small businesses to corporate clients. We help manage their money and make sure access is simple and easy. We have some high-net-worth clients for whom we provide wealth management as well. Apart from the core business, we have partnerships with fintech startups to help bring innovative technologies to our customers. We also provide financial education for free at our learning portals.

Glenn: This appears to be a wide portfolio of services you are offering. I know that you like investing in modern disruptive technology for the betterment of the bank. Are there any disruptive technology-based projects you have in your pipeline that are close to your heart?

Mr. Smith: Yeah, of course. The one I am most keen to see the light of day is the artificial intelligence–based chatbot for which David is an executive sponsor. I had recently been to one of their demonstrations, where it almost felt like I was talking to a person, except that it was entirely an artificial intelligence chatbot.

Mr. Chen: I am waiting to see how our customers enjoy those services.

Mr. Smith: AI-based chatbots have a lot of potential in today's market. Another interesting one is extending our bank's digital platform to investment tools as well so that the customer has one portal to access all their wealth details.

Glenn: Great. Looks like the bank has a lot of projects lined up for quite some time. From what we understand, you are now a bit concerned with the overall security posture of the enterprise on the whole with so many moving initiatives already in motion. Could you tell us a bit more about this?

Mr. Smith: Yeah, essentially that is why we wanted to have a chat with Prolink Solutions. I think we all know about the Nexus Bank incident. Their data breach was an eye opener for most of the banking sector here locally. When we were briefed on the actual attack, I felt the entire attack pattern was an interesting lesson to learn for us. As a non-technical business owner, it was interesting how a weak link in another partner enterprise led to a massive one million customer data loss. This is not something I want to happen to us. That was when I started thinking about security augmentation with long conversations with Christopher, but I never really got around to looking at overall strategy as my team highlighted that we are fine with our current security controls. I was told that we are under low risk of an attack and hence we should be fine. Later in the year, COVID-19 happened, and our workforce started accessing all resources remotely, and I saw a huge decline in general operational efficiency. Samantha was firefighting customers almost daily, all the while dealing with decreased sales, lesser initiatives, and delayed project timelines.

[Ms. Lee nods pensively.]

After a couple of months, we had an outage in our data center and our VPN was operationally down for almost a day until the operations team realized that the devices were not sized enough. I still remember Samantha did not go home those three days when this incident occurred.

Mr. Eaton: I was fairly confident that our security controls were sufficient, but after this incident I began to wonder if our enterprise is actually resilient enough to a change in environment. I began to think what happens to us if we get attacked. Are we ready? I spoke to Jed, my technical director, and his team about their confidence in the efficacy of our security controls, and they were confident that we could handle the increase in workload and other possible attacks. I wanted this efficacy validated externally. I got an external expert to validate the security posture of the data center for us, and they painted a very different picture. My team was concerned as well at how vulnerable we were, and subsequently we started looking at all the deployment recommendations.

Ms. Lee: Some recommendations included optimizing the firewall rules, but essentially made it more operationally complex. My operations team is constantly loaded with large volumes of rules to be created, and now recommendations are to add more context to the rules. I am concerned about the impact it will have on customers and employees and their overall experience. Eventually I do need to support Christopher to maintain compliance to the regulatory bodies that govern the bank. Changing design and increasing rule bases, etc., were some other recommendations we received. We even got recommendations to add new firewalls. For me, this seemed like overkill, and my focus is to make sure the overall IT operations run smoothly. I could not see the justification of adding in new firewalls. Christopher and his team then started exploring Zero Trust as a side thought.

Mr. Eaton: The key messaging of Zero Trust is not clear, with all the vendors saying they can do Zero Trust. Since you are a neutral entity and our trusted partner in managing our infrastructure, we wanted to understand what you think about this entire concept and leverage your expertise in gauging our current state.

Glenn: We had understood that this was a key top-of-mind topic for you. Have you or your team evaluated Zero Trust as a strategy for your enterprise?

Mr. Smith: One of my team members was looking for VPN enhancement and best practices on the Internet after the downtime we had, and he saw an article about how Zero Trust was doing away with VPN. He seemed to like the overall concept, and he mentioned it to me, and I liked it, too. He then presented some aspects of Zero Trust that we felt were relevant to our bank environment and got us interested enough to pursue.

Glenn: Could you maybe give us some insights on what makes you think Zero Trust is a good way forward?

Mr. Smith: First of all, one of the bank's main visions is to make access seamless to our customers, and today we have to create so many paths for a user to access their own information and money. If a user wants to withdraw money, they need to be validated, authenticated at multiple checkpoints, and then authorized by the bank personnel, and it's a hassle. The same applies to our employees. They need to go through so many approvals if they need access to a specific system. There are processes in place to make sure there is no fraud and the money is safe. If Zero Trust can ease this process, that is something I am interested in pursuing. VPN, of course, is a personal as well as enterprise concern due to the recent outage. How much more can we scale our VPN concentrator? If we can move the dependency from there and make it simple for users, then why not? Don't you agree? *[Glenn nods]* I am looking to transform our IT infrastructure and make it more agile, scalable, flexible and of course cost effective. I want it to be secure and functional at the same time.

Glenn: Indeed, and as part of this entire exercise, we will try and explain how Zero Trust aligns to this vision. We will look at your entire network and see how we can incorporate Zero Trust and why, in reality, Zero Trust is something that is usually part of the overall security strategy without being explicitly labeled as Zero Trust. It is likely that you have already been practicing Zero Trust tenets in your current network, and with additional enhancements to your existing infrastructure, the goal is to take your enterprise to the next maturity level. If you don't mind, I can brief you on what Zero Trust constitutes, and we can go into some aspects of the strategy itself.

[Glenn proceeds to explain what Zero Trust really is. The next two chapters will be a detailed explanation of the Zero Trust concept and how it fits into overall IT and security strategies.]

Endnote

1. The x in CxO is symbolic of a variable. This has been used to include all-executive-level leadership in an enterprise and should not be confused with a CXO (chief experience officer), which is a subset of the representation.

Chapter 2

The Zero Trust Kaleidoscope

Zero Trust is not a new approach to security, but it does need a bolder mindset to tackle. If you survey the entire enterprise landscape, you will find that there are deployments and architectures with various maturities. Some of them are designed with complex threat modeling. Others have simple dot1x to provide network access. These deployments are polar opposites of each other, but what they have in common is static rule-based access control. Zero Trust essentially shifts the focus from this common aspect of all security mechanisms. It can be identified as an amalgamation of most security concepts we already know. An enterprise with an established security portfolio may find the adoption of Zero Trust moot. However, enterprises that have an aggressive compliance strategy or have been exposed to a breach (of their own assets or those of other enterprises in the same business vertical) are more likely to be open to adopt Zero Trust because the risk reduction after Zero Trust adoption is evident. This does not mean that an enterprise needs to be breached to start conversations about Zero Trust; however, when tangible loss impact has been witnessed, risk discussions no longer take a back seat. There are other deterrents to adoption of Zero Trust, but the overall conversation will most likely not be as simple as convincing a network owner to install a firewall. It takes substantial understanding of the business flows to be able to factor Zero Trust into the enterprise security road map. This is a critical aspect you will observe in conversations you have with enterprise leadership because a majority of enterprises have applications deployed decades ago and no one has updated application flows and requirements. Currently, if you ask an application owner for flows relevant to an application, you will get only a list of TCP/UDP ports allowed.

Complacence is killer as well. A common theme across small- or medium-scale enterprises is that the enterprise itself has not been attacked, so why should the organization spend time and money changing the direction of security when it seems to be working. Lack of an attack is considered a performance indicator, which is fundamentally wrong. Lower attack counts to your enterprise could simply mean that you do not have enough visibility into your network to gauge if an attack has actually happened. It is likely that there are already some elements of advanced persistent threats as well as reconnaissance artifacts in the enterprise network and endpoints. To those customers, your approach should simply be, "It is not a matter of *if*, but *when*." You cannot consider security a reactive measure to an attack. Risk is no longer processed based on just the occurrence of an exploitation attempt. With a shift in risk perception and focus on identity-centric proactive security, policies and control systems consider the asset being protected and the potential threat to specific types of assets. The flip side of the coin involves security-focused enterprises that have heard of the concept of Zero Trust and see the value it can provide if associated with overall sales of their product. A direct correlation is then inferred between the act of adopting Zero Trust and selling secure products. The truth is that a vast majority of enterprises have already embarked on the Zero Trust journey without realizing it or have implemented basic Zero Trust tenets in their network as part of overall security best practices. The product is just a means to the end.

Considering that concepts like network segmentation are an integral part of Zero Trust and have been in most security conversations for more than a decade, why is it that interest in Zero Trust, as a viable enterprise security strategy, just picked up a few years ago? Some justification for Zero Trust uptick could be attributed to how sophisticated attacks have evolved and how much they rely on lateral movement. In reality, it doesn't matter how well you can infiltrate a network. If you cannot move laterally, you will not be able to extract a lot of data and will be stuck in a specific sandbox. Of course, there are other deterrent factors, such as market competition, which existed long before Zero Trust was standardized. Market jargon essentially led to multiple vendors claiming that their own product could provide a Zero Trust solution, thus confusing the common enterprise owner in the process.

Organizations cannot live in the past within a castle surrounded by moats. Fully depending on perimeter-based security alone is not a desirable security strategy, and a software-defined perimeter is already being considered as a viable option to enhance the existing strategy itself. Re-evaluation at each segment is the critical aspect that propagates dynamic risk-based control. In the next section, we will look at some of the reasons why Zero Trust took much more time to get adopted.

Delay in Adoption

I have been fascinated with how such a simple strategy like Zero Trust took almost a decade to get adopted into general security architectures and frameworks. After conversations with multiple experts in the field, I have listed some factors that will help you understand why Zero Trust finally became a journey that the entire security community was comfortable embarking upon. The sections that follow explore these factors in more detail.

Mindset and Perception

The following are some of the common deterrents that are stemmed from the way people perceive Zero Trust:

- **Misinterpretation of purpose:** Over the course of my research, I found unconventional reasons why Zero Trust was almost shunned. People took Zero Trust literally.[1] Enterprises stopped trusting people and completely denied access to actual resources, claiming that if employees really needed them, they could ask for permission to access. Zero Trust is not an excuse to justify a toxic work culture. This was one of the reasons why the topic of Zero Trust adoption was never raised. The concept itself was not considered "enterprise friendly." Employers took pride in trusting their employees and assumed that a Zero Trust architecture was not needed.

- **Implicit trust complacency:** Another aspect of why Zero Trust was slow to pick up was the complacency of security teams implementing security controls. It was almost always assumed that with the right firewalls and IPS technology, internal networks were safe. This, of course, has changed over time, with most breaches being root-caused to social engineering and human negligence. Add another pinch of lateral movement and you have a witches' brew of advance persistent threats and data exfiltration. Security is only as strong as your weakest link, and rightfully Forrester changed its viewpoint on Zero Trust by shifting its focus to data. Data now sits at the center of the Zero Trust Extended Framework,[2] with People, Networks, Devices, and Workload focusing on protecting the data. Currently, all aspects of security are being incorporated into the information security policy, which paves the way to a cleaner adoption and implementation of Zero Trust.

- **Reluctance to install posture and Zero Trust agents:** Users are not keen on getting their devices postured. Some perceive this as invasive, and others complain about performance issues. Additionally, organizations such as government agencies need the endpoint agent to have a minimum level of security certification, which generally takes time to achieve. Hence, different enterprises take different approaches to solve this roadblock. One such solution is to bring the workforce to the office. The statement, simply put, reads, "If you want to avoid installing agents on your computer, make sure you're in the office." This breeds implicit trust, and it can be inferred that if you are in the office, then your laptop and you can be trusted. This is the implicit trust we want to move away from, which inevitably also paved the way for many enterprises to start exploring and adopting Zero Trust. The implicit trust model is not feasible given the constantly evolving threat landscape, and the remote work connectivity necessitated by the COVID-19 pandemic hollowed it out even more. An enterprise needs to be sure that the end host connecting to the network is not compromised, and it is secure enough to be given access by continuously validating against a policy engine at each entry point. Posturing and Zero Trust agents have changed the way enterprises and security practitioners perceive security, and with better security posture measurement, we are slowly seeing traction towards cloud access and subsequently Zero Trust architectures and principles.

■ **Full access privilege (convenience over security):** Enterprise employees do not like the concept of restricted access. I recall mentioning to my friend from college that I take pride in the fact that my company gives me unfettered Internet access. What it took for my company to make sure I was not a threat became clear to me five minutes into the Zero Trust architecture journey. Users want full access, and on the flip side security practitioners and custodians have been advocating for need-to-know access. This, then, forces enterprises to reach a middle ground where users are given access but based on whether they need to access the resource (principle of least privilege) and whether their profile has a minimum secure context to access the desired resource. For example, the policy engine might not be very strict in giving an employee access to a local FTP server but will check if the same employee has a well-secured endpoint before giving access to a production Active Directory server and will deny access if that employee is not an Active Directory administrator, even with all of the preceding conditions satisfied. There is no more implicitly secure server. Cisco as an enterprise does not give access to Box or SharePoint from a device that is not in the trusted device management category, but not all enterprises implement context-based controls. Many enterprises that are just starting out believe they have bigger issues to fix, such as increasing sales and augmenting their products. Security takes a back seat, and leaders choose convenience over security.

■ **Reluctance to use the cloud for security controls:** Enterprises are still not comfortable exposing their applications and databases to the Internet. Forrester extended Zero Trust to data centers, but also brought in a concept called Zero Trust Edge,[3] the primary function of which is to provide protected access for users to their workload over the Internet by using cloud-based security and networking. It is referred to as Secure Private Access (SPA) under the Cisco Secure Access capability list. Because the cloud was a more recent network development than hardware networking or security, security is more amalgamated with the networking infrastructure; hence, the implementation of context-based dynamic control becomes simpler and more effective. Yet, even today, it is unlikely that an enterprise will willingly expose its databases to the cloud, even if we set up the best context-based controls, because it is the mindset that needs to change, not the technology itself. You will explore server-to-server communication via the cloud in detail in Chapter 8, "Building a Zero Trust Architecture."

When security mechanisms like cloud firewalls are available today, enterprises shy away due to similar fears of exposure over the Internet. Even when cloud firewalls provide the capability to scale at will, the fact that traffic needs to traverse the Internet, be inspected by a firewall (which might not be part of the enterprise), and then return back to the data center seems to be a daunting prospect for many enterprises. Data Governance is another aspect where the enterprise cannot send local data outside of the country if the data processor resides beyond the country's boundaries. This stops a lot of enterprises from allowing private access to data center applications, which then leads to design decisions such as NAT and Virtual Desktop Infrastructures. These technologies, however, bring in the baggage of performance and traffic bottlenecks.

■ **Work from anywhere, access anything:** We've had our fair share of life-changing events over the past few years, but nothing has had as dramatic an impact as COVID-19. The work-from-home paradigm has forced enterprises to change the way they look at their workforce. The entire company may be working out of 14 countries, each with a different access requirement. Location just becomes a context setter rather than a fixed security parameter. To facilitate employees to access resources from wherever they are, enterprises are now beginning to look at Zero Trust as a viable access model, but not before exhausting all options with their VPN. VPNs have always been an inhibiting factor for Zero Trust, until people started seeing issues with the technology itself. The original ideology was that as long as there is a strong encryption tunnel to protect data in motion, there's no need to move to Zero Trust. With recent VPN scalability issues and overall deployment complexity, Zero Trust has started gaining traction. After heavy loss of production hours and growing numbers of VPN users, CISOs are considering the power of context-based access over limited-bandwidth VPN solutions. However, seasoned executives and infrastructure leads choose to stay with a proven solution rather than exploring a new efficient method.

Security Architecture and Technology

The following deterrents are related to an overall technology and architecture gap in the enterprise that prevents organizations from considering Zero Trust:

■ **Scalability of hardware:** When one is designing next-generation firewalls for customers, there is always a dilemma of creating enclaves for the customer. An enclave is a logical boundary that segments different functions in the network. For example, a campus enclave will segment the campus LAN from an Internet of Things (IoT) enclave, which segments the IoT devices. An extreme viewpoint would be that every application or function must ideally be a functional enclave, but scale becomes the villain, and customers have around a thousand applications with subfunctions. How will a next-generation firewall scale enclaves and subsequent sub-enclaves? And even if it does, how can one perform network access control (NAC) for all of these segments without affecting the hardware scale of devices servicing the network? These are common questions that most architects will have, and Zero Trust completely levels the playing field. You will learn about some of these approaches in Chapter 8 and Chapter 9, "Critical Security Mechanisms for Zero Trust." Essentially, Zero Trust will make you ask the question, "Isn't this how we should have secured the networks in the first place?"

This is easier said than done, however. Due to the inherent lack of information regarding applications, their functions, and subsequent lack of macro-segmentation, operation teams deploy wider crude enclaves, which are easier to operate and deploy. This makes the current setup much more attractive to operation teams, and they question the need to segment further and follow a new Zero Trust strategy, as this would lead to more application discovery and rule creation. This is not attractive to both network and security operations.

■ **ICS and IoT restricted architecture:** IoT (including CCTVs, IP phones, and similar headend devices) as well as operational technology (OT) endpoints now need to access the Internet to be able to provide real-time information for processing. Industrial control systems (ICSs) were traditionally locally managed and had all their communications over a Layer 2 domain. With more versatile placements of sensors, it becomes increasingly difficult to extend a Layer 2 domain. Extending Layer 2 across any network is generally frowned upon, which then leads OT architects to start using the Internet as a medium to send data to cloud-based processing and analytics. Certain aspects of real-time feedback and monitoring, which are time sensitive, greatly benefit from the movement of data processing to the cloud. Attacks on ICSs now push enterprises to extend their perimeter and subsequently move to Zero Trust architectures, which they experience on all cloud platforms. Architects must consider the type of endpoints, the traffic they process, and the subjects that will access these OT endpoints, and then design security access around that context. This adds complexity to a cookie-cutter design that most architects are accustomed to, and hence people tend to stay away from Zero Trust.

■ **Dynamically changing perimeter of a distributed enterprise:** With the changing security perimeter, protection is no longer placed only at the edge. Gone are the days you could buy a firewall, place it at the Internet perimeter, and relax. Security has changed its focus. Threat modeling is now one of the many inputs to a larger architectural shift, which is asset-based contextual security. Some people think that asset-based protection and threat-based protection are different. In reality, they are complementary to each other. Assets, especially data, do not always reside on-premises. The cloud bubble has burst, and now everyone has a piece of their pie in the cloud. In comparison to on-premises devices and data, the cloud is flexible and scalable. You might have multiple workloads in availability zones you did not expect. How would you scale your security architecture when your workloads are scaling leaps and bounds? That's when we realize that rather than securing the perimeter against threats, we need to secure the assets. Protect the data rather than protecting the perimeter. Security must be deployed closer to the assets rather than their perceived perimeter. Of course, by extension, assets can be anywhere, and hence security must scale and follow data. This becomes a challenge in the current security landscape. A distributed enterprise needs a more flexible security strategy, and that is where Zero Trust gets lost in translation, because enterprises assume that Zero Trust is more restrictive. On the contrary, Zero Trust as a concept is more liberating if done right and gives more power to the consumer and the custodians.

■ **Rigid application architecture:** If you ask customers to give a full application dependency and port mapping for applications in the network, very few of them will even think about providing an answer. Applications are so vast and beyond an operational scale that customers consider application architecture only when they are decommissioning an application or installing a new one. Applications also are rarely changed to fit the microservice architecture, which means that even if we do get a monolithic application in place, it is unlikely it will be segmented. For example, monolithic applications would include a front-end web service, the actual application, and a back-end database all in one physical server. Compromise of a web server would essentially compromise the entire application. Enterprises are not open to any changes on an application because they do not want to impact the business. "If something is working, let it work—why break it?" is the common philosophy.

Why is this a roadblock to Zero Trust? The reason is that without knowing the application dependency, which is the "object" aspect of the equation, it will be increasingly difficult to provide the right context because you do not know what level of security you need to provide for the object, especially without taking data classification into consideration. Accessing a database server's administrator details is different from accessing the database itself. Due to application complexity and reluctancy to change application architecture, Zero Trust remains elusive for some.

■ **Legacy infrastructure:** A majority of enterprises shy away from Zero Trust because they are not confident enough that their large number of legacy devices can fit the new strategy. These devices might be considered "legacy" for of a lot of reasons, including the following:

■ Lack of support for modern segmentation methodologies (VRF, group-based policy)

■ Lack of support for enabling a Zero Trust agent

■ Capabilities to perform advanced networking and security functions(active directory integration)

The enterprise might be in the process of migrating these devices or might just not be able to remove or decommission them due to critical applications being installed on them. Either way, enterprises might not see value in Zero Trust if a large percentage of their infrastructure is legacy hardware and outdated software.

Personnel and Related Skillsets

Although it might at first seem unlikely, the lack of the right knowledge of the implementation of various security technologies is a key factor why Zero Trust has not been adopted. Not all enterprises put in the effort to improve the security workforce, and due to the lack of right skillset, senior leadership is not comfortable with moving to Zero Trust, especially when a lot of effort needs to be spent in configuring and maintaining a centralized policy. You will learn the importance of showcasing strength in the Zero Trust team selected to strategize and deploy the solution in Chapter 7, "Zero Trust Avengers, Assemble!"

Consider the following example of a simple enterprise policy requirement: An enterprise wants to block Facebook chat across all headquarters and branches. The enterprise tries to block the chat itself and cannot find an expert in the product. This leads the enterprise to block the application entirely. Users get agitated and complain that they want to access Facebook, and the enterprise replies sternly that it is not part of the acceptable usage plan. Users then download a proxy application and surf Facebook. Not only has the organization repelled its own users, but it has also motivated the users to find alternative and often dangerous paths to browse the Internet. The enterprise tries to block the proxy application as well but fails to do so. The specific enforcement point does not have the technical capability to effectively block proxy applications. When the policy cannot be implemented, it becomes a detractor for the operations engineers. They then move to configure a more contextless policy and end up blocking Facebook for everyone. The overall lack of hardware support and personnel skillset could lead to multiple back doors in the enterprise. This type of static rule-based access then becomes detrimental to the user experience, impacting all aspects of the overall business, which leaders do not like and tend to avoid.

If the deployed technology is complex to learn, it is unlikely there will be efforts spent to understand it. The bare minimum knowledge will be acquired by personnel to perform just the basic operations. The highlight of this segment is not to bring about general reluctance to learn. The point here is that your Zero Trust strategy might be simple, but the technology selected to implement the strategy might not. When the deployment and operations engineers do not have the right skillset to translate your Zero Trust strategy to security mechanisms, they choose to avoid it altogether.

Process and Maturity

The following deterrents listed consider reluctance to change existing processes and focus on overall lack of orchestration and automation:

- **Orchestration and automation maturity:** A common deterrent to Zero Trust adoption is the lack of automation and security orchestration possibilities. Currently, there are all forms of behavior analysis, threat modeling, information processing, and automation and orchestration that can serve the business to provide an accurate view of the network. Many years ago, before risk-based talks and network visibility were considered viable for enterprise security, "SOAR" referred only to birds flying in the sky, not *security orchestration, automation, and response.* With greater automation came greater visibility and context building. Network access control (NAC) became much simpler to monitor, and as threats increased, Zero Trust began to look more enticing with its dynamic access control and constant monitoring. Yet, when various enterprises were asked as to why Zero Trust was still not being considered, their answer was that the concept itself was attractive but they did not have the kind of visibility needed in the network to effectively deploy and build policies. When you look at Chapter 6, "Understanding Zero Trust Maturity," you will learn that automation and orchestration are not mandatory to begin the Zero Trust journey. The most difficult step in Zero Trust is to begin. Overall, the lack of SOAR maturity inhibits Zero Trust adoption.

- **Nonuniform policy and lack of integration:** Not all enterprises have taken visibility and flow monitoring seriously, which indicates that decisions are made based on very broad conditions such as IP and geolocation. Similarly, for enterprises to really make use of Zero Trust, they need to have an entity that integrates with all the information sources, which is currently lacking in most enterprises. It is of late that open source intelligence (OSINT) is being considered as a key information source for security intelligence.

Currently, some of the leading banks do not implement an effective security operations center (SOC). Neither is there a basic NetFlow ingestion mechanism. There is negligible orchestration, which means each device is its own kingdom. Although a firewall might allow a VPN user at the edge, the same user might be blocked at the application level. Distributed decision points lead to multiple policing, which can be detrimental to the overall flow. A simple analogy to highlight this is getting caught by the police at various checkpoints. Imagine you're traveling on a road and you forget to wear your seatbelt. You get caught at a checkpoint and the police issue you a ticket for US$50. You travel a little further and you reach another checkpoint and that police officer issues you a ticket for US$100. You contest the charges saying that you've been fined already, but the police officer at this checkpoint tells you that this checkpoint has its own fine-grained policies and the penalties are separate. This is essentially how an end-to-end flow suffers when policed at different siloed security enforcement points.

■ **Lack of asset inventory:** You cannot design a trust flow if you do not know what you need to protect. Similarly, you cannot have a Zero Trust model if you do not know who can access the objects in your network. This is another key reason why a lot of customers do not move to a Zero Trust model. The asset inventory is just not in place, and setting one up could take months or even years to reach a usable point.

Of course, this should not be a deterrent because Zero Trust is not all about asset inventory. Asset inventory is important when creating Zero Trust policies, but is not a prerequisite.

The truth is that many leaders perceive what they lack as a reason to avoid embarking on the Zero Trust journey because they foresee a roadblock due to these missing controls. This is, however, not true, as you will learn in Chapter 8, "Building a Zero Trust Architecture". Zero Trust essentially has minimal technical prerequisites to start deployment, but it does need a robust strategy and a road map to direct your efforts. What you also need is to identify your business and technical catalysts, showcase to your promoters the return on investment (ROI), and prove to your detractors that the strategy can align to all business needs. You also need to identify a starting point for the enterprise as you build the tactical and strategic road map.

Segmentation is needed to be able to isolate various segments in the network and build Zero Trust policy. That is our final goal, but is it a prerequisite to begin exploring cloud firewalls ? No. As long as you have a waypoint on your road map, all paths lead to the same destination, which is to create and manage an effective Zero Trust policy to control access of a subject over an object.

Peeling the Zero Trust Onion

The true intent behind Zero Trust is fairly simple; however, due to the market jargon and essentially infinite interpretations of its definition, the concept of Zero Trust has truly become a kaleidoscope. The more you twist the scope, the more varied images you see, but the subject being viewed is the same.

In general, there are multiple ways to approach a specific subject. You could define it at its core and then expand its interpretations. A metaphor to understand this imagery is the spreading of ink on a napkin. Once the drop hits the napkin, it is not easy to control the direction in which the ink spreads, and we lose the overall intention of where we started. This book begins exploring Zero Trust from the outside-in, by essentially looking at various perspectives of security leaders on Zero Trust Solutions and picking the right keywords to define the true intent of the narrative. The analogy is entering a maze from any direction and slowly making your way to the prize. A maze analogy is apt in the interpretation of Zero Trust because there is a plethora of information on what Zero Trust represents, but far less information on what it really takes to design and achieve a Zero Trust policy.

Similar to the maze analogy is the "peeling of an onion" analogy, which is also an outside-to-inside strategy. You peel off layers of the onion to reach the core of the bulb where the true taste of the onion lies. Based on this approach, you will explore market interpretations; see how the key terms have been identified by Forrester, Gartner, and the NIST documentation; and then finally be able to grasp the core intent of Zero Trust as a concept. This will also help reinforce certain tenets of Zero Trust, which we must keep in mind when designing Zero Trust architectures. Peeling off various layers of the Zero Trust onion will help you understand what actually lies at the heart of the concept. Figure 2-1 illustrates this concept.

Zero Trust Market Interpretations

This layer represents all the market jargon around Zero Trust.

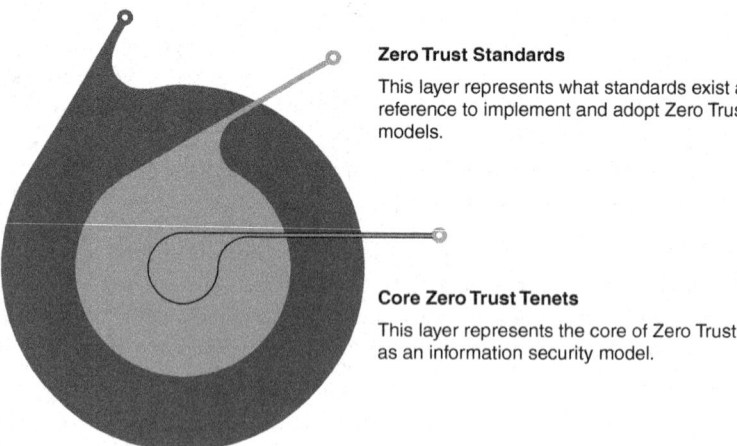

Zero Trust Standards

This layer represents what standards exist as reference to implement and adopt Zero Trust models.

Core Zero Trust Tenets

This layer represents the core of Zero Trust as an information security model.

Figure 2-1   *The Outer Layer of the Zero Trust Onion*

Beginning at the outer layer of the onion, you will analyze the product and vendor interpretations of the framework. Without going into details of specific product sales talk, each Zero Trust leader's definition of Zero Trust will be unfolded. Some of these enterprises have created their own interpretations of Zero Trust based on their strengths, while others have adopted common standards like NIST and Forrester. Each enterprise has strived to maintain the core tenets of Zero Trust with great finesse. In the end, the tenets remain the same but are hidden from plain sight. Our goal is to truly see what Zero Trust is.

Referring to the "Zero Trust eXtended Ecosystem Platform Providers"[4] report from Forrester, we will use some of the key vendors listed as a starting point, and you will explore their perspective on Zero Trust, their products, and how they align the overall Zero Trust tenets to their product itself. Please note that the report referenced is a point-in-time statistic to capture some leaders and their challengers in the Zero Trust ecosystem and observe their interpretation of Zero Trust. Please refer to the latest "Zero Trust eXtended Ecosystem Platform Providers" or its equivalent report for the relevant details.

Illumio's Take on Zero Trust

Illumio focuses on *segmentation* and an *assume breach mentality*. It defines the need as follows:

> "Effective containment begins with an 'assume breach' mindset, which, in turn, drives a least-privilege approach to building security controls—this is Zero Trust security."[5]

Illumio's strategy includes the following:

- Visibility everywhere
- Least privilege access
- Adaptability and consistency to provide uniform policy irrespective of location
- Proactive posture

CrowdStrike's Take on Zero Trust

CrowdStrike's definition of Zero Trust focuses on *de-perimeterization* and then builds on NIST 800-207 with more focus on *continuous verification*.[6] CrowdStrike defines Zero Trust as follows:

> "Zero Trust assumes that there is no traditional network edge. Networks can be local, in the cloud, or a combination or hybrid with resources anywhere as well as workers in any location."

The three key focus points are as follows:

- Verify access and continuously validate.

- Segment the network and minimize the blast radius.

- Automate context creation so that it is dynamic and not a point-in-time activity.

Cisco's Take on Zero Trust

Cisco approaches Zero Trust with the following principles:[7]

- Establishing trust in every access request, independent of location

- Securing and enforcing trust-based access across all applications and networks

- Continuously verifying trust

- Responding to change in trust

- Extending security across a distributed workforce and workload

They access systems, regardless of location. A Network and Cloud pillar focuses on secure access to the network for any device based on the security posture of the device. This pillar also considers multicloud architectures and the controls needed to maintain a uniform policy across all cloud platforms. Cisco also has an Application and Data pillar that focuses on preventing unauthorized access within application environments irrespective of where they are hosted, along with a key focus on data classification and protection.

As an extension, Cisco focuses on offering integrated analytics, automated decision-making, and segmentation controls across the entire infrastructure.

Palo Alto's Take on Zero Trust

Palo Alto approaches Zero Trust with a focus on *continuous validation*.[8] It secures the network as follows:

> "... eliminating implicit trust and continuously validating every stage of a digital interaction. The prime focus being always verify an actor and never trust irrespective of where it is coming from. It gives weightage to strong authentication methods, leveraging network segmentation, preventing lateral movement, and simplifying granular, least access policies."

Palo Alto approaches Zero Trust with three aspects, along with strong asset validation:

- Device integrity and the principle of least privilege

- Removing implicit trust

- Continuously monitoring at runtime to validate the right access

All devices and users must be covered under the zero-trust approach. You should not have segments that are unaccounted for. There are Zero Trust models to support legacy deployments as long as they are in the enterprise inventory.

The focus is to continuously monitor access and scan for malicious behavior.

Zscaler's Take on Zero Trust

One of the largest market presences of Zero Trust, Zscaler defines Zero Trust by considering the "any resource by anyone from anywhere concept."[9] Zscaler defines its Zero Trust Exchange as follows:

> "… enables fast, secure connections and allows your employees to work from anywhere using the internet as the corporate network. Based on the zero trust principle of least-privileged access, it provides comprehensive security using context-based identity and policy enforcement."

Zscaler extends the concept by focusing on the fact that no user or application should be inherently trusted and that every entity attempting to access the network is malicious. Trust is established based on user identity and its extended attributes. Services and applications are expected to be protected, even if they are across networks that are untrusted, such as the Internet. Zero Trust facilitates secure connectivity for users, their devices, and desired applications using contextual policies over any network.

Zscaler considers three aspects when it comes to Zero Trust:

- Attribute-based access control
- Flexible agent-based/non-agent-based deployment with senior leadership support
- Ensuring a process is in place to be able to continuously monitor the access provided for each actor via security information and event management (SIEM) and endpoint security

Forcepoint's Take on Zero Trust

Forcepoint approaches Zero Trust by replacing implicit trust with explicit permission every time.[10] The aspects that Forcepoint focuses on are as follows:

- Preventing lateral movement of threats
- Controlling access and usage of data
- Continuously monitoring for change in risk profile

Forcepoint has multiple deployments as well, which may include agent or agentless deployments, but the core selling point for Forcepoint is to get rid of the VPN with its Zero Trust Network Access.

Cloudflare's Take on Zero Trust

Cloudflare highlights *de-perimeterization* as well as *identity verification*. The definition states the following:

> "... strict identity verification for every person and device trying to access resources on a private network, regardless of whether they are sitting within or outside of the network perimeter."[11]

Microsoft's Take on Zero Trust

Microsoft extends explicit verification to include threat factors as well as general enterprise governance as context factors into the policy engine.[12] It focuses on the following:

- Explicit verification

- The use of multiple data points to build context

- The use of least-privileged access

- Limiting user access with just-in-time and just-enough-access (JIT/JEA), risk-based adaptive policies, and data protection to help secure both data and productivity

- Assuming a breach, minimizing the blast radius and segment access

HashiCorp's Take on Zero Trust

HashiCorp mandates identification of machines as well as users so that controlled access can be established from users to machines. HashiCorp's approach is more *multicloud-centric*[13] and begins with "enabling scalable, dynamic security across clouds." HashiCorp focuses on the following:

- Improving the enterprise security posture

- Reducing the likelihood of a breach

- Accelerating secure multicloud adoption

A Common Pattern Emerges

By now, you should be seeing a pattern with some common themes. Whereas different vendors sell various products, their Zero Trust definition is tailored to the product they sell, but every definition has common attributes that come down to the core Zero Trust principles, which we will discuss in detail when we look at National Institute of Standards and Technology (NIST) definitions in Chapter 3, "Defining Zero Trust." There might be certain modifications, but the essence remains the same.

If one had to visualize the entire set of Zero Trust keywords as a word cloud, some key topics will stand out broadly, and these topics will inadvertently be the core of Zero

Trust, as we will observe in Chapter 3, "Defining Zero Trust.". Figure 2-2 shows a visualization of this concept.

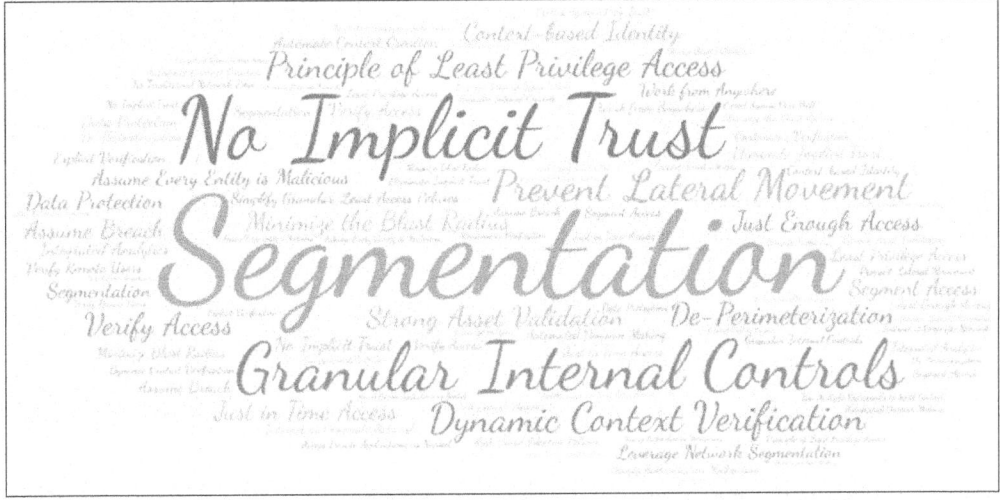

Figure 2-2   *Visualization for Zero Trust Concepts*

At this point, if a list of key Zero Trust terms from all vendors was created, the following would be the top five:

- Continuous visibility
- Context-based access
- Segmentation
- Breach assumption
- Removal of implicit trust

These will eventually be the key drivers to enable enterprises to begin their Zero Trust journey. In Chapter 3, "Defining Zero Trust", we will peel back the inner layers of the onion and finally reach the core to see what exactly Zero Trust is all about.

The Interview

[Glenn stops explaining Zero Trust from a vendor's perspective and takes a pause to make sure his audience has grasped what he is trying to explain.]

Glenn: With this introduction to existing Zero Trust solution providers and how they view Zero Trust, you should be able to gauge some key aspects of what we perceive as the right approach to understanding Zero Trust.

In summary:

- Zero Trust has been in the market for a long time in various formats.

- Multiple reasons have slowed down Zero Trust adoption. A major chunk of these causes being mindset, architectural and technical challenges, skillset concerns, and lack of mature processes.

- Our approach to security has been reactive, where breaches lead to security augmentation. This is contrary to strategy being driven by risk and assets.

- Multiple vendors provide technology that can achieve a Zero Trust architecture, but there are key topics common across all product and service offerings related to Zero Trust. Some of them include:

  - Continuous monitoring and verification of access

  - Dynamic granular controls based on context derived identity

  - Need for network and application (macro and micro) segmentation to reduce the blast radius

  - Getting away from the "implicit trust" paradigm

  - Freedom of movement with controlled tailored access over restricted movement with standard access

  - De-perimeterization of the enterprise

  - Application, flow, and operational visibility

Mr. Chen: For a concept that is just picking up, I see that a lot of vendors have adopted it into their products. I see the pattern, and I understand where the security community is moving toward, but I am still not clear on what really is the value-add of Zero Trust, especially, how it augments my strategy and infrastructure transformation vision. Why can I not just pick my current compliance reports and just implement the security controls? When William, our security lead, brought up Zero Trust, it was difficult to wade across this sea of jargon available on this topic. How is Zero Trust any different from other strategies available for us to adopt? What exactly makes Zero Trust unique in its tenets?

Glenn: I'll be glad to walk you through some of the key tenets of Zero Trust so you have a better picture of what Zero Trust really is at its core.

Endnotes

1. "Zero Trust Shouldn't Mean Zero Trust in Employees," https://www.darkreading.com/endpoint/zero-trust-shouldn-t-mean-zero-trust-in-employees

2. "The Zero Trust eXtended (ZTX) Ecosystem," https://www.forrester.com/report/the-zero-trust-extended-ztx-ecosystem/RES137210

3. "Take Security To The Zero Trust Edge," https://www.forrester.com/blogs/take-security-to-the-zero-trust-edge/

4. "The Forrester Wave: Zero Trust eXtended Ecosystem Platform Providers, Q3 2020," https://www.forrester.com/report/The-Forrester-Wave-Zero-Trust-eXtended-Ecosystem-Platform-Providers-Q3-2020/RES157494

5. "Zero Trust—The security paradigm for the modern organization," https://www.illumio.com/solutions/zero-trust

6. "Zero Trust Security Explained: Principles Of The Zero Trust Model," https://www.crowdstrike.com/cybersecurity-101/zero-trust-security/

7. "Cisco Zero Trust Architecture Guide," https://www.cisco.com/c/en/us/solutions/collateral/enterprise/design-zone-security/zt-ag.html

8. "Zero Trust with Zero Exceptions," https://www.paloaltonetworks.com/zero-trust#:~:text=What%20is%20Zero%20Trust%3F,stage%20of%20a%20digital%20interaction

9. "Work From Anywhere," https://www.zscaler.com/solutions/modern-workplace/work-from-anywhere

10. "Continuous Zero Trust Security," https://www.forcepoint.com/use-case/zero-trust-security

11. "Zero Trust Security | What is a Zero Trust Network?" https://www.cloudflare.com/en-gb/learning/security/glossary/what-is-zero-trust/

12. "Embrace proactive security with Zero Trust," https://www.microsoft.com/en-us/security/business/zero-trust

13. "Zero Trust Security," https://www.hashicorp.com/solutions/zero-trust-security

Chapter 3

Defining Zero Trust

In this chapter we will peel back more layers of the onion to get to the core of what Zero Trust is and, more importantly, what it is not. An unconventional but effective approach to learning a concept that could change our perception about a well-established design is to spend time understanding what that concept is *not*. Over many talks and discussions with colleagues, I have noticed that changing a presumption about a subject is almost impossible because every corner of the conversation ends up with one of us telling the other "that's not true" or "that's not it." Approaching the discussion by prefacing it with "what it is not" helps strengthen the perception of the concept being discussed.

Figure 3-1 illustrates the layer that focuses on amalgamating all the core tenets into a reusable reference architecture framework. You are now one step closer to understanding the core of what Zero Trust is.

Zero Trust Market Interpretations

This layer represents all the market jargon around Zero Trust.

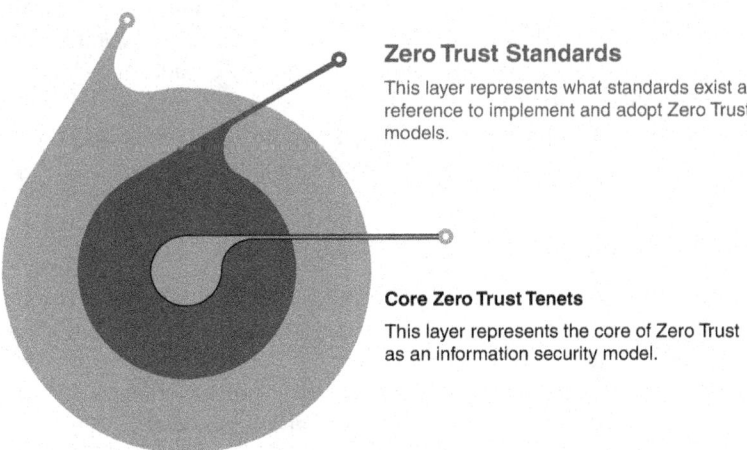

Zero Trust Standards

This layer represents what standards exist as reference to implement and adopt Zero Trust models.

Core Zero Trust Tenets

This layer represents the core of Zero Trust as an information security model.

Figure 3-1 *Peeling the Middle Layer of the Zero Trust Onion*

In Chapter 2, "The Zero Trust Kaleidoscope," we explored how vendors consider and incorporate Zero Trust concepts into their products. In the next section, you will understand what Zero Trust is *not*. This is a key step in peeling the onion and reaching its core.

Zero Trust Is Not...

A Product

Attempting to sell Zero Trust is like trying to sell the Biba model, which is an information security construct used to make sure that integrity is maintained when data flows across various security levels. It is a well-established concept, not a marketable product or software. Products incorporate this concept to make sure the overarching architectural principle of maintaining integrity is achieved. Zero Trust is no different. If a vendor tries to sell you Zero Trust, they are not, because Zero Trust in all aspects of its definition cannot be sold. Zero Trust is not a product. Essentially, what vendors sell are products or security controls that will help you achieve a Zero Trust network architecture (ZTNA) that provides Zero Trust Network Access (also ZTNA). Zero Trust network architecture is a network design concept that will help provide Zero Trust Network Access, which Forrester considers as a technology or mechanism. For example, Cisco sells next-generation firewalls that can be placed to protect the network, but the policy to configure the firewall must come from the Zero Trust policy. To achieve the Zero Trust policy, a Zero Trust architecture must be created, which in turn requires a senior leadership mindset change. Vendors such as Cisco have expertise and a deep understanding of Zero Trust to help customers transform their networks and begin this Zero Trust mindset change. Remember that even if you have the best technology that enables Zero Trust, you still need skilled personnel to create the right policy and implement effective processes to enhance the created policies with feedback and continuous monitoring. Otherwise, the offer is still a product and not a solution.

To summarize, while technology and mindset are interrelated, Zero Trust itself is not any of the above and is more of an **information security model** and a **philosophy**. It needs a lot of coordinated effort from the IT infrastructure, security, and governance teams in the enterprise and must be strongly supported by senior leadership. To achieve Zero Trust, multiple solutions must be selected to align with the business strategy, which is why understanding the business use cases is pivotal. Chapter 4, "Always Start with 'Why,'" emphasizes the importance of alignment.

A New Technology Fad

As discussed earlier, Zero Trust is not new. All tenets of Zero Trust were inherently part of a larger security framework, hidden in plain sight. However, these controls were siloed and had lesser impact when implemented separately. For example, asset management on its own is an important activity; however, unless the asset inventory is fed into a micro-segmentation program, all you will be left with is a perfect inventory. Segmentation and multifactor authentication (MFA) have been the primary focus in the security domain for many years. Unfortunately, they have been either a reactive measure to achieve compliance or the sole methodology to enhance identity and role-based control. When adopted together, they have more meaning and their purpose aligns to what the business needs. Zero Trust is the glue that joins all the disjoint security siloes.

Achievable Only with the Best Hardware

By extension, since Zero Trust is not a product, it cannot be achieved by buying a lot of hardware or even the best hardware. Hardware can provide security control to support your strategy, but hardware alone cannot help an enterprise achieve Zero Trust. Zero Trust is an amalgamation of multiple technologies and solutions. It is a change in the way you perceive access control. It means choosing and understanding the people who enter your home rather than installing a security camera prepopulated with a static list of names of people who can enter.

Achievable with Just Identity Management or Multifactor Authentication

Whereas identity management is key for implementing Zero Trust Network Access, it is not the final activity. In all likelihood, identity and access management (IAM) is already implemented or is on a road map to be implemented in the enterprise. Similarly, multifactor authentication is one of the most common identity-validating strategies deployed by most enterprises, but it does not mean that Zero Trust is achieved or is close to being achieved with the implementation of just MFA. Zero Trust needs constant visibility and feedback into the policy engine. Enhancements such as posture checks and device health must feed into the overall identity context. Similarly, without segmented applications with the right security controls, MFA is just one part of a larger vision.

An Outcome of Segmentation

Zero Trust is rarely just about segmentation. It is usually impractical to implement a strong segmentation policy without dynamic identity-based access and constant monitoring. This would end up becoming an operational nightmare because operation engineers would need to manually validate false positives and negatives based on how stringent the policy is. It is usually much more effective to make sure that multiple solutions can fit together to provide complete Zero Trust.

A Security Project

Two things are combined here that you must consider. It is unlikely to complete a Zero Trust implementation with a standard design project. It is a journey to take, a destination to arrive at, and an entire framework in itself to build and maintain. Parking it as a security team's headache is usually a bad idea and results in more inter-team clashes. A Zero Trust initiative is a joint venture by multiple teams and team leads, which you will learn about in Chapter 7, "Zero Trust Avengers, Assemble!" The important takeaway is to validate that Zero Trust is not one team's project responsibility. It is an entire group of teams' effort at maintaining security without giving up convenience.

A Restrictive Regiment

Many consider Zero Trust as a means for the enterprise to tighten the screws on security and make life difficult for its employees. That is far from the truth. In reality, Zero Trust is liberating. It allows you to give more freedom to the subjects without compromising on security.

Now that you know what Zero Trust is not, let's begin defining what it is.

Zero Trust Standardization

The discussion in this section will relate to how Zero Trust as a concept has been standardized. CISA, NIST, and Forrester have standards created for Zero Trust that are reference points for most enterprises embarking on the journey.

Forrester refers to Zero Trust as an information security model. According to the Forrester paper "The definition of modern Zero Trust,"[1] the definition of Zero Trust can be summarized as "an information security model that denies access to applications and data by default. Zero Trust advocates these three core principles: all entities are untrusted by default, least privilege access is enforced, and comprehensive security monitoring is implemented."

According to NIST Special Publication 800-207,[2] Zero Trust is a model that "provides a collection of concepts and ideas designed to minimize uncertainty in enforcing accurate, least privilege per-request access decisions in information systems and services in the face of a network viewed as compromised."

These are two major definitions based on standards. Notice how certain themes keep repeating themselves. You should also observe how most enterprises have extracted certain key themes from the Zero Trust definitions to create their Zero Trust product strategy. Forrester focuses on denying access to applications and providing access only based on context (attribute-based control). NIST focuses on assuming the network is compromised and restricting access per request (assume breach). This means that if the same subject requests access to the same object, the access must be validated again by a policy engine.

Forrester also extended its original Zero Trust definition with the Zero Trust eXtended Framework,[3] or ZTX, which puts data at the center of protection, surrounded by people, workload, network, and devices. Here are some key points to note:

- Data is the crown jewel that needs protecting. Data sits at the center of the systems that will implement Zero Trust.

- The Workloads pillar includes resources that process or essentially handle the data. These should be unique functions in the customer's environment. For example, a web server workload is different from a database workload. These functions should drive the business and segmenting them is key.

- The Networks pillar focuses on segmentation at the network level and covers ubiquitous access from multicloud architectures.

- The People pillar covers accurately segmenting users and managing them to provide accurate identity insights. Implementing MFA or SAML is considered within the People pillar.

- The final pillar is the Devices segment, which focuses on identifying the assets in the environment and isolating them.

Cisco implements its vision of Zero Trust with similar pillars: User and Devices, Network and Cloud, and Application and Data, which align with Forrester's five pillars, as illustrated in Figure 3-2.

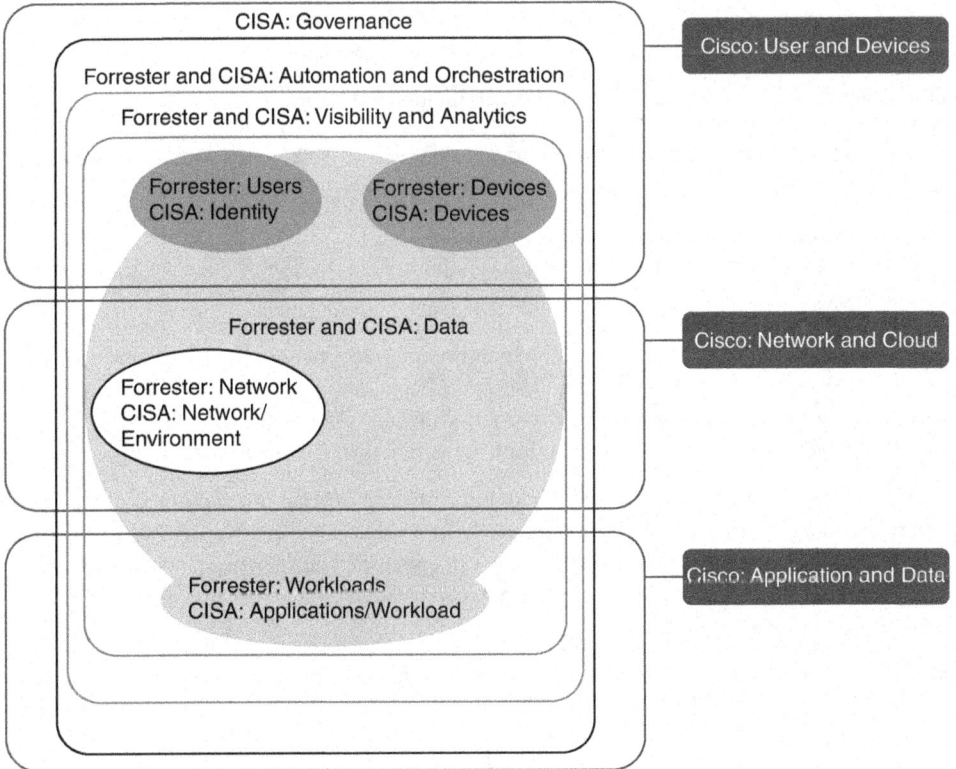

Figure 3-2   *Cisco's Overlay of Zero Trust*

When the Cisco Zero Trust overlay[4] is considered, it fits right in with Forrester's ZTX ecosystem as well as CISA's Maturity Model v2. A subject's context is defined by the user attributes as well as the user's endpoint device. Cisco considers that an enterprise's strength and its weakest link is the workforce and hence combines a user's attributes with their device information. The network depicts the environment or transports the subject needs to use. It depicts the critical communication channel between the subject and the object and maps to the access methods (VPN, LAN, branch, cloud) that a subject uses to access their workload. Essentially, the Network and Cloud pillar encompasses the virtual or physical networks or environments from where the subject accesses the object. The requirement for the networks pillar is to maintain a secure communication channel between the subject and the object.

We must appreciate the fact that the phrase "enter the enterprise boundary" is not used anymore because an enterprise boundary is no longer a fixed enclosure. The enterprise is where its data and assets reside. This could be on the branch, on the cloud, or even on other endpoints. The Application and Data pillar represents the objects or applications that are accessed by the subjects and the data processed and stored by these applications. Similar to the enterprise boundaries, workloads are no longer restricted to only a data center or server farm. They exist where the business needs them to be and must be secured with the same access control that would be implemented in a server farm.

Other tenets of the Forrester ZTX framework include automation and visibility:

- Visibility is getting as much information as needed to make better security decisions and provide the right contextual access.

- Automating the existing manual process of evaluating context for access is another security mechanism within Zero Trust. Each implementation of security policy, incidents, and the response should be automated.

CISA adds a final layer of Governance, which is sometimes overlooked but is even more important because it is critical to align the Zero Trust strategy to business. The topics discussed are some common tenets of Zero Trust when taking into account the various standards. We will discuss in detail some of these tenets in the next section.

Our focus here is not to reinvent the wheel but rather to step on the shoulders of giants and try to simplify the core aspects of Zero Trust so that when anyone wishes to refer to the standards and policy documents, the concepts make more sense, and they are in a better position to enable and support the enterprise.

All standards covered in this section greatly help to answer the question of "how" to implement a robust Zero Trust architecture; however, this is not the whole story. Before you can decide how to implement Zero Trust, you need to understand various aspects of Zero Trust as an advocate and practitioner. You need to convince a larger set of enterprise business groups to make sure they align with your vision as well. Through this book, you will attempt to answer the questions of "Why should you?" and "What will it take?" Let's peel back the last layer of the onion.

The Core of the Zero Trust Onion

This is it. The core of the onion, the center of the maze. This is where the book will bridge the gap between knowing and understanding. After all the resources you have considered until now, this section will help define Zero Trust at its core. To make it simple to understand, I have included a "travel size" version to explain to someone without a security background and a more detailed version for seasoned practitioners. Based on the security acumen of your audience, you can choose to start with the smaller version and build up to the larger more detailed definition. Figure 3-3 illustrates the inner layer of the Zero Trust onion.

Zero Trust Market Interpretations

This layer represents all the market jargon around Zero Trust.

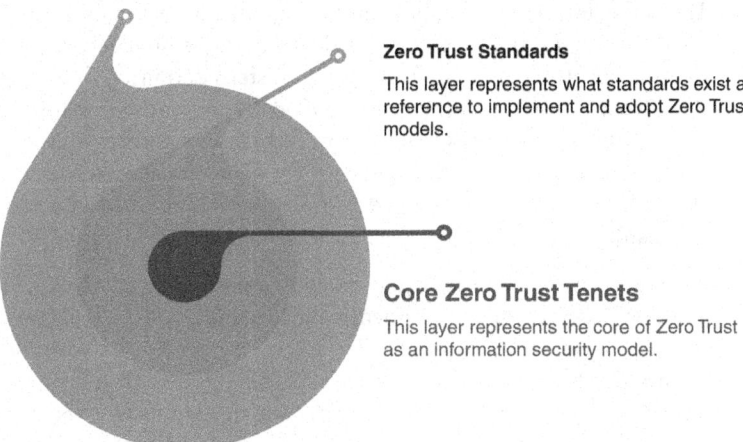

Zero Trust Standards

This layer represents what standards exist as reference to implement and adopt Zero Trust models.

Core Zero Trust Tenets

This layer represents the core of Zero Trust as an information security model.

Figure 3-3   *The Inner Layer of the Zero Trust Onion*

> **At its core, Zero Trust is all about continuously securing how a subject, at any location, from any device, securely accesses any enterprise-owned object and making sure the subject is authorized to access the target object over a secure communication channel.**

This would be a good start on how to explain Zero Trust to someone who is not security savvy. Zero Trust is practiced by everyone every day. Your friends and a stranger do not get the same access to your house. The same applies when your friend becomes a stranger. We, as humans, perform contextual access control in every aspect of our lives. The concept is simple and everywhere around you, but from an organization's perspective, it is important to detail each definition so that the organization can understand how to prioritize the required activities to achieve the final vision. Based on the multiple aspects of the Zero Trust paradigm discussed so far, here is a detailed definition of Zero Trust to consider:

> **Zero Trust is an information security model that provides an authenticated subject need-to-know, least privilege access to an object within the enterprise asset inventory scope. The access provided is dynamically generated and validated at the time of an access request based on granular parameters that add to the overall context of the access and include, but are not limited to, identity, location, risk profile, governance, and compliance. Granularity in policy is facilitated with effective user, network, and application segmentation, and a resource access verdict is facilitated by a policy engine that ingests information from multiple information sources that are continuously monitoring contextual changes to make sure the access requested is authorized just in time.**

The preceding definition covers all the tenets of a Zero Trust model, and the goal is to dissect this definition to make sure the tenets are embedded into your understanding of the concept. To begin with, consider Forrester's definition of Zero Trust and validate that Zero Trust is an information security model, similar to the Bell-LaPadula model or Biba model, which are security models covering various aspects of security, such as confidentiality and integrity, respectively. Zero Trust encompasses a subject accessing an object over a secure channel, but the key differentiator here is the understanding of an authenticated and authorized subject and an asset that is accounted for.

As an enterprise, it would be unwise to be considering any subject that is not authenticated. Zero Trust extends the authentication boundary and floats the idea that authentication alone is not enough to provide access. You need to be an authenticated subject who is authorized to access a specific object. The object must not be an entity you do not have visibility into because if the enterprise does not know the object, it likely does not know the extent to which access must be controlled to the object. For example, if you are not aware of a web server in your environment, you would not know the type of data it handles or the type of authentication that the web server must provide. You also will not know which applications or users access this web server and its function in the network. By extension, you would not be able to segment it into the right zone. Hence, asset inventory is an important aspect of Zero Trust because it helps you build the initial scope to which your Zero Trust architecture will apply.

The second aspect relates to the policy engine. The policy engine must ideally be a centrally managed logical policy entity that can process information from various sources and make a contextual decision to control access to resources within the asset scope. It must possess the capability to dynamically validate access each time a request is made.

The final key factor is to continuously monitor the security posture of the subjects and objects. Active Directory group memberships might change, and objects might start processing highly confidential information, making them more susceptible to attacks and thus requiring stricter control. A continuous monitoring infrastructure coupled with security automation and orchestration is key to ensuring any change in security context is captured and fed into the policy decision-making infrastructure automatically.

Automated responses to authorization or posture changes must be in place to ensure access is processed just in time.

Effective implementation of Zero Trust increases your general security posture, as you are now changing your protection strategy from static perimeters to dynamic boundaries and focusing on providing access based on identity rather than location. This will force enterprises to strengthen their security posture with stronger identity management, effective segmentation, and, in the end, a secure environment to promote and enhance the business.

Are We Crying After Peeling the Onion?

Now that we have peeled the Zero Trust onion, it is time to understand that an onion in its purest form is still only a vegetable, similar to how the Zero Trust model is still only a model and framework. Onions must be cooked with other food items to bring in better flavor. It is the base for many dishes and an additive that brings value to the main dish. Similarly, Zero Trust must be used in concert with other security best practices to make the best use of its principles. Essentially, there is no single interpretation of using Zero Trust; it must always be tailor-made for the enterprise that is consuming its benefits.

Unlike an onion, which makes us cry more as we go deeper, getting at the core of Zero Trust will give you more understanding of what your enterprise is lacking in the form of security controls, skillsets, and processes. It is not a complex set of principles to follow or a difficult technology to master. It is a simple concept whose interpretation completely depends on how the enterprise wants to proceed.

Rethink Security: A Common Breach Scenario

Security has been supporting networks forever, but it has gained traction recently because of the increase in the complexity of breaches and larger focus on security as a business driver than a compliance initiative. Exfiltration is no longer about getting access to the file that stores the passwords. Attacks range from persistence in the network for years to bringing down nuclear reactors. It is almost always the human factor that essentially leads to a breach, but there are other factors as well. Why is it that if you gain access to one credential you essentially get the keys to the kingdom? Let's consider these common real-life scenarios:

- Access to one user gives access to all users, some of which might be privileged. There is no segregation between privileged and unprivileged users.

- Assuming a privileged user is compromised, this compromised user has access to all services provided by a system, not just the services for which they are administrator. There is no segregation at the service level.

- Assuming that the attacker does get network access to an application, this privileged user imposter has full access to other applications at an application level, too, because the attacker is now part of a privileged group, and control on the other nonrelevant applications is local to those applications rather than a common decision point.

Imagine you are a security vendor, and your network gets attacked. How would that impact your reputation? These are risks that executive leadership needs to consider. Data stored in the data centers and server farms is not just enterprise data. It includes data of the customers who use enterprise services as well, so the responsibility is not just for locally owned data but also for customer data. From a data security perspective, the focus has always been hiding data with large perimeter security. In general, practitioners do not consider granular controls like making data unavailable to unauthorized users or restricting editing rights to specific users. With more enterprises distancing themselves from perimeter-based security and increasing focus on context and identity, overall IT and security strategy needs to change.

In the next sections you will learn about some of these core tenets, and in Chapter 9, "Critical Security Mechanisms for Zero Trust," you will explore some implementation strategies with greater detail.

Concepts and Tenets of Zero Trust

This section covers some of the tenets of Zero Trust that all enterprises must keep in mind when designing architectures and selecting solutions to augment their Zero Trust model. These tenets must also be considered when an enterprise is just beginning to consider implementing Zero Trust. A gap analysis will showcase how certain targeted solutions will help achieve all the relevant tenets.

Zero Trust is not about distrusting everything. This is a common misconception that prevents a lot of companies from adopting Zero Trust. It is a data-centric and identity-centric model that hugely impacts secure workload access. It is a response to declining and ineffective perimeter-based security. With new workload definitions and larger subject scope, the network-centric approach is not effective. Zero Trust is an effective access model for cloud-native or distributed workloads, and it adds value to the current perimeter-based networks as well. Enterprises setting up greenfield data centers or starting with applications that are cloud-native will greatly benefit from this model.

Zero Trust adds flexibility and scalability to the current infrastructure. The reason is that, in general, enterprises are no longer interested in maintaining or tracking physical perimeters. There are very limited entities in the equation for Zero Trust: the subject, the object, and the context. If we can identify our critical assets (personal health information(PHI), personal identifiable information(PII), critical information infrastructure(CII), trade secrets, IP, and so on) and provide the right protection for them, Zero Trust becomes very streamlined and helps provide much more granularity from a policy creation and governance perspective. Contrary to the belief that more granularity can hinder the business, more granularity (if done correctly) can exponentially augment the business and secure it. It is preferred to keep all the protection close to the assets.

Avoid Creating an Implicit Trust

If you revisit the definition of Zero Trust, there is a statement that alludes to controlling access to an object in the enterprise asset inventory irrespective of location. This is one of the most basic requirements of Zero Trust. Initial security designs focused on identifying the most secure location for storing the crown jewels of the enterprise and then securing that location with everything the enterprise can offer. There have been definitions of attack continuum, where identity and access management (IAM) has been considered a mitigating factor *before* the attack. Security controls *during* and *after* the attack were attributed to solutions like firewalls, IPS, and endpoint security. This is a fairly traditional viewpoint to security and has changed over time. It is only recently that IAM has been considered a continuous process and not just an entry-validation mechanism. This means you can allocate access based on your persona (what you have and what you know by extension) rather than where you are. Consider your current access in your workplace. You are allowed to enter your company premises because you are an employee, but once on the premises, you are not given full access to all rooms. You are allowed to access all buildings, but not all rooms. Access to server rooms, hosted data centers, and so on is not granted to all employees. Trust should not be assigned to a location. Of course, practically speaking, it is impossible to completely remove trust from a network segment. That would mean assigning a separate segment per user or application, which is neither feasible nor scalable with traditional on-premises infrastructure. The practical approach should be to reduce the implicit trust zone to an atomic level such that segmenting it further would adversely impact business. Location becomes an attribute rather than a mandatory access control (MAC) criterion. This tenet aligns with the "All communication is secured regardless of network location" tenet according to the NIST definition and "All entities are untrusted by default" tenet according to the Forrester definition.

Another aspect to this tenet is the inclusion of asset inventory and its definition. Assets in the context of this tenet are not just servers and users but all types of end devices that communicate with each other. CCTVs, OT clients, mobile devices, and so on are all considered within the scope of the assets and hence part of the Zero Trust scope. This essentially does away with an already shrinking enterprise perimeter. It also aligns with the "All data sources and computing services are considered resources" tenet according to the NIST definition. Any device that connects to the network and utilizes network resources is also part of the "subject accessing object" equation and hence becomes a part of the overall Zero Trust strategy.

Access Control Based on Dynamic Risk and Context

This tenet is the second and most versatile tenet of Zero Trust. The focus of this tenet is about constantly changing access requirements for the same subject. Imagine the company is going through a critical crisis and has planned to let go of a large number of employees on the same day. The same employees who clock in at 9 a.m. will no longer be employees at 10 a.m. How would the enterprise weed out or cater to a potential disgruntled employee who still has access? Let's say Sally logged in to her servers in the morning and then got fired, but in spite of her Active Directory account getting shut down (which rarely occurs so quickly), the application she uses still provides her access based on her previous authenticated identity. That means a non-employee (possibly disgruntled due to the recent firing) gets full access to an employee asset. This is where Zero Trust brings value by enforcing dynamic access control based on the context of access. A user not in Active Directory loses all enterprise context to access an enterprise resource immediately.

I recall having conversations with several esteemed colleagues about complete removal of location as a decision criterion. The overall consensus has always been that removal of location as a deciding factor is not practical. What is practical is basically using location as additional context. A user in the corporate network should ideally have the same access while accessing resources via a VPN but might have additional posture checks when accessing the same resources from a private endpoint rather than a corporate endpoint. Location becomes a factor, along with the type of endpoint or asset the subject is accessing from. Users from any location must be given access to a specific resource only if they are authorized. The principle of least privilege must be followed, and subjects must be provided access dynamically at the time of request rather than based on historical approval. Strong identity and access management must be implemented along with MFA to make sure that context is tracked dynamically.

Another factor that contributes to the context is the change in posture of the object or service that the subject wants to access. This can be impacted by the data or transactions being processed. Application control also becomes critical, and this must be fed into the policy engine so that the next time a request comes in, the context generated is accurate and up to date. Of course, the subject could be service accounts, machines, IoT devices, or a variety of devices that can request access over an IP network.

Other factors can be queries such as software version, open ports, placement in the network, and history of attacks. Anomalous behavior may also feed into the policy engine to provide an accurate context to be validated continuously. Checking for endpoint posture also adds to the context of access of the subject. The subject may be a combination of a user and a machine with a valid posture. Posture checking must include strict rules like mandating an antivirus software and endpoint detection and response (EDR) software if available. Checking domain membership, blocking USBs, or checking for hard disk encryption like BitLocker are some of the common posturing conditions when posturing devices.

This tenet aligns to "Access to resources is determined by dynamic policy—including the observable state of client identity, application/service, and the requesting asset—and may include other behavioral and environmental attributes" and "The enterprise monitors and measures the integrity and security posture of all owned and associated assets" according to the NIST definition, and it aligns to "Least privileged access is enforced" according to the Forrester definition. The focus is the changing context and in turn the access of the subject.

Authorized Once Does Not Mean Authorized Forever

Access must not be broad static access and must be dynamically extracted from a policy engine each time a request comes through. This is important because one can have all the segmentation and security controls in the network; however, if the user is not constantly challenged to provide identification, you are essentially going back to implicit trust. You are accepting that if a user previously authenticated, then that user is implicitly allowed in the network at all other times, which is wrong on all levels. This Zero Trust tenet basically states that a subject may have a dynamically changing identity and a subject's context is not fixed. Essentially validate if the identity is the same at all stages of the end to end access request. To facilitate this requirement, the context must be monitored and validated each time access is requested. An employee today could be an ex-employee in half an hour. The policy engine must take multiple information points into consideration, preferably in real-time or near real-time. The focus of this tenet is the avoidance of any implicit trust, which we discussed in the previous tenet, by continuously checking the changing context of the subject.

Continuing with the example discussed in the previous tenet, but in a slightly different scenario, let's say Sarah logs in to her HR application using MFA to check for any pending cases for the day. Sally, her friend, asks her for some help. Sally's phone was acting up and she got denied three times by the authentication server. Now her account is locked and she needs an employee to open a case for her. Sarah moves to Sally's workspace and logs in with her credentials and opens a case for Sally. Later, Sally's manager calls her in and lets her know that the company was not doing well and they have to let her go. Sally comes back to her workplace and sits for a while, contemplating how she spent seven years working hard for this company and has been let go like she was nothing. She then realizes that she was unable to log in because she had been removed from the Active Directory. Rage fills her up and in a split-second she decides to wipe out the entire HR database. She logs in to her HR server as Sarah, removes all the entries, and walks off.

This might not be a common scenario, but the catch here is that when Sally attempted to log in to her server again, it did not ask her again if she was authorized to access the server. The session did not time out. This static authorization has been the death of many enterprises, and this is what Zero Trust aims to kill. Dynamic access based on context is now key to what you can access. Of course, there is less value and an operational nightmare waiting for us if we dynamically check for access change without automation and orchestration. Manually checking access for each request is close to impossible. The focal concept in this tenet is the constant and dynamic checking of access.

Access must be constantly validated against an always changing context. This aligns to "Access to individual enterprise resources is granted on a per-session basis" and "All resource authentication and authorization are dynamic and strictly enforced before access is allowed" according to the NIST definition and the "Comprehensive security monitoring is implemented" according to the Forrester definition of Zero Trust.

More Information Is Good Information

Policy enforcement points are key to control data flows, but the core tenet of Zero Trust is to process as much information as possible. Information should be assimilated from local sources such as a security information and event management (SIEM) solution, correlated by a network behavior analysis solution, or extracted from external sources such as open source intelligence (OSINT). The bottom line is that since access is now a privilege, the more accurate a context is, the more granular the policies can be. For example, HR can access HR systems as long as they are authenticated with MFA and access is requested from a postured device. This is an example of a context-based policy. This can be further granularized based on the following criteria:

- **User segmentation:** HR admins
- **Target asset:** HR application server (not a database, not a web application)
- **Time of access:** Work hours (9 a.m. to 5 p.m.)
- **Device posture:** Compliant
- **Authentication:** Via MFA
- **Activity requested:** Adding a new admin user, sending data to an Internet IP

Hence, the policy engine will allow an HR admin to access an HR application server at 2 p.m. from a security posture–compliant endpoint and will allow the same administrator to create an admin user. After 5 p.m., the same user will be able to only read from the application server and not create an administrator account. This is the advantage of information. Multiple sources can provide the needed information to differentiate between a valid flow and an invalid flow, which would otherwise have been implicitly allowed under a broader policy. This is aligned to the tenet "The enterprise collects as much information as possible about the current state of assets, network infrastructure, and communications and uses it to improve its security posture" according to the NIST definition. It is important to remember that data without context doesn't support the business and is no longer a security or business driver. Hence, more information adds context and is always useful. Remember, you cannot protect what you cannot see.

Automate Procedures and Orchestrate Simple Processes

As part of Zero Trust maturity, an enterprise must have automation involved in log collection and ingestion. In the example discussed previously, when Sally is removed from the Active Directory, an automated task should be triggered to remove any existing sessions for Sally so that any subsequent access requests are immediately reauthenticated. This is a proactive approach to maintaining a dynamic context rather than using an expiry-based reauthentication. Traditionally, most security deployments would use a time-based context, and a reauthentication would occur typically after 30 minutes. This is reactive and suboptimal. A more proactive approach is simply to use exposed APIs for applications and keep feeding in new information as and when it changes rather than waiting for the policy engine to ask for detailed information.

Security must be deeply embedded into the entire ecosystem to be able to add dynamic context for all subjects and objects involved. There must be a means to proactively share information either via an API or via other integrations. User lifecycle management has some factors that can directly trigger information exchange between policy information points and policy decision points. Having one centralized logical policy decision point is a key factor so that there is one big security viewpoint of the entire environment. As we increase integration with multiple components, the context building becomes stronger, and by extension granular policies will become more effective. Manual intervention from, say, a security operations center (SOC) analyst should also be a trigger or even an information point for the policy engine.

Always Start with Why

Chapter 4 will revisit in detail the importance of "why," but even though this is the last tenet covered in this segment, it is the most important one and the one most commonly overlooked. You could deploy and transform the entire environment of a customer to Zero Trust perfectly, but if it cannot achieve the enterprise vision, then the entire process is moot. Whenever you start having conversations with senior leadership of an enterprise, it is important to understand why they want to move toward a Zero Trust model. As architects of their network, we need to understand what their side of the story is. Maybe an enterprise wishes to move to the Zero Trust model without really knowing what it is signing up for. It's possible that the enterprise is not aware of the architectural changes that will need to be completed to achieve the desired Zero Trust architecture. It is important to make sure that even with changes in senior management that the general direction and strategy still remain and the vision is not lost. The user experience might change, and people might not be happy with a security agent being installed on their mobile phones, but these topics need to be discussed with senior leadership, and they must be aware of these changes and their impact on the business as well as employees. Senior leadership owns each aspect of the enterprise, which includes employee experience, security architecture, critical network infrastructure, and so on. Therefore, they need to be on board with all the changes that will be introduced. Employees must be consulted and feedback must be extracted to make sure the transition to Zero Trust processes is a smooth one.

With any enterprise embarking on the Zero Trust journey, there will be strategic and tactical goals. It is important to align the business drivers with the Zero Trust vision. When a road map is being formulated, some of these tactical goals being achieved will further enhance the Zero Trust strategy within the enterprise and help strengthen the vision of the enterprise. That is why I cannot stress enough how important it is to understand why the enterprise wants to move to a Zero Trust model. It is a big change, and it needs more than just money, technology, or "know how." It needs a broader mindset and deeper understanding of the enterprise vision.

Zero Trust Catalysts

In this section, we will explore some of the common Zero Trust catalysts needed to implement a successful Zero Trust strategy. There is a reason they are catalysts—most enterprises have already embarked on a Zero Trust journey by augmenting or initiating some of these activities. Spending time and money on some of these endeavors goes a long way in speeding up the Zero Trust journey. It is important to highlight that none of the catalysts covered in the subsequent sections can singularly achieve Zero Trust. Recall that the siloed security initiatives have already been in the security space for quite some time. Each initiative is in itself extremely useful to the enterprise, but when deployed as standalone initiatives, they cannot achieve the broader Zero Trust vision. Zero Trust removes silos and needs all initiatives to follow the same design principles and tenets. The catalysts covered in the following sections must be used in conjunction with each other to be able to get the right Zero Trust effect.

Asset Inventory

Assets are the key to an enterprise, and this makes asset management a primary focus in an enterprise. Its importance to Zero Trust is in defining the scope of the Zero Trust strategy. You must know what types of assets are in your enterprise to augment the context of the asset's identity. Applications and devices must be able to utilize and process this context. For example, on the same physical server you might have a syslog server and an HTTP server. Access to both services must be controlled separately, and they must be considered separate assets. Similarly, if the same server has services processing information of different classification, then the access to the services must be evaluated separately. Providing access to just the physical server is not enough, and considerations must be taken to separate these services physically or logically via microservices.

Assets don't always mean applications. Any entity that needs to interact with your enterprise is an asset. Subjects and objects alike are both assets. Users, devices, applications, networks, and more importantly data are all assets. Contrary to focusing on protecting assets within a fixed perimeter, Zero Trust places the protection around the assets themselves. The reason fixed perimeter protection is flawed is because an asset's value doesn't reside in a specific location. Assets are always dynamic. A user can access an application from anywhere. This application might have business attributes in various systems. The asset value is key, and it is an important factor in Zero Trust.

Does this mean we need to start with data asset inventory and classification immediately and ignore the rest of the assets? The answer is "yes and no." Of course, data classification is important to begin with, but security controls for other pillars should run in parallel as well. As a matter of fact, security controls for network, devices, users, and workloads will have common intersections with data security, which is why asset inventory is not just about data but about all assets. Controls placed for other pillars will support the data controls. Performing data asset inventory is not easy, so there should be a plan to start somewhere.

A larger concern when it comes to asset inventory is how well it is implemented in an enterprise. Automation in asset inventory is not commonly seen in enterprises, and most enterprises have manual or spreadsheet-based automation, which is rudimentary. User assets are maintained in the Active Directory and are therefore much more mature when it comes to automation. The point here is that assets are a cumulative whole, not just users and servers yet protection for various types of assets may vary based on the value and nature of the asset (user, data, or application).

Whereas using Active Directory is the most common method of maintaining a user asset inventory, products like Cisco Duo and Cisco Identity Services Engine (ISE) can be coupled with Active Directory to provide an accurate view of users and devices accessing the network. Products like Cisco Application Centric Infrastructure (ACI) and Cisco Secure Workload can provide valuable asset visibility into the workloads in the server farms.

Identity and Access Management

Users are key assets in the environment and the original weak link in any enterprise. A large number of attacks begin with the people because they can always be engineered to provide information. Hence, understanding all the subjects, their roles, and their access requirements is important. User control typically is extended to devices as well, and this leads to utilization of user and endpoint behavior analysis (UEBA). UEBA helps map out whether certain flows from users are expected. Users and devices commonly exhibit various behaviors, and this is something that must be captured by a visibility and analysis solution. For example, Cisco Secure Analytics Cloud helps provide UEBA to applications in the cloud environment.[5] By extension of the current definition of an asset, vendors and subsidiaries, for example, are also considered assets because they will need to access resources and applications in the enterprise. UEBA combined with products like Cisco ISE can provide key insights into anomalous users, devices, and their access patterns.

Whereas the user segment remains the same, devices have changed by leaps and bounds. We now have employees with phones and tablets that connect to your network. Fridges and coffee machines now have access to the network and the Internet. The relevance of "need to know" access to the network is much more when it pertains to endpoint devices.

A strong IAM system is key to the implementation of a Zero Trust solution. Strong multifactor authentication and a robust policy engine is key to implement the granular policy derived based on the context of the identity. End-user machine authentication and subsequent security posture checks are another important aspect in making sure that the machines that users connect from are secure enough. This is a key addition to the context of the identity. Continuing from the previous example, Sally accessing the network via her corporate laptop and Sally accessing the network via her mobile phone would be similar in the traditional networking paradigm, but the endpoint now provides more context and hence provides more scope to create granular rules.

Segmentation

Segmentation is an important accelerant for overall adoption of Zero Trust. Irrespective of the direction the enterprise wants to consider, be it macro-segmentation or micro-segmentation, the very outcome of breaking the enterprise (user, network, application, data) into relevant segments leads to a more streamlined path to building a Zero Trust policy. You will learn how macro-segmentation and micro-segmentation lead to faster adoption in the subsequent subsections.

Networks (Macro-Segmentation)

Networks are the core of communication in an enterprise. Whereas networks are critical in the environment, what we are trying to change is the implicit trust that most networks bring today. Isolation and segmentation are therefore key. As with any cybersecurity kill chain, lateral movement is almost always needed to get access to higher value assets, and the motive of segmentation is to reduce and control such movements. All technology that helps perform isolation is part of the network segment. Firewalls, VLANs, virtualization, containers, and TrustSec are all network segmentation controls. Chapter 9, "Critical Security Mechanisms for Zero Trust Architectures" will cover isolation and network architecture elements in more detail. The important aspect to remember here is that a network is an asset but not a mandatory policy enforcement control. The subject must have access to the object irrespective of the network it comes from. The architecture must understand and control a business flow, not just a network flow. Solutions like Cisco Software-Defined Access (SDA) help provide administrators with an accurate understanding of the network. With segmentation built into the solution, SDA incorporates the key Zero Trust tenet of network segmentation.

Workload (Micro-Segmentation)

Isolation is not just at the network level but also at the workload level. Workload typically depicts an application stack, generally a web, application, and database that can perform a specific business task and help support the business and the customers. Occasionally, some of these stacks might not be microservice architecture-based and they do support general workload segmentation, but the isolation extends much more than a stack-level function. For example, file share services must be segregated from critical information infrastructure servers. This seems to be a simple principle, but it is not new to see enterprises having a large implicit trust boundary in the server farm. Containers, VMs, legacy servers, even physical servers all are workloads and must be isolated to avoid lateral movement. Products like Cisco Secure Workload can provide both asset visibility and flow visibility for enterprise workloads. With flow visibility, it is easier to incorporate micro-segmentation as part of the IT and software strategy.

A core tenet of Zero Trust is dynamic context, and to be able to maintain granular policies, the users, network, and applications must be segmented so that an end-to-end context of a subject accessing an object can be policed accurately. With the right segmentation in place, security can focus on supporting and even augmenting business by helping the organization make the right decisions and security control placement. Cisco Secure Workload, Illumio, and Akamai Guardicore are solutions that help monitor flows and constantly provide feedback to the administrator on new flows that might be anomalous. This then helps the administrator or the decision point to analyze and understand whether the context of a flow has changed.

An Open Mindset

As someone who is going to lead the customer to achieve their Zero Trust vision or as a leader who is taking the enterprise on a Zero Trust journey, you must begin with an open mindset. Leaders must be able to understand that the entire process cannot be finished within three or four months and might take years to fully achieve based on the company's current maturity level and available skillset. Hence, the journey is more important than the destination. Changes in the workforce, processes, and technologies are inevitable, and while Zero Trust tries to fit into all architectures, it can be strict compared to existing norms when it comes to policy enforcement. If the right enforcement points are missing, this will add to the capital expenditure of the enterprise.

Achieving Uniform Policy Enforcement

With traditional perimeter-based control, a firewall on the Internet does not need to know if a fridge needs to communicate over the Internet or whether it's safe that a coffee machine is accessing YouTube. The server farm firewall will check for a specific set of policies that will not be the same as the Internet firewall, and this becomes a difference in enforcement for the same business flow. The concern with this approach is the distribution of decision capabilities. With a common logical decision point, the policies can be implemented uniformly across multiple technologies. This decision point must be fed with constant information from the network, and that's where visibility and analysis become relevant. With greater maturity of visibility and analysis, more detailed context as well as segmentation can be performed, which will in turn help feed the common decision point and maintain uniform policy. A common policy decision-maker, such as Cisco ISE or the Security Services Edge (SSE), deployed in customer networks or in the public cloud can perform a wide range of decisions based on multiple integrations with both Cisco and third-party information points.

A similar scenario can be seen when the same flow passes through both on-prem and cloud infrastructures sequentially or simultaneously. Consider a scenario where currently the enterprise has direct Internet access, and outbound Internet is enforced on the cloud with security controls like URL filtering and IPS via a cloud firewall infrastructure such as Cisco Umbrella or SSE. If the Internet link goes down, the enterprise backhauls the traffic to its data center or uses another transport path. When it takes an alternate path, the traffic gets subject to the WAN firewall's security control as well as the central hub's Internet firewall access control, which may not have URL filtering or may be allowing specific URLs. This is a common scenario, and the minor setback here is the lack of uniformity in policy control. With the right social engineering skills and the right targets, all one needs to do is bring the local network down to be able to bypass URL filtering. This is against Zero Trust principles, and a major prerequisite for consistent enforcement is to make sure all the enforced policies are uniform across all enforcement points, including multicloud architectures.

It is possible that a policy information point might be used directly at the application level. Granular user-level enforcement is generally owned by the application and more coarse-grained enforcement is performed at a network level. A combination of both of these enforcement approaches is covered under the tactical initiative of uniform security enforcement. The decision points may be at different locations in the network, but as long as they are synced with each other, the overall business flow control is retained.

Integrating Effective Information Points

It is imperative to have a mature SOC to be able to ingest multiple sources of information and provide valuable inputs to the policy engine in order to dynamically enforce policy. Automation around proactive event ingestion is an important part of generating the right information needed to identify users, devices, and their respective context. Hence initiating or designing a SOC helps have a single source of information or at least processed logs for the policy engine to consume. A SOC need not be the only source of information, but if it is, there will be multiple sources integrating with the SOC where the SOC collates all the information together. Strong SOC and SIEM integration along with mature logging services ensures accurate information processing. The decision point has sufficient data to make an accurate risk-based decision. Products such as Cisco Splunk are commonly used by enterprises to integrate multiple modules to create a mature SOC.

Implementing Effective Automation and Orchestration

With great SOC maturity comes the need for better orchestration and automation. An effective task (like analyzing a false positive) adds value tactically, but strategically the effectiveness of the task will always be measured by how well it can be orchestrated with other tasks and scaled out (AI-based identification of false positives). A simple example showcasing the need for Orchestration can be observed when responding to incidents, where orchestration can help make incident response much more streamlined with the right playbooks. Products such as Cisco XDR can help provide a common orchestration platform and a single plane of glass for a majority of orchestration requirements. Using its large integration portfolio, an administrator can configure multiple playbooks to fit the orchestration need of identified tasks. A SOC analyst is best equipped to understand and isolate tasks and processes that can be automated. The playbooks will introduce proactive orchestration so that analysts can focus on more important and complex tasks. Orchestration makes the path to mature processes much simpler, thus speeding up the Zero Trust initiatives.

Software-Defined "Anything"

The biggest advantage of deploying any solution based on software is the added capability of automation and orchestration. Software-defined networks support underlay provisioning and automation, whereas software-defined policies allow for dynamic policy creation based on flows and context. Software-defined visibility provides enhanced visibility and monitoring capabilities that support Zero Trust tenets, whereas software-defined wide area networks (SD-WANs) provide secure access for branches to their headquarters. By leveraging programmable assets, organizations can gain granular insights into network traffic, user behavior, and device activities. This visibility helps identify anomalies, detect potential threats, and respond quickly to security incidents, aligning with the Zero Trust principle of continuous monitoring and adaptive security.

SDx (software-defined anything) also helps abstract some of the functionalities that fit above technology. For example, routing decisions, security policies, traffic doctoring, and so on are all configuration in current networks that need to be manually replicated across devices. With the need for uniform policies and dynamic evaluation of context, SDx becomes a key catalyst for enterprises to move toward Zero Trust.

The Interview

[Glenn pauses after the explanation of the core Zero Trust principles and then continues.]

Glenn: Essentially, Zero Trust is simple. You give the user only the access the user is supposed to have to objects that the user needs access to. Everything else adds additional protection and security layers over this concept.

In summary:

- Zero Trust is not a technology or a service but an information security model.

- Everything that provides a service, computes information, or in any way handles data is considered a resource or an object.

- Anyone who requests access (including services) is considered a subject.

- Irrespective of where the end user is and what is being accessed, the access and the communication channel must be secured.

- Access is provided only just in time and just as much as is needed. Hence, if access is requested again, the context must be specified and validated again.

- Access must be provided based on at least identity, endpoint, and the application desired to be accessed. The more data we can collect, the more context we can add, and the more granular the policy can be.

- The enterprise must make sure that owned assets have continuous posture checks to validate compliance.

- Visibility into the flows, assets, and their communication is key to defining the Zero Trust policy.

- Segmentation at the user, application, data, and network levels reduces the blast radius of an attack. It avoids larger losses of data and makes incident handling simple.

- Orchestration helps coordinate mundane tasks so that analysts can focus on complex activities. Automation helps provide a repetitive procedure for the same task.

- The core tenets of Zero Trust are as follows:

  - Avoid creating implicit trust.

  - Assume Breach.

  - Control access based on dynamic risk and context.

  - Proactively reevaluate risk and context.

  - Continuously monitor and build risk-based profiles.

  - Use automation procedures and orchestrate simple processes.

  - Align with the business.

- Some Zero Trust catalysts include:

  - Asset inventory projects

  - Identity and access management augmentations

  - Segmentation initiatives

  - An open mindset to take on new risks to reduce existing ones

  - Uniform policy enforcement strategies

  - Rich and effective information sources

  - Effective automation and orchestration

Mr. Smith: Wow, now that I think about it, like you mentioned we are following this concept almost every day, but it is easier said than done in an enterprise scenario. How does one change the trust landscape of the enterprise so quickly? How have other enterprises and their employees reacted when they reduce the trust they have in the existing workforce?

Glenn: It's not about never trusting. It's about reducing the implicit trust circle, and the key when it comes to Zero Trust is that it is not a point-in-time task or activity. It is a journey. It's not a one-way ticket; instead, it's essentially like visiting your grandma every year. You know you will get cookies at the end of the trip that you enjoy all the way back home.

Mr. Eaton: Alright, then my question would be, why I should adopt Zero Trust? How will it benefit me?

Glenn: Well, basically with a better idea of what you want and a clearer idea of what it takes to reach your goal, your enterprise will benefit with stronger security policies and effective security controls. Zero Trust provides you a framework to achieve the best security capability with technical augmentations, effective processes and skilled personnel. The bigger question is, are you mentally prepared for the changes? If you are, then the next step is....

Endnotes

1. "The Definition of Modern Zero Trust," https://www.forrester.com/blogs/the-definition-of-modern-zero-trust/

2. NIST 800-207: Zero Trust Architecture, https://doi.org/10.6028/NIST.SP.800-207 and https://csrc.nist.gov/publications/detail/sp/800-207/final

3. "The Zero Trust eXtended (ZTX) Ecosystem," https://www.forrester.com/report/the-zero-trust-extended-ztx-ecosystem/RES137210

4. "Zero Trust Maturity Model," https://www.cisa.gov/zero-trust-maturity-model

5. "User and Entity Behavior Analysis," https://www.cisco.com/c/en/us/td/docs/wireless/controller/ewc/17-8/config-guide/ewc_cg_17_8/m_ueba.pdf

Align to the Business Vision and Mission and Craft Metrics for Success

This phase is all about aligning the Zero Trust strategy to the enterprise's vision and mission. The focus will be on identifying the common reasons why enterprises choose to move to a Zero Trust model and how you can create a Zero Trust vision and mission aligned to the overall business vision and mission. This is often overlooked and is an important phase before beginning any strategy or implementation conversations.

In this phase, Zenith Trust Bank will also go through the critical step of identifying how to measure the success of the Zero Trust initiative. This phase will take into consideration feedback from all stakeholders to make sure everyone is able to measure the value of the Zero Trust initiative through their own lens. The following chapter is covered in this phase:

Chapter 4   Always Start with "Why"

Chapter 5   Measuring Zero Trust Success

Chapter 6   Understanding Zero Trust Maturity

Always Start with "Why"

When it comes to technology, the tendency of most teams is to jump straight to product selection or solutioning without considering other factors in the enterprise that could influence the decision. This, in turns, leads to organizational misalignment and possibly a total reversal of the idea itself. In most cases, Zero Trust conversations for enterprises begin with engineers and mid-level managers who then pitch the concept to leadership to see if it makes sense to adopt. However, inevitably, the conversations always steer towards "how" the Zero Trust architecture is implemented and what changes would be needed. When a new concept is incepted, the common goal should be to understand the inner workings of the concept. It is rare that one would consider why the framework was incepted in the first place. Does the enterprise even need such a paradigm shift?

The interesting part is that once the technology manifests itself in its true form, we accept it and move on to the next technology. Many enterprises embark on the Zero Trust journey because they can, rather than examining why they wish to utilize Zero Trust. As transformational partners for enterprises, we need to keep our implementation-focused mindset aside and consider a business perspective. The question you should be aiming to answer is, " What is Zero Trust bringing to the table for the enterprise?" Sure, it sounds cool to implement Zero Trust on an AWS cloud, but how will it impact capital expenditure and operational expenditure for the enterprise. Does the cloud workload justify a strategic movement? Like Simon Sinek once rightfully said, "People do not buy what you sell but why you are selling it."

This phase of the book (all covered in this chapter) keeps in mind this one specific aspect—alignment to business vision. It does not matter if you are the owner of the enterprise or a consultant supporting it, the vision must be known and clearly document-ed. A vision statement encompasses a broad final state the enterprise wants to achieve. If we had to formulate a Zero Trust Vision, it would be similar to the following:

> **"As an enterprise, we wish to adopt the Zero Trust information security model to provide any entity (including non-enterprise subjects) with convenient but autho-rized, secure access to any resources owned by us. We would like to achieve this access model in two years in incremental stages."**

This is a broad statement that covers the overall intention of Zero Trust. Note that this is not a business vision. This is a Zero Trust vision that is aligned to the business vision. For a bank, a business vision could be to be the best banking partner, but the Zero Trust vision must align with this statement by making sure data is secure yet accessible. The Zero Trust vision must therefore facilitate the business vision. The enterprise wants to make access convenient but not remove the security aspect from it. This is what every enterprise wants to move toward. Enterprises wish to make sure that they make the experience as simple for the user as possible without losing the security aspect. One such example would be to remove passwords and perform password-less authentication. Users are tired of typing passwords and then using multifactor authentication to log in. Even with single sign-on (SSO), the process gets cumbersome. To reduce friction, password-less authentication with biometrics included will be the key to solving this inconvenience and still maintaining security for the enterprise.

Along the same lines, enterprises have various reasons to decide to embark on the Zero Trust journey, and as consultants and trusted advisors, we must understand why enterprises want to adopt Zero Trust. It is from these "business objectives" that the mission statement will be derived, along with their technology road map as well as their overall security strategy. Some of the common use cases that enterprises focus on when they choose to move toward a Zero Trust architecture are discussed in the "Aligning Zero Trust with the Enterprise Mission" section.

Take the Time to Ask Why: Understanding the Vision

The first step of any Zero Trust engagement is to speak to leadership and understand what they want to achieve by moving toward a Zero Trust access model. Zero Trust as a framework is not demanded by most regulations because Zero Trust is abstract. Regulations focus on protection of assets with tangible security controls. It is possible that audits will demand certain security controls, but those are in response to a specific security deficit identified by a gap. Regulators do not, however, dictate what access control model is used. For example, PCI-DSS would direct an enterprise to protect credit card details. It could even direct the enterprise to use AES-256 encryption when storing personally identifiable information (PII); however, it does not ask the enterprise to utilize a Zero Trust architecture. This is because compliance typically concerns itself with the result and is less concerned in how it is achieved. For an enterprise, though, how a security strategy optimizes costs and security is an important metric to craft.

Many security models can achieve the preceding requirement, but the advantage of Zero Trust is that it covers more ground. Even though it is categorized as an information security model, it covers all assets of an enterprise, including people, technology, and process. It's not just about the data. Zero Trust models preach that protecting the environment surrounding data automatically protects the data. Zero Trust is a logical means to achieving the overarching vision, which is "convenient and secure access to the enterprise assets." There is no need to choose between convenience and security. Zero Trust facilitates both. Even though Zero Trust has many advantages, it is not as easily adopted among enterprises due to the many reasons discussed previously in Chapter 2, "The Zero Trust Kaleidoscope."

It is precisely due to the many delays in adoption that it is imperative to speak to senior management and the technical leads of an enterprise to understand and conclude why Zero Trust as a model fits their requirements. Security is no longer an add-on to the business but is more of an innovation engine that strives to support the business. It is a flexible, versatile gel that glues all the business units of an enterprise together. As part of overall governance and risk conversations, the business alignment of the information security model must be identified and etched in stone. Of course, the means of achieving the vision might vary, but the overall business goal must be the same.

In this case, when we consider Zenith Trust Bank's vision statement, the overall security vision can be summed up as "convenient access without reducing the security posture." This vision is crafted based on the initial conversation that Glenn from Prolink Solutions had with Zenith Trust Bank. The CIO had highlighted two key top-of-mind topics. One was the VPN downtime, and the second was the attack on a peer bank. When you speak to leadership, you need to pick key top-of-mind topics that are important to them because those keywords will drive the overall vision, and Zero Trust must align with those key-words. A VPN caused resources to be unavailable and impacted business as usual, thus affecting performance metrics, and the peer bank's attack rattled him enough to do an audit and validate his own posture, which was very low. This translates to a risk metric. Therefore, the vision of "convenient access without reducing the security posture" is accurate and aligns with the overall security vision. Note this is not the overall business vision, but it's expected to facilitate the achievement of the overall business vision. Write down this vision statement as you will repeat this back to leadership once your conversation is done.

Key Attribute

Security Vision: Convenient access from anywhere without reducing the security posture.

Aligning Zero Trust with the Enterprise Mission

This section will answer the question "Zero Trust aligns with my business strategy because…." The focus of the section is to understand what are the key drivers that lead an enterprise to support the adoption of the Zero Trust model.

A CISO's Viewpoint on Zero Trust Adoption

A vision statement is a broader final state that an enterprise strives to reach. To illus-trate this further, take the simple example of travel. "I want to travel all over the world" is a vision. A mission statement is more specific and relates to how we want to achieve the vision. Utilizing the same example, a mission statement would be "I want to reach Singapore in the shortest duration of time with the least layovers and most luxury." It is more specific and hints as to certain aspects of the trip that will be made that are not covered in the vision. You could have multiple mission statements as well. For example, another mission statement would be "I want to travel to France and make sure I visit the Eiffel Tower by pre-booking tickets if needed." The example simplifies the efforts needed to extract the mission statement, but that is the next step. When the vision is clear, you need to iron out what you are planning to do to achieve the vision.

It is important consider how senior leadership looks at Zero Trust and some of the aspects of what architects need to consider when identifying mission statements. This section will focus on some of the common Zero Trust business drivers and what the CISOs think about Zero Trust. The statistics and information have been captured by the survey "CISO Perspectives and Progress in Deploying Zero Trust," conducted by Cloud Security Alliance and published on March 6, 2022.[1] It is strongly recommended to go through this research paper to get detailed insights along with any updates to the material as they are provided.

General Top of Mind Based on Existing Security Landscape

If you zero in on the responses for common drivers for leadership to move toward Zero Trust, the highest priority (35% of responses) is to reduce the attack surface. This is justified with the large number of breaches and the constant inability of traditional security to support the drastically changing threat landscape.

The second priority is to simplify the user experience (a close 34% of responses). Pause and compare this with the initial vision extracted for Zenith Trust Bank, which is "convenient yet secure and authorized access of the network." These two topics together cover close to 70% of most leadership's top of mind. Enterprises are no longer just focusing on the crown jewels. Users are generally considered as a weak link, and to strengthen that link, enterprises need to make sure the end-to-end access to the network and workload is seamless and less complex. The more complex it is, the more chances that the workforce will find a workaround and open a potential backdoor.

Finally, the third priority is to improve security risk posture and resilience. This is in line with most security visions, and the risk-based access control that governs most Zero Trust policies is a key mission statement. Risk-based drivers are covered in the subsequent sections, but to summarize, risk-based access control takes into consideration the risk profile of an end-to-end flow rather than just an asset. This approach will thus consider the subject, the object, and the overall context of the flow. Whereas enforcing principle of least privilege and governance enhancements are other top of mind topics for leadership, the overall thought process is fairly clear. Make security easy for end users so that they can appreciate and contribute to enhancing the overall security posture of the enterprise.

Keep this in mind and frame a mission statement for Zenith Trust Bank. Write down the statement to present to leadership.

Key Attribute

Mission Statements: *The mission is to provide a secure and reliable network infrastructure that enables our clients and workforce to access their applications and workloads with ease. We achieve this by:*

- *Implementing risk-based profiles to identify and mitigate potential security threats*

- *Reducing the attack surface by implementing industry-standard security measures*

- *Making access to the network and workload simple and streamlined*

- *Implementing robust security measures and ensuring simple and secure access from anywhere to the network and workload through streamlined authentication and authorization processes*

Enhanced Data Handling

Another important consideration for leadership, especially CIOs, is the data that is handled by the enterprise. Certain risks and threats come with certain types of enterprises. Hospitals and most organizational human resource departments or third-party human resource services, for example, have to store personal health information (PHI) or personal identifiable information (PII), and this puts a target on their backs for information brokers and malicious threats. A media company, on the other hand, might have sensitive information about people or a social occurrence that might even be a security risk to the country. Today, enterprises are interested in how unauthorized access and modification of information can be reduced. Zero Trust provides a simple access model to do so. From a CxO perspective, data handling and the impact of all other pillars on data are key when it comes to Zero Trust, and this topic remains top of mind when it comes to Zero Trust adoption.

Keep this in mind and frame a mission statement for Zenith Trust Bank. Write down the statement to present to leadership.

Key Attribute

Mission Statements: *Our mission is to provide high-quality services while maintaining the privacy of our clients' data. We are committed to upholding the highest standards of security, confidentiality, and privacy for all the personal and confidential information that we handle.*

- *Our aim is to be a trustworthy partner for our clients by ensuring that their personal data is collected, processed, stored securely and in compliance with all relevant data protection laws and regulations.*

- *We strive to promote transparency in all aspects of our operations and to provide clear and concise information to our clients about how their data is used.*

- *We are dedicated to continuously improving our data protection practices by regularly reviewing our policies and procedures, identifying areas for improvement, and implementing new measures to safeguard against potential data loss risks.*

- *We believe that protecting the privacy and security of personal data is not only an ethical imperative but also a fundamental aspect of our business operations.*

Business Drivers

Business drivers motivate or influence a specific activity or vision that supports the overall business. These drivers are usually to augment a larger business goal or strategy and are not technology-focused. Some common business drivers that benefit from Zero Trust are as follows:

- Overall user experience

- Cloud migration of workload

- Risk-based access control implementation

These drivers will be explored in more detail as more mission statements are deduced for Zenith Trust Bank.

Enhance Overall User Experience

Enhancing user experience constitutes providing more power to the user. Essentially, the core driver here is to include only the necessary "middlemen" in the end-to-end flow and let security handle the rest of the access control. This will not only make movement of users easy across the enterprise or the cloud, it will also keep the access model unified across multiple locations. In modern enterprise security architectures, a common fallacy is to represent all security mechanisms with a separate hardware or software product extension. This not only adds to the operational overhead of the network, but also the latency of the end-to-end flow. Typically, when a user accesses a service there are a plethora of firewalls, load balancers and application gateways performing siloed controls. The mission statement, however, is to make the user experience seamless and simple. If an enterprise can trust the system enough to let users access internal websites over the Internet, true Zero Trust has been achieved.

Create Optimized User Access to Applications

As part of user experience (UX), it is critical to constantly authenticate and authorize the user, device, and application attempting to access resources within the network, and this is a core tenet of Zero Trust models. This can result in a more complex and time-consuming login and access process for users, which can negatively impact the UX. Imagine telling users that the enterprise is adopting a more empowering model but then they have to authenticate every hour and they need to key in their username and password as well as their MFA token. Users will not take kindly to misinterpretation of facts. Therefore, it is essential to ensure that Zero Trust security measures are implemented in a way that does not overly burden users and disrupt their workflow. Some examples of how this can be achieved include use of streamlined and intuitive authentication methods, such as biometric authentication or single sign-on or by implementing context-aware access controls that can automatically grant or deny access based on a user's role, location, and device.

Maintaining transparency with users and explaining the need for Zero Trust and how it improves their workflow is an important step in end-user stakeholder management. Without seeing the impact on their own workflows, end users will not fully understand the extent of how Zero Trust enhances their experience.

Enhance End-to-End User-to-Workload Access

End-to-end UX is a key focus of most enterprises. Consider the example of an external user accessing an internal server. In modern network access designs, when a user has to access an internal server, some of the common deployment options include either to use a jump host in the DMZ or to implement access-restricted VPN or per-application VPN tunnels. When a user must access a jump host, the user must authenticate to the jump host machine and from the jump host machine perform another remote access login into the server (see Figure 4-1).

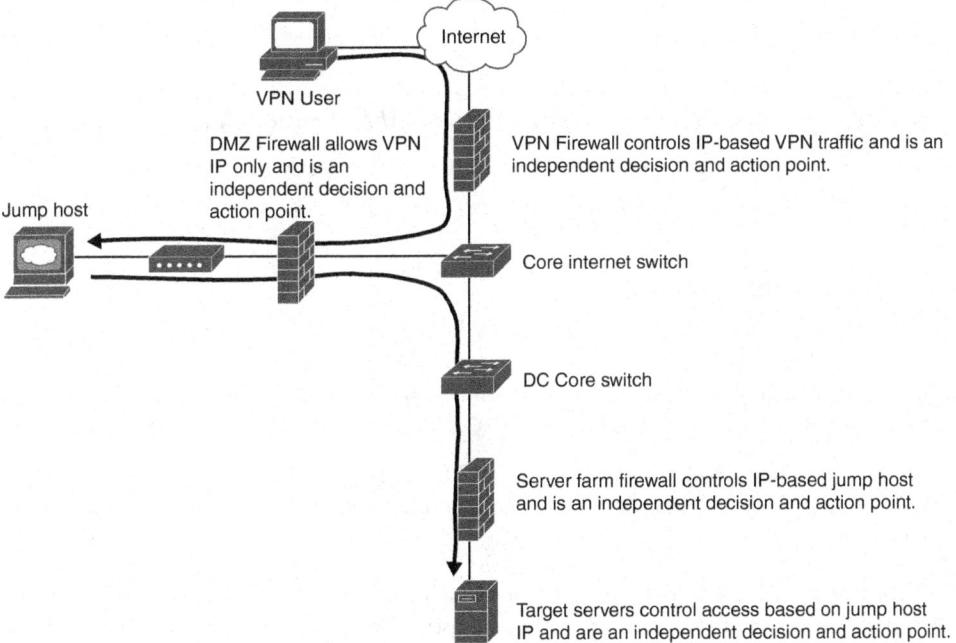

Figure 4-1   *A Typical Jump Host Flow*

There are some inherent problems with this entire flow:

- This format of access separates the accountability of various custodians. For example, when there is a technology refresh in the server farm network, no one is looking at the demilitarized zone (DMZ) jump host or the overall user access. The server farm devices and networks are refreshed to bigger devices, but the user still needs to remote in multiple times to access the workload. The same applies to software. Services get their software upgraded but jump hosts are considered lower priority which exposes them as a potential attack surface.

- The overall experience is a hassle and not very user friendly. There are enterprises where even the jump host is not accessible from the Internet and must be accessed over VPN. If we compare this scenario to real life, we could take an Apple showroom building entry. To enter the Apple showroom in Building 1, you need to access their partner showroom in Building 2, which is five miles away and the entrance to the Apple showroom is only via the partner building.

- This access model is sourced from a perimeter-based mindset and continues to propagate the same mentality. Secure the crown jewels by controlling access from specific locations.

This, however, is a roadblock that Zero Trust access control aims to do away with.

Key Attribute

Mission Statements:

- *We aim to empower our users with efficient and intuitive access to the information and resources they need, while ensuring their privacy and security.*

- *We aim to provide the apt security to a user's flow with the right context and risk awareness rather than provide siloed security controls.*

Support Cloud Expansion

Cloud migrations are a major game changer when it comes to Zero Trust adoption. Security in the cloud is not the same as security controls in traditional networking. With the changing responsibilities and accountabilities in various cloud models (IaaS, PaaS, SaaS), a standard next-generation firewall (NGFW) at the perimeter is no longer valid.

As a reminder, SaaS, PaaS, and IaaS are three different types of cloud computing service models that provide different levels of infrastructure, platform, and software resources to users:

- **SaaS (Software as a Service):** This model allows users to access and use software applications over the Internet on a subscription basis. The software is hosted and managed by the service provider, who is responsible for maintaining the infrastructure and security of the software. Examples of SaaS include Salesforce, Dropbox, and Google Apps.

- **PaaS (Platform as a Service):** This model provides a platform or environment for users to develop, test, and deploy their own software applications without having to worry about the underlying infrastructure. PaaS provides users with a set of tools, programming languages, and APIs that they can use to build and run their applications. Examples of PaaS include Heroku, Google App Engine, and Microsoft Azure.

- **IaaS (Infrastructure as a Service):** This model provides users with access to virtualized infrastructure resources, such as servers, storage, and networking, over the Internet. Users can deploy and run their own software applications on this infrastructure, which is hosted and managed by the service provider. Users have full control over the operating systems, applications, and data that are running on the infrastructure. Examples of IaaS include Amazon Web Services (AWS) Elastic Cloud 2 (EC2), Microsoft Azure VM, and Google Cloud's Compute Engine.

Cloud security and on-premises security are two distinct security paradigms, but they are not mutually exclusive. The way in which cloud security impacts the on-premises security strategy will greatly depend on how an enterprise uses the cloud and what security measures it has in place already as part of the existing infrastructure.

Provide Location Agnostic Security

With cloud computing, an organization's data and applications might be hosted in a remote data center that the organization does not manage. This means that data is transmitted over the Internet and is vulnerable to interception or unauthorized access. Organizations need to have robust network security measures in place to secure their on-premises networks and cloud connections. This eventually leads them to Zero Trust because it provides an access model that can support location-agnostic access control. Leased lines or a direct VPN is a lesser preferred option, as the cost per megabyte (MB) for data transfer will soon overcome the cost per MB the enterprise is willing to pay. Zero Trust access control suits the cloud deployment model well because cloud deployment models are built on principles of freedom and ubiquitous access.

Enhanced Data Security

Cloud providers generally have robust data security measures in place to protect customer data; however, organizations need to ensure that their sensitive data is encrypted and that access to the data is tightly controlled. This can involve implementing on-premises data security measures such as access controls, encryption, and data loss prevention (DLP) solutions to mimic the cloud security model. It is ironic that due to the general fear of movement to the cloud, the security measures in place to protect workloads on the cloud have made them even more secure. Cloud Security has Zero Trust tenets built in due to the overall nature of how data is accessed and processed. This is a huge paradigm shift when it comes to on-premises security mechanisms. Cloud migrations or deployments allow a consultant to support an enterprise to grasp the subtle differences and enhance security on the premises. It will be simpler to move to a Zero Trust model to the cloud if the basic principles have been implemented on-premises as well.

Unify and Augment Identity and Access Management Across Locations

Cloud computing has a huge impact on the organization's identity and access management (IAM) strategy. IAM is a core for all Zero Trust architectures. As organizations deploy multicloud workloads, they need to ensure that their IAM policies and procedures are consistent across their on-premises and cloud environments. They also need to ensure that access to cloud resources is tightly controlled and that their IAM systems can integrate with cloud providers. This means local AD and a cloud-based AD must be in sync to make sure the decision is being made based on the same data set of information. This aspect makes cloud migration a huge driver toward approaching Zero Trust for both cloud and on-premises networks.

With more adoption of cloud solutions, a cloud access security broker (CASB) is common when considering identity management in the cloud. CASB and on-premises IAM serve different purposes, but both are important for ensuring the security of an organization's digital assets. CASB is a security solution that helps organizations secure their cloud-based assets by providing visibility into cloud applications and enforcing security policies. CASB solutions are deployed as a cloud-based service, and they can help organizations to monitor and control the use of cloud services by their employees, partners, and customers. CASB solutions can also provide data loss prevention (DLP) and threat protection for cloud services. CASB is an important aspect of the Security Services Engine capability that is part of the overall Zero Trust Network Access concept; an example would be the Cisco Security Service Edge solution, which provides the CASB feature along with other overall security controls like Firewall as a Service, DNS security controls, DLP, and web security controls. Together, these controls provide the overarching SSE capability. On the other hand, on-premises IAM is a security solution that is deployed locally within an organization's network. It is designed to manage user access and authentication to local network resources, such as servers, applications, and data. An on-premises IAM system is usually integrated with the organization's network infrastructure and provides centralized management of user accounts, authentication, and authorization policies. An example of one such product that provides IAM capabilities is Cisco Identity Services Engine (ISE). While both CASB and on-premises IAM focus on security, CASB is specifically designed to secure cloud-based assets, while on-premises IAM is designed to manage access to local network resources. Access control on the cloud or a public SaaS via a CASB is very different from accessing an application on the Internet, and with varying cloud deployment models, enterprises wish to move to cloud and have a common security control model. Products like Cisco Multicloud Defense and Aviatrix gateways provide access control and orchestration across multiple cloud providers Their access model is different and the use cases they cater to are different. Zero Trust unifies the deployment and allows for more robust IAM and control strategies.

Easier Path to Unified Compliance

Compliance requirements could differ between on-premises and cloud environments. For example, cloud environments need to showcase their efficacy with SOC2 and SOC3 reports. An on-premises network has different compliance requirements. Nevertheless, compliance applies to the organization as a whole, and therefore principles governing the compliance initiatives must be applicable to both on-premises and cloud networks. When enterprises do not follow a common access model, the compliance teams need to maintain a separate set of requirements for both on-premises and cloud networks. Organizations generally wish to ensure that their compliance strategies are consistent across their multicloud architectures. This involves implementing common on-premises and cloud security measures that meet regulatory requirements for data privacy and security. This is where Zero Trust shines. Cloud security has Zero Trust principles built in, and striving to achieve a common security model can drive an enterprise to adopt Zero Trust on its premises as well. As the organization is exploring more cloud-native applications and moving workload to the cloud, migrating the security and compliance requirements is much simpler if the on-premises workload was secured with the same underlying guidelines. This becomes a huge driver for Zero Trust Adoption and subsequently migration to the cloud.

Overall, Zero Trust facilitates cloud movement and makes applying automated security control much simpler. It is in the enterprise's best interest from a governance and compliance perspective as well to utilize Zero Trust tenets. Implementing an enterprise-wide governance plan with a common access control policy is an attractive proposal for leadership.

Keep this in mind and frame a mission statement for Zenith Trust Bank. Write down the statement to present to leadership.

Key Attributes

Mission Statements:

- *Our mission is to provide comprehensive multicloud security solutions that enable organizations to operate securely in a diverse hybrid environment while maintaining strict security compliance standards.*

- *Our team is dedicated to constantly evolving and adapting to the ever-changing threat landscape to ensure the highest level of security for our clients.*

- *We believe in a proactive approach to security and aim to provide the best state-of-the-art security controls for both on-premises and cloud-based workloads.*

- *We aim to bring more convenience to the customer by expanding services and applications to the cloud with the appropriate security controls.*

- *We aim to simplify compliance initiatives by unifying the requirements across all networks.*

Risk-Based Access Control Adoption

Most organizations perceive the changing threat landscape with impending fear and would like to build effective security against zero-day threats. It is a commonly accepted corollary that if you do not provide access to an object by default (default deny), a zero-day is most likely going to be blocked and the object is protected. This is of course far from the truth, but it is a theme that is attractive to most enterprises because it is easier to implement rules and then forget about the control in place. The repercussions of this form of access control are far worse. By reducing the entire worth of an asset to just a rule, you are depriving it the security control it actually needs. A critical information infrastructure server needs much more controls than a set of NGFW rules to support them.

A common means to overcome the general lack of risk consideration is to have layers of security controls for an entire flow. Most firewalls, for example, have a static security level assigned to a zone or segment, and each zone is implemented with a specific set of security controls relevant to that zone. Thus, as information traverses zones, it gets controlled based on the risk associated with the zone; however, this also extends the perimeter-based access control that the Zero Trust paradigm does not consider. In contrast, a better option is to combine a Zero Trust solution like the Security Service Edge with additional security controls like endpoint-based security providing endpoint detection and response (EDR). The concept of layered defense still prevails; however, it is more asset- or function-specific rather than location-specific.

Risk-based access control in a Zero Trust architecture grants or restricts access to resources based on an entire flow's risk profile. It is not restricted just based on the subject but the object and other attributes as well. Threat modeling becomes key here, as it gives a lot of insight into various threats that can be identified for specific objects. However, assigning risk to flows requires a better understanding of overall business flows in the network, hence, most risk-based initiatives begin with flow discovery in the network. In risk-based access control, potential risks associated with a user's identity, device, location, behavior, and other contextual factors relating to the object as well, are evaluated before access is granted to it. This approach is particularly useful in enhancing the Zero Trust security strategy. This approach, however, cannot be standardized because the risk appetite of an enterprise varies and hence the profile definition cannot be made the same. This is why there are no resources claiming to showcase a standard risk profile template. Chapter 8, "Building a Zero Trust Architecture," delves more into some basic profiles that can be used to identify risk for subjects or objects. This will be a baseline to build on and create subsequent custom risk profiles tailored for each enterprise. For example, social engineering might be a low risk for Company A and high risk for Company B.

Additionally, a large chunk of breaches have occurred because a remote user was connected to a corporate network and clicked a malicious link, or more recently due to MFA push phishing. Enterprises today also do not take security endpoint posturing as seriously as they should and therefore end up having an extremely secure enterprise network edge but allow the propagation of a much larger attack surface via VPN users. Essentially, VPN users are an important pivot when it comes to movement toward Zero Trust. Subsequent sections will showcase some pitfalls of the VPN solution, but essentially once enterprises start seeing risks with a VPN solution, the shift to Zero Trust is fairly expedited.

It must be noted that it is not as simple to move to risk-based access control, which we also observe in practical scenarios. Several challenges are associated with its implementation:

- **Deployment complexity:** Implementing a risk-based access control system can be complex, requiring a deep understanding of the organization's assets, risks, and user roles. This complexity can make it difficult to design and implement an effective system.

- **Lack of visibility:** Visibility into flows may not be mature enough to explore a more advanced risk-based access control.

- **Data quality:** Risk-based access control systems require accurate and up-to-date data on user roles, privileges, and access rights. Poor data quality or incomplete data inventory can undermine the effectiveness of the system.

- **Policy consistency:** Risk-based access control requires consistent security policies deployments across all users and systems. Inconsistent application security deployments can lead to security vulnerabilities and compliance risks.

- **Scalability:** Risk-based access control systems need to be scalable to support growth and changes within the organization. This can be challenging, particularly for organizations with complex IT environments or rapidly changing business needs.

- **Privacy concerns:** Risk-based access control systems rely on collecting and analyzing data on user behavior and access patterns. This raises privacy concerns and requires careful management of data collection, storage, and use.

- **User acceptance:** Risk-based access control systems can be perceived as intrusive or unfair by users who feel that they are being unfairly restricted in their access to resources. This can lead to resistance to the system and reduced adoption rates.

All things considered, risk-based access control is an important business driver and a key pivot to moving to Zero Trust access control models.

Keep this in mind and frame a mission statement for Zenith Trust Bank. Write down the statement to present to leadership.

Key Attribute

Mission Statements:

- *We aim to provide apt security to a user's end-to-end flow with the right context and taking into consideration the object being accessed.*

- *We strive to incorporate risk and enhance the overall security posture of each flow rather than provide risk-averse siloed security controls.*

Technology Drivers

Technology drivers are the factors that influence and shape the design, implementation, and maintenance of enterprise security controls and processes. It is important to understand the differences between a business driver and a technology driver. A business driver influences a business decision, and a technology driver helps influence a technical decision. For example, a business driver definition would be "to provide strong access control for users," and a technology driver definition would be "to make sure the identity of a user is continuously validated multiple times using multifactor authentication when access is requested." In the business driver, the method of access control is inconsequential as long as the overarching business need is achieved. A technology driver is usually a security mechanism like encryption, multifactor authentication, Direct Internet Access, SD-WAN, and so on. The key here is that these technology drivers can be adopted without having a Zero Trust vision in mind. As these technologies get adopted independently, the drive for Zero Trust is accelerated. For example, as SD-WAN implementation progresses, so does the need for segmentation, which leads to independent segmentation initiatives in the enterprise. Eventually, they will converge to end-to-end segmentation and uniform policy access across various branches. With technical capabilities in place to create a Zero Trust policy, enterprises get more confident in embarking on the journey.

Encrypted Traffic Visibility and Control

Encrypted traffic is a key concern when it comes to inspection in modern networks. It is possible to inspect encrypted traffic; however, to do so requires decrypting the flow using an SSL decryptor. Using SSL decryption has its limitations, with the main one being performance. Not many enterprises today can perform SSL decryption and still maintain the line rate of the interface. Additionally, most products clearly state a massive impact on traffic when decrypting SSL. Hence, the simplest alternative is to use the Server Name Indication (SNI), or more recently Encrypted Traffic Analytics (ETA). These are technological capabilities that are introduced into the network to be able to perform specific functions.

What if we could just let the access control model decide if the access is authorized and let the traffic stay encrypted? If an HR user needs to access an HR server and it is from a branch network, technologies like SD-WAN facilitate data-in-motion encryption, and traffic control is performed based on identity, MFA, PAM, and other security controls without the need for decryption. Zero Trust access control helps to remove dependency on decryption and focuses on the contextual access control aspect. Encrypted traffic flows get subjected to more granular policy-based controls without giving up on performance. This means that access to encrypted traffic can be restricted to only the specific users, devices, and applications that need it, reducing the attack surface and potential for data breaches. Mechanisms like Encrypted Traffic Analytics[2] provide valuable insights into the traffic for a fraction of the time needed to decrypt and inspect. Encrypted Visibility Engine[3] serves a similar purpose, which is to be able to control traffic based on its metadata instead of payload. Both ETA and EVE greatly help push forward an overall Zero Trust agenda. When primary pain points regarding encrypted traffic are considered, it is the reluctance to decrypt and impact performance that prevents most enterprises from adopting Zero Trust. Enterprises do not decrypt and hence do not incorporate security controls deemed necessary by a Zero Trust strategy. With capabilities like ETA and EVE, enterprises get more details about traffic without having to decrypt it by using flow metadata. This then allows for more information to be available, which supports decision-making and eventually a timely action.

A well-planned Zero Trust model should use advanced security technologies such as machine learning and behavioral analytics to detect anomalous behaviors in encrypted traffic flows. This allows for faster detection and response to potential security threats. With end-to-end encryption, when a subject tries to access an object, all that needs to be done is to deny the access and check for policy. It's easier said than done, but essentially that is what Zero Trust is all about. It's simple to understand and implement if the primary use cases are clear. When an enterprise wants to enhance security but not compromise on the performance aspect, Zero Trust access models become a strong technical driver. With almost 99% of traffic being encrypted, the access model must be able to control the flow, and Zero Trust can facilitate such control without impacting performance. The driver here is capability to police encrypted traffic.

Keep this in mind and frame a mission statement for Zenith Trust Bank. Write down the statement to present to leadership.

Key Drivers

Mission Statements:

- *Protect all client and employment data at rest and when it's in motion and inspect the traffic for anomalies efficiently without forcing decryption. Aim to achieve this by augmenting existing security controls as part of the Zero Trust strategy.*

- *Utilize next-generation technology like machine learning to intelligently identify anomalies.*

Implementation or Enhancement of End-to-End Flow Visibility

Traditional rule-based security is no longer effective in today's security context. Traffic flows are often allowed based on predefined access control rules that assume trust between different parts of the network. This implicit trust makes it difficult to monitor traffic flows and identify potential security threats or anomalies and becomes a technical driver to move to Zero Trust because a Zero Trust architecture needs detailed understanding of flows and the capability to monitor flows continuously. For example, with 99% or more traffic being encrypted, one might lose visibility into most payloads; however, with the right context of the subject and object, a flow can still be created and monitored.

Data loss or data exfiltration is another driver that closely relates to flow visibility. A common scenario for an attacker to exfiltrate traffic is by getting a foothold in the network and requesting for a command and control (C2) server for more information or a script to copy data across SSH or some other "firewall-safe" protocol. An Internet-bound flow (traffic outbound) is a great example of how Zero Trust helps mitigate this kind of attack and how many enterprises see value with adopting Zero Trust specifically for this use case. Zero Trust access models typically include more detailed monitoring and logging capabilities, which provide better visibility into network traffic flows. For example, network administrators clearly see who is accessing in-scope resources, when they are accessing them, and how much data they are transferring. This further helps identify potential security incidents or compliance issues. Zero Trust is extremely compliance friendly and almost all initiatives under Zero Trust will help the organization achieve the required compliance. Additionally, enterprises are not entirely confident with the effectiveness of on-prem DLP solutions and are looking to move toward the Zero Trust cloud inspection model. End-to-end flow visibility is a larger umbrella, which includes looking at server-to-server as well as user-to-server flows in detail. It is not sufficient just to log siloed data to a centralized collector. There must be analytics happening on all these logs to stitch the context of an entire flow, which is easier to achieve as part of the Zero Trust strategy because there are basic initiatives like enhanced asset inventory and unified logging that augment this capability.

It is important to highlight the gray areas that most technologies bring. Flow visibility is both a technical and business driver. This is a good example to truly understand what the difference is between a technology and business driver. Augmenting network visibility can help augment the overall business functions by providing the right visibility into all aspects of network, user, and workload traffic. Because it is supporting the broader vision of flow visibility, it becomes a business driver. At the same time, designing and implementing specific technology-focused like a security operations center (SOC) or implementing application visibility leads to higher visibility into the network and thus enhances the Zero Trust access control policy with better and more granular control. Essentially what this means is that when an enterprise starts using technologies like unified SOC to get more visibility into the network, and when enterprises see more traffic metadata than was previously possible, they then get the required baseline and technical strength to start embarking on the Zero Trust journey.

Keep this in mind and frame a mission statement for Zenith Trust Bank. Write down the statement to present to leadership.

Key Attribute

Mission Statement:

- *We aim to provide the required security to a flow with the right context and risk awareness rather than provide siloed security controls.*

Implementing Identity and Access Management Enhancements

Strengthening one's identity and access management (IAM) strategy is both a prerequisite and a technological driver. Whereas some enterprises might have a moderately mature IAM, Zero Trust demands a highly mature and automated IAM. With a high-level strategy to move to Zero Trust, enterprises now see that augmenting their IAM solution inevitably takes them down the road of Zero Trust. The information and attributes provided in a Zero Trust environment have much better value to the decision-maker in the network. Similar to visibility, IAM and identity-centric policies form a core of Zero Trust architectures, and when the cloud is part of the enterprise strategy, IAM on the cloud will drive the change in the overall security strategy, thus leading the enterprise to explore Zero Trust.

Keep this in mind and frame a mission statement for Zenith Trust Bank. Write down the statement to present to leadership.

Key Attributes

Mission Objectives:

- *Improve user's access experience to their workload with tailor-made enterprise-specific identity and access management strategy.*

- *Strive to achieve an enterprise-wide risk-based identity-centric policy that considers real-time changes in identity posture and manages access with just-in-time and need-to-know session management.*

In summary, leadership wants to transform their enterprise user experience with the same or better security posture, and Zero Trust is a means to achieve their vision. Leadership wants to align with whatever their current business is and will consider their Zero Trust journey a success when the desired overarching business driver has been achieved. To achieve the vision, mission statements are crafted to provide a set of more specific goals that can all converge to the one common and critical business vision identified as part of initial discussions. Zero Trust, as an access model and security framework, is a strategic adoption of achieving the mission statements. It is also an important task to convert these mission statements to key risk indicators and key performance indicators, which you will explore in Chapter 5, "Measuring Zero Trust Success," and Chapter 6, "Understanding Zero Trust Maturity."

Common Drivers for Zero Trust Adoption

Apart from business and technology drivers, there could be specific conditions or requirements that will drive adoption of Zero Trust. It is not always that a specific technology or business requirement will lead to Zero Trust considerations. It is common for an enterprise to approach Zero Trust in response to a specific risk indicator or a specific performance concern in its existing network. In this section, the discussion will revolve around some of the common use cases that greatly benefit from Zero Trust. Some of these are fairly common, while others are more niche.

VPN-Less Access of Enterprise Services

VPN technology was innovative when first introduced (and still is), but as of today there have been too many concerns with the growing home-based workforce. It is a bottleneck in the network, where bandwidth is limited and control is not easy. The access control must be done at a NGFW or VPN concentrator, and there is a dependency on subnets and VPN pools, which leads to much more complex VPN designs. Add that to a constantly growing workforce, and we have a technology that was not meant for scale trying to support network access to a diverse and large set of end users.

Another aspect of VPN is the complacence people gain when they implement the solution. For example, if an enterprise has around 500 cameras and they connect to their camera hub via a site-to-site VPN, then is the camera network secure? Yes, the *transport* between the camera and the camera hub is secure, but is the camera hub constantly authenticating the cameras? Does the camera hub allow the camera to perform actions it doesn't need? Or does the camera hub at least profile the cameras to make sure they are still cameras and not endpoints emulating cameras? And let's not even get into scalability issues. What if our camera workload doubles? Will the camera hub be able to handle the increase in traffic? Couldn't we just have used HTTPS with the right authentication and authorization mechanism?

With Zero Trust, all that is needed is a web portal to provide the user access. With the right MFA and authentication deployment, an accurate access control decision point is an important consideration to either allow or deny access. This is a fairly straightforward and simple method of providing access to the internal resources and is much more preferred by users but avoided by operation engineers. VPN-less access of services is a mission statement that can be achieved with the myriad of security service edge providers in the market today.

Shifting away from VPN and adopting a more ZTNA-based approach remains as the primary use case for many enterprises to move to Zero Trust. Not everyone is happy with using VPNs due to its erratic issues with remote access clients and scalability concerns with site-to-site VPN. Even with technology like DMVPN, FlexVPNs, and many other VPNs, which claim to support scalability and ease of use, the design and deployment are cumbersome needing many hours of thought spent in making sure the design is risk free. Exposing internal servers to the Internet and controlling ubiquitous access to services is by far the most common driver to steer away from VPN and move toward Zero Trust. Enterprises are either keenly looking to get rid of the existing VPN solution and move to a robust Zero Trust model or are implementing non-VPN solutions already like Virtual Desktop Infrastructure (VDI), per application VPNs or inbound SSE solutions. As we have observed while we were peeling the onion, some of the leading vendors in security showcase their own Zero Trust solution as a means to do away with VPNs. This is another reason why enterprises that have a specific security vendor may choose to adopt the vendor's Zero Trust principles, and these principles might involve movement away from VPN.

Figure 4-2 showcases what the traditional VPN design looks like.

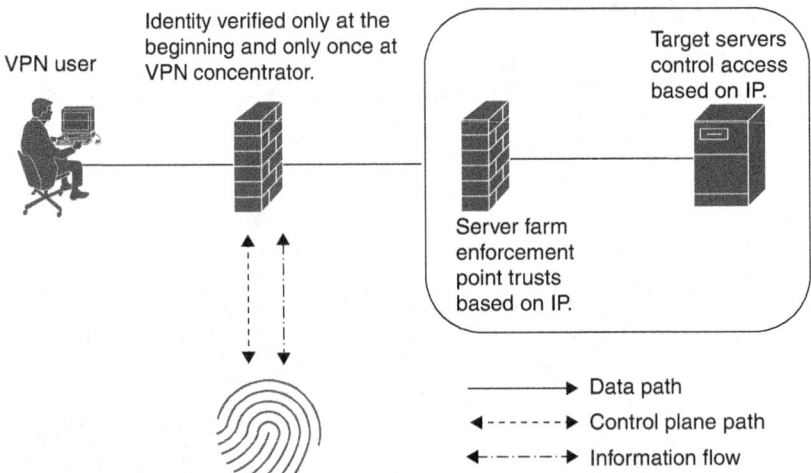

Figure 4-2   *A Traditional VPN Use Case*

Do not worry if you see unfamiliar terms like N-PEP and A-PEP. You will deep dive into these terms in detail when you explore Zero Trust Designs in Chapter 8, "Building a Zero Trust Architecture" and Chapter 9, "Critical Security Mechanisms for Zero Trust." An N-PEP is a network enforcement point like a firewall or next-generation firewall. An N-PEP has the capability to control traffic inline (allow, deny). An A-PEP is an application enforcement point that resides local to the application. Consider modern VPN solutions based on the illustration in Figure 4-2. VPNs extend the concept of a secure perimeter. The previous conversation of complacency becomes more relevant here. Because the user accesses the network via a VPN, the assumption is engrained into both the user and operational engineers that the access is secure as long as the user is authenticated at the VPN gateway. Control on the server farm firewall is loose because there is an implicit assumption that if traffic has reached the server farm segment, it is more trusted than external traffic. This is what Zero Trust aims to do away with.

VPN is definitely not dead and in no way are we reducing the impact of having a VPN solution built into the network or even using an existing one. There are setbacks, however, that we need to address to truly understand why moving away from VPN (especially remote access VPN) is becoming a key driver to adopt Zero Trust. There are certainly use cases where site-to-site VPNs or direct access VPNs stay relevant as part of the hybrid workplace use case; however, the primary use case of "access from anywhere" will no longer see a need to depend on remote access VPN. Similarly, we cannot do away with IPsec VPNs; however, implementing them as part of a DMVPN over MPLS solution has lesser impact than implementing it as part of a more modern SD-WAN solution. VPN is both a business and technology driver, similar to flow visibility. An enterprise may aim to improve user experience, which supports the overall business. At the same time, an enterprise might see multiple issues due to scale and technology, which could inevitably force leadership to look for alternate methods of providing access to the workload. This will make departure from VPN a technology driver. It is important to understand some of these setbacks before we understand how Zero Trust can solve these problems.

VPN Setbacks

VPN as a technology cannot be entirely eradicated; however, there are several challenges that enterprises face with VPN. The sections that follow highlight the setbacks encountered with modern VPN implementations.

Perimeter Extension

One of the main setbacks of VPN technology is the concept of perimeter extension. VPNs are essentially a perimeter extension, and it needs the service being accessed to be connected to the corporate network. Imagine a scenario where the enterprise is fairly strict with their network access and all resources are essentially not exposed to the Internet. If this enterprise has workload running in the cloud, either they have to spend time and effort in setting up a remote access VPN on the cloud or connect a direct connection line from the cloud to on-premises devices and then access it via the VPN. This is a site-to-site private line VPN and is cumbersome and impractical yet a fairly common deployment scenario in today's hybrid multicloud data centers. Rather than extend the perimeter, the best course would just be to assume the resource as its own security perimeter and just control access to it with the right security controls. This will also allow achieving end-to-end encryption directly to the application rather than depending on the network. Zero Trust solutions simplify the access model by removing the dependency of location as well as securing traffic across the Internet. Access is given to a resource as long as the enforcement points can communicate to a decision-maker in the network to request for an access request verdict.

Trust Extension

Another aspect of VPN is the *extension of trust*. Traditionally, a VPN user is considered trusted, and enterprises do not focus on adding in security controls to devices from which VPN users connect from, but as recent breaches have revealed, a VPN user is all an attacker needs to perform push phishing and then access on-premises servers. If a VPN user is given full access to the server farm based on its subnet rather than the user's role and context, this opens a new attack surface for malicious actors.

Scale

No discussion about VPN would be complete without discussing work from home. COVID-19 taught enterprises a valuable lesson, which is productivity is not location specific. Similar to Zero Trust, it doesn't matter where the employee is as long as the work can be done. This led to new limitations on the number of people connecting to a VPN concentrator. Most enterprises were not ready for this drastic and permanent increase. Many enterprises looked to expand licenses for VPN devices, but some chose to understand the underlying issue. Whereas expanding the scale of the VPN users or redesigning VPN solution to scale users was a viable temporary solution, a more permanent solution would be to remove the dependency of scale altogether. As long as the application can maintain an encrypted path to the end client and as long as the end device was postured according to the enterprise policy, there was no need for concern regarding external on-path (previously known as man-in-the-middle) attacks or traffic sniffing.

Location-Specific VPN

There are VPN solutions deployed where different sets of resources are placed at various locations like a primary server farm data center(DC) and a backup disaster recovery server farm site (DR). To cater to an active/active application profiles, users need to switch VPNs to access the various workloads. For example, a user can access their remote server network on the corporate VPN but the DMZ workload via a DMZ VPN. This will provide isolation of networks but will affect user experience. Similarly, the same workload needs a separate VPN tunnel for various contractors, end users, employees, and so on. Rather than simplifying and unifying the access for all users by utilizing role-based access controls, VPN complicates the solution.

Limited Inspection Before Decryption

Finally, VPNs hugely restrict the bandwidth capabilities of the solution and have limited inspection capabilities on encrypted traffic. The traffic must be decrypted at the VPN headend and can be inspected only after decryption, which adds to the already increasing latency. Any security devices in the path before decryption cannot inspect inner payload and hence provide limited security capabilities for that traffic.

The Zero Trust Advantage

When enterprises move to Zero Trust, the dependency on maintenance of scale is removed completely. Zero Trust only needs to have validation mechanisms for a specific subject's access to a specific resource under certain conditions. Consider Figure 4-3 as an example.

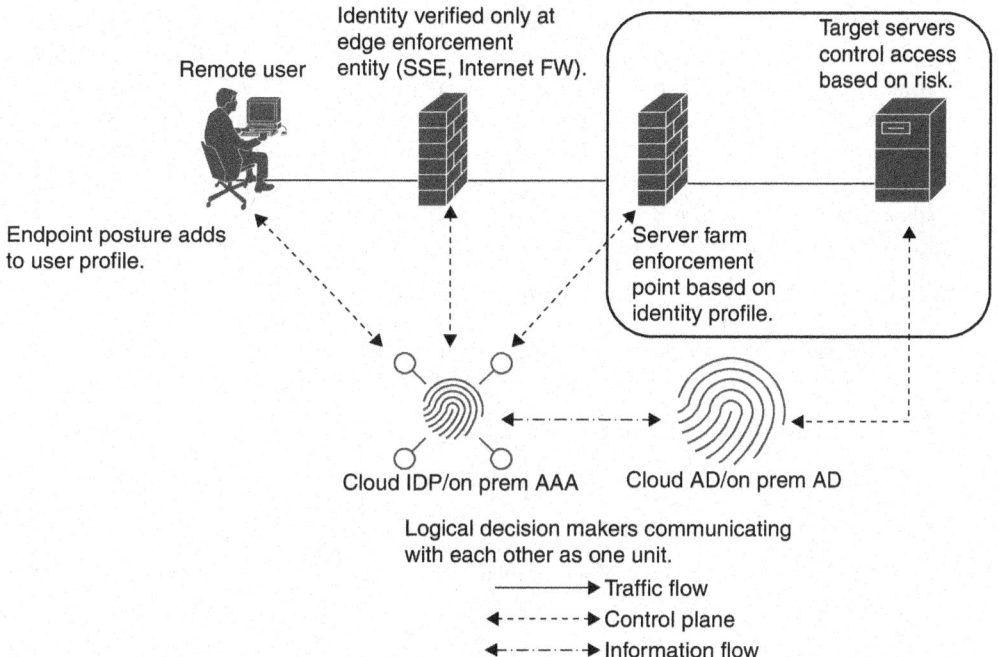

Figure 4-3 *Zero Trust Network Access Use Case*

Zero Trust solutions are simple to implement but could be complex to design depending on what technology is already deployed. There are multiple enforcement points that basically get decisions on whether the traffic needs to be allowed or not from a common logical decision point. The core of the solution is to centralize policies, which drives home the software-defined policies paradigm. There is no concern about traffic being encrypted because user-to-server communication will be encrypted end to end at an application layer (HTTPS, SFTP, and so on). As long as the enforcement point gets access to the decision point and gets a verdict based on the information processed by multiple sources of information, the access is extended at each junction of the network. This also propagates the "access anything from anywhere by anyone as long as it is allowed" paradigm. Operations become simpler, and it is much easier to track who accesses what than to check which access rule was hit on a firewall. Enterprises, however, might struggle with legacy applications that cannot provide any form of application-level encryption. The architecture must cater to placing controls protecting such legacy setup so that the mechanisms deployed can provide the required encryption; however, there will have to be some acceptable level of implicit trust. It is not ideal but it is unavoidable.

Note the addition of multiple enforcement points, including an endpoint protection agent, which is most likely a posture agent or Zero Trust agent. Ideally, there will be some level of fine-grain decision-making capability at the application itself, but the idea is to remove the decision-making aspect from data path as much as possible. For example, a database activity monitoring (DAM) solution should be enabled on most database applications, as this is an application-specific control. The DAM solution will get context from a common entity like an Active Directory server or an AAA server, and the decision to act on the built context will be local to the DAM. On the contrary, a solution like network access control or basic firewalling could be controlled programmatically to edit policies based on flows or identity. A common AAA server or Active Directory would evaluate the context of the flow and provide a verdict action that the network device can adopt and implement without performing any decision-making on its own.

The advantages of this approach are manifold, including the following:

- Adding new subjects doesn't need a configuration change on the enforcement points. The subject just needs to be added to the right group to gauge the right risk profile.

- Users access their applications from anywhere with the right security controls.

- Resources across multiple locations don't need a switch in VPN, and so on.

- There is no need to decrypt encrypted traffic to enforce policies.

- There are no more identity silos. All identities are synced to the centralized decision-maker.

- There are no bandwidth or user count bottlenecks due to VPN concentrator hardware limitation.

- There are no concerns of simultaneous logins for VPN.

Moving away from remote access VPNs gradually is generally a good place to start for Zero Trust deployments. The messaging is definitely not to completely do away with VPNs, and generally most enterprises will deploy a hybrid setup where both remote access VPN agents (more likely VPN as a Service) and Zero Trust agents exist together. The starting point for most initiatives driven by Zero Trust Network Access (ZTNA) is enhancing the user segmentation in the enterprise. Once the user segmentation and Active Directory augmentations are complete, the move toward network and application Zero Trust begins. Since the focus of this driver is to enhance the way users access resources, the focus is also on segmenting them according to policy and controlling access as granularly as possible based on identity.

It is important to highlight that when we discuss VPNs, we are considering both site-to-site and remote access VPNs as part of the VPN-less access use case; however, the primary of course in this case is remote access VPN for end users. There are use cases relevant to workloads accessing other workloads across S2S VPNs or SD-WAN, and, in all likelihood, it will be increasingly difficult to incorporate non-VPN-based access for workloads. Another aspect utilizing the SD-WAN capability is branch-to-headquarter communication. SD-WAN, which is a key secure access service edge component, has IPsec tunnels to propagate data between branches and headquarters across various transport media such as the Internet and MPLS. SD-WAN, along with SSE, builds the secure access service edge (SASE) architecture. Hence, it is important to note that we cannot do away with VPNs. However, the direction of the technology is to merge security capabilities (SSE, NGIPS, NGFW) with the underlying transport encryption capability (SDWAN with IPsec) to create one unified secure access service edge architecture. The key driver for typically most sales pitches when it comes to Zero Trust solutions is to move away from remote access VPNs (though most pitches use just "VPN" as a general term). However, site-to-site VPNs and Direct Connect links will likely be persistent, and a hybrid environment will be the most practical approach to achieve Zero Trust Network Access.

Keep this in mind and frame a mission statement for Zenith Trust Bank. Write down the statement to present to leadership.

Key Attribute

Mission Statements:

- *We strive to make sure that access control is simplified for users with innovative means to identify, authenticate, and authorize them.*

- *Utilize technology that scales with an increasing user base and expanding workload with a chance to cause minimal downtime.*

- *All users will access only their applications from anywhere across any transport with the right need to know access control.*

Uniform Policy Enforcement Across Locations

An enterprise having multiple branches and network initiatives running parallelly might realize that managing the policy at each location is not simple. On the contrary, it is an operational overhead to maintain policy at different locations. Custodians across various locations have no synchronization between each other, and essentially policies tend to become local and location specific. The second consideration is the enforcement of least privilege. Without having uniform policy for an end-to-end flow, different enforcement points will enforce traffic based on the administrator's configuration and technical capabilities of the local enforcement point. Hence, some enforcement points might not effectively implement least privilege, which then violates basic Zero Trust prerequisites.

This leads some enterprises to evaluate Zero Trust so that they can implement a common unified policy across all locations and make sure the principle of least privilege is applied across an entire flow and not at each security enforcement point. Policy is not restricted to just operational access rules but also tactical policies such as endpoint policies and segmentation policies.

Concerns with Lack of Uniform Policy

This section covers some common pain points relating to non-uniform policy enforcement. Note that most of these pain points are more relevant to a large and mature enterprise with many employees and locations. It is rare to finds small enterprises strategizing uniform policy deployment as a priority.

Loss of Contextual Detail Across Locations

A common concern observed when designing segmentation strategy is that the enterprise itself has multiple locations with various functions. Enterprises spend time and effort building a strong segmentation strategy with the help of vendors like Cisco, but the implementation of such a strategy is done only at a specific segment like the headquarters. This headquarters, or the central hub, is where all traffic must be routed to. For example, the headquarters has multiple virtual router forwarding (VRF) instances for wired and wireless employees, and for each employee there are segments based on their departments. When the branch network is validated for the same routing design, within the same enterprise, it is observed that there is just one user group assigned, which is a branch user.

Alternatively, when the traffic reaches the branches from headquarters (HQ) or other branches, the context or attributes assigned from the headquarters is lost because there is no way to propagate the context assigned to specific traffic. To illustrate the meaning of this concern, if the HQ monitors SSH traffic from the branch, and the incident response (IR) team considers the activity anomalous, how would the IR team identify which endpoint or user has been targeted across a branch of around 200 users? How could they isolate this SSH traffic from interacting with other server farms or critical information infrastructure (CII)? Figure 4-4 illustrates the scenario in a standard server-farm-to-Internet flow use case. A compromised server can be protected only if the C2 URL it reaches out to is part of known malware.

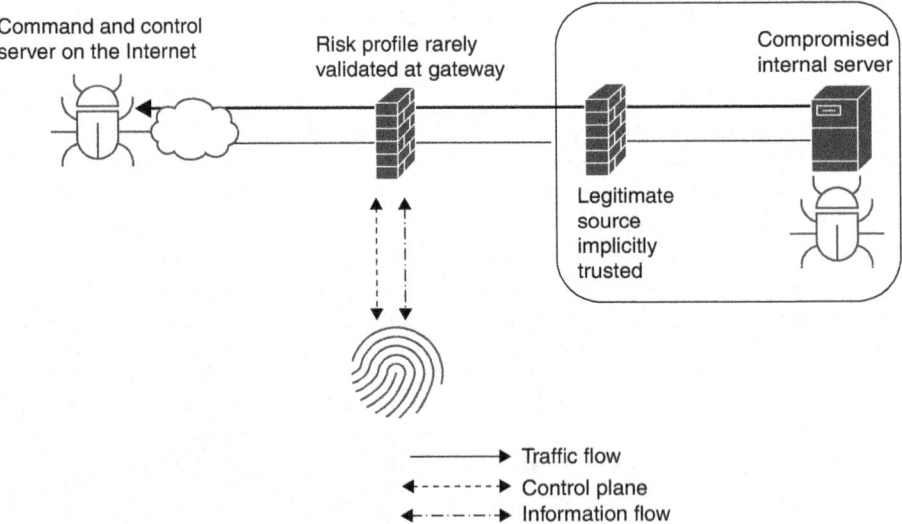

Figure 4-4   *A Malicious CNC Exfiltrating Traffic from the Branch via SSH*

Another example to consider is if a remote user accesses the VPN located in the HQ and then tries to access the branch IP phone for voice communication with a branch user, as illustrated in Figure 4-5. The only way for an enforcement point in the branch to understand if the user is VPN or not is by its IP information, which alone is not sufficient to provide context-based access.

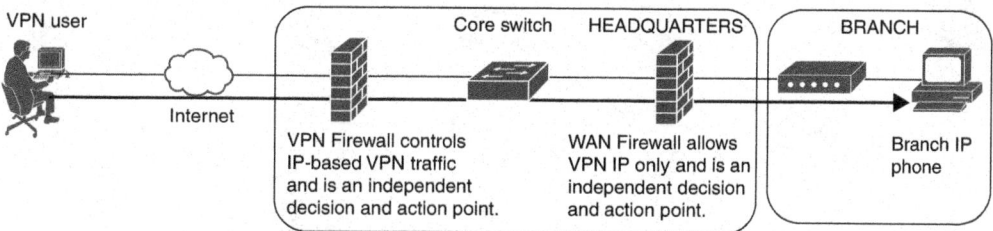

Figure 4-5 *Branch IP Phone Access by VPN User*

Finally, let's consider the common practice backhauling of all Internet traffic to the headquarters (HQ). Figure 4-6 illustrates that though traffic from an employee's device originating from the branch enters HQ, the traffic itself will have no identity attribute allocated to it and hence gets enforced based on the local enforcement point's policy which is likely a firewall. The most common concern with this form of access is the lack of control. Employee A is Employee A when accessing the Internet from campus but is VPN User A when accessing it via a VPN and is Branch User A when accessing the Internet from the branch. Irrespective of whether differentiated Internet access is desired, an employee needs to wear multiple hats when accessing common resources, and those hats are most likely tied to their location (IP address). This is counterproductive to the overall identity-based access control paradigm.

Lack of Unified Security Controls and Mechanisms (Hardware Capabilities)

Most enterprises do not implement the same security controls at branches. Some consider them as an extension and others consider them as just cold sites for Internet access. Either way, branches do not get the same segmentation or security treatment as head offices, and this is a huge concern, as illustrated further in Figure 4-6. A supply chain attack originating from a branch becomes more real than ever. A branch user whose posture is not checked is now a threat to HQ.

Increase in Bandwidth Needs and Total Cost of Ownership

An employee surfs the Internet at the campus and gets a fixed acceptable usage policy (AUP). When this same employee goes to a branch, there is most likely no direct Internet access set up. Due to this technical limitation, Internet access is backhauled via the DC server farm to the headquarters. At a branch, the employee's endpoint posture and user credentials are not effectively verified, and hence the branch becomes a backdoor for malicious activities toward the headquarters.

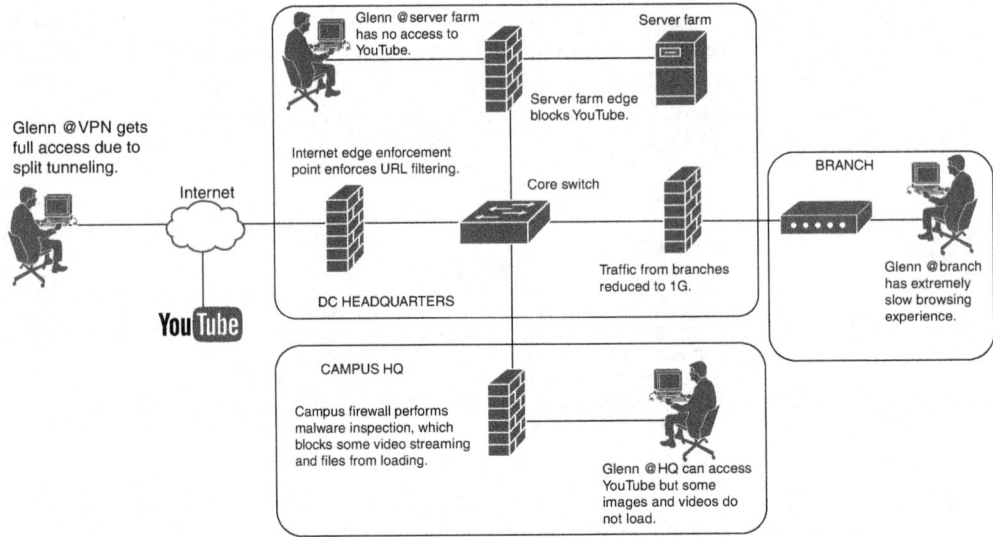

Figure 4-6   *The Same User Accessing Internet as Multiple User Roles*

Backhauling in general is considered a common practice when Internet is provided via the existing server farm. It increases latency and impacts overall user experience. This also overloads the enforcement points (firewalls, IPS) due to the additional load it needs to cater to. Alternatively, a direct Internet link and a firewall per branch might be considered too costly for some enterprises. The backhauling concept, however, is no longer a feasible solution because of impact to the HQ firewalls and the usability of the Internet leg in HQ. Couple this with VPN access, and we could easily be looking at a slow VPN/Internet link unless its upgraded.

Disjointed Access Policy in a Hybrid Environment (Enforcement)

Secure Access Service Edge (SASE) has cloud-based Security as a Service (SecaaS) as a key technology driver in the form of Security Services Edge (SSE). SASE is a network architecture model that consolidates network security and cloud services into a single, unified offering. The term was coined by research and advisory firm Gartner in 2019. The components of SASE include the following:

- **Network security:** This includes firewalls, intrusion prevention systems (IPSs), secure web gateways (SWGs), cloud access security brokers (CASBs), Data Loss Prevention (DLP) and more. These security components are typically deployed as cloud-based services.

- **WAN connectivity (workplace):** This includes software-defined wide area network (SD-WAN) and other networking components that provide connectivity to cloud-based services.

- **Cloud services (workload):** This includes cloud-based applications such as SaaS, IaaS, and PaaS offerings.

- **Identity and access management (workforce):** This includes authentication and authorization components such as MFA and IAM systems.

- **Analytics (visibility and automation):** This includes components that provide real-time visibility into network and security events, as well as the ability to identify and respond to security threats.

When adoption of SASE is considered, a branch user with direct Internet access will be subject to specific policies that are not the same as a campus user who is subject to an on-premises firewall. If there is no common policy validator or there is no integration between on-premises firewalls and the SASE provider, various flows get enforced differently for the same subject. If an enterprise goes with the SSE option and provides a direct Internet access to some branches without point of presence (POP) resiliency, they end up using a cloud-based security provider for primary Internet access and the on-premises server farm for secondary Internet access. The capabilities in the cloud are very different from the capabilities of the on-premises firewall. Consider now if the primary Internet link goes down: the policy that was enforced on the cloud is now enforced via a completely different policy set and less-scalable on-premises hardware. This is a strong motivator for malicious actors to DoS your Internet links in the branch. These are common challenges with a hybrid approach to edge access hence the need for a "unified" SASE architecture.

Similarly, observe that the Active Directory is a key **centralized decision-maker** as well as information point to a firewall, which is also a **localized decision-maker.** This will be the common design factor at multiple locations with different functional firewalls. For example, a campus user accessing a server workload needs to pass through multiple enforcement points, each with its own policies created by independent decision-makers. This flow, however, will not be enforced in the same way as a VPN user accessing the same workload because the enforcement points and decision-makers are different.

The Zero Trust Advantage

The primary business driver here is not only to simplify user access with common access control but also to make security operations more streamlined and easier to manage. A technical driver is to propagate the identity context to all locations in an enterprise. Consider the SASE architecture illustrated in Figure 4-7.

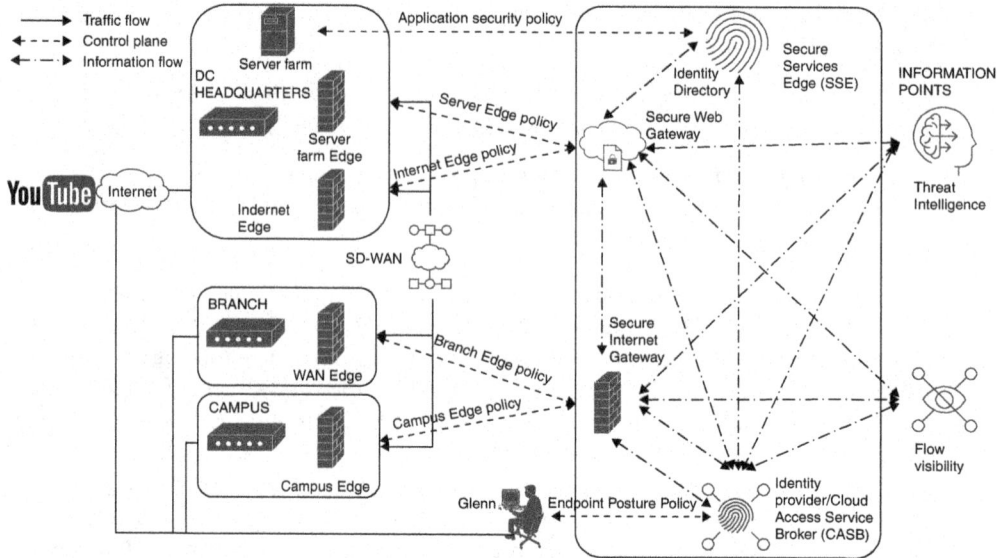

Figure 4-7   *Zero Trust Architecture for Uniform Policy*

An important aspect of this illustration is the presence of a common decision-maker who manages the entire policy. Networks following the software-defined networking paradigm have a common controller, which is the brain of the network. Similarly, in software-defined policing, a common decision-maker controls access to various segments of the network. In a Zero Trust deployment, it doesn't matter that the end user is at a branch or a different location. The context and grouping will remain the same, thus preserving the segmentation strategy. Similarly, irrespective of whether access to an Internet segment is via a cloud-based security service edge(SSE) solution or via an on-premises device like a firewall, as long as the enforcement points(SSE or firewall) get their verdict from a common logical decision point, the implementation of that verdict may be different, but the policy would still be retained. Enterprises may consider this as a strong business driver to move to Zero Trust and remove the dependency on the location or underlying network. An extension of this driver is to propagate the identity and context of access via technologies like SD-WAN coupled with Security Service Edge (SSE), which together is the SASE architecture provided by solutions like Cisco+ Secure Connect. Zero Trust helps streamline Internet access as well as overall segmentation strategy implementation for all users, irrespective of branch or HQ, and makes sure the same policy is applied across all locations.

This business driver is usually adopted by enterprises that wish to work on network readiness as an initial step toward Zero Trust Architectures and aim to make sure policy enforcement is uniform across all locations. Keep this in mind and frame a mission statement for Zenith Trust Bank. Write down the statement to present to leadership.

Key Attribute

Mission Statements:

- *Provide the same level of contextual security control to users irrespective of their location with effective segmentation and continuous monitoring for changes in security posture.*

- *Adapt effectively to increasing users and workloads by expanding to a multicloud architecture and architecting secure access across all networks.*

Secure Third-Party or Vendor Access

It used to be a common scenario in large multivendor projects for an enterprise to create temporary VPN profiles for vendors to access workloads and network devices remotely. The deployment may be done granularly with an Active Directory group and selective control, or it may be deployed temporarily as a local user on the VPN concentrator. Times have changed, and no enterprise is keen on providing access even via Active Directory to a vendor or third party—the biggest fear being supply chain attacks. A supply chain attack is a type of cyberattack that targets the software supply chain, which is the system of organizations, people, activities, information, and resources involved in creating and distributing software (in some cases hardware) products. In a supply chain attack, a cybercriminal targets a third-party vendor and injects malicious code or malware into their software product or its update. This is done in order to compromise the security of the end users who rely on that software product, as the malware is then distributed to all those who download and use the affected software. The end goal of a supply chain attack is usually to gain access to sensitive information, steal intellectual property, or install ransomware or other types of malware. Supply chain attacks are particularly dangerous because they can affect a large number of users and organizations, and they are often difficult to detect and mitigate. This is because the malicious code is hidden in legitimate software, making it difficult for security systems to identify. Imagine securing the main house with all the right controls but your barn already has a broken door that leads to the forest. Irrespective of how well you have secured the house, people will break through the broken barn door to get into your house because you have not secured the path from the barn to the house. Supply chains follow the same analogy.

Setbacks of Current Third-Party Access Methodologies

Keeping security concerns aside, managing external users and their user lifecycle is also an operational overhead for most operations teams. The sections that follow address some setbacks with the current design setup for third-party user access.

Operational Overhead

Restricting access to vendors or third-party entities is viewed as a good practice considering the enterprise is not extending implicit trust to another enterprise, but it ends up being counterproductive to completing work. Essentially, a third party's access scope is very limited. Consider a use case where a vendor has ten employees working for a customer. The customer has to provision users on the Active Directory under the vendor OU. This user then needs to have the right access to the network via a unique, isolated routing domain called virtual router forwarding (VRF). Routing needs to be configured for this vendor, and access control must be in place at all independent enforcement points like switches and firewalls. Firewall ports need to be identified, and rules must be created. Operations teams need to perform multiple tasks and complete paperwork just to add one user as a vendor and provide access to one device.

Because the resource that needs to be accessed is known, most enterprises tend to mandate being onsite before creating a temporary user on the Active Directory and allocating a separate VRF to vendors or third parties. With user and VRF segmentation, vendor traffic is expected to be isolated from the legitimate campus traffic. The concern here is that the VRF segmentation is not maintained end to end, and what ends up happening is that vendor traffic merges with corporate traffic at a fusion point like a core switch or a firewall.

Operationally, management of the user lifecycle, creation of a network adaptive to the vendor workload, and the cleaning up of users after projects are some activities that most operations teams avoid taking on, if possible, due to the large operational work involved.

As we have come to know, Zero Trust networks might also need a Zero Trust agent installed, especially if access needs to be provided to a private IP. If the server is maintained with a public IP, then all that needs to be implemented is access control to the server. Installing agents, however, is not considered feasible by many enterprises. Some of the reasons include the following:

- **Lack of awareness or understanding:** Many organizations are not aware of Zero Trust agents or do not fully understand how they work. This leads to resistance to change and general complacence. What you do not understand, you fear.

- **Complexity:** Implementing and managing Zero Trust agents are perceived to be complex and time-consuming tasks, requiring specialized knowledge and expertise.

- **Cost:** Some organizations view the cost of implementing Zero Trust agents as too high, especially if they are already using other security measures. Mass-deploying Zero Trust agents requires time and might need to be done as a project in itself.

- **Compatibility issues:** Zero Trust agents may not be compatible with all types of software or systems in the enterprise, especially legacy software, which can limit their effectiveness. Operational administrators will avoid these changes because they will need to maintain two systems in a hybrid environment.

- **Compliance issues:** Installing an agent may be prohibited as part of compliance.

- **Performance impact:** Zero Trust agents can introduce performance overhead, especially if the agent is installed on every endpoint, leading to potential issues with user experience.

If installing an agent, irrespective of whether it's a posture agent or Zero Trust agent, is not feasible, there are still alternatives where vendors are given access via Virtual Desktop Infrastructure (VDI) or cloud desktop model where these virtual devices could have the

Zero Trust agent installed in them, and we could just access these VDI devices over the Internet. Jump hosts are another alternative, where the vendor is given access to a specific jump host. Jump hosts are rarely exposed directly to the Internet, as a VDI infrastructure is much more mature (from deployment as well as operational perspective) to support the shared resource access model. Either way, these solutions are not permanent and will eventually get crippled with the scale of users, as was discussed with the jump host access example in the previous section.

In summary, operation takes a hit, and there are challenges that block the overall adoption of Zero Trust in spite of having the potential to make operations much simpler in the long run.

Vendors Are Treated as Guests

A vendor performing network activities in the environment cannot be provided with the same access requirements as a guest user. A guest does not need access to a network device to perform configuration. A vendor essentially is treated as a guest when they access the network because they cannot use corporate access. This is because there will be specific policies in place on the proxy or other enforcement points for vendors, guests, and corporate users based on IP address. Vendors and guests usually share the same IP pool because the overall perceived risk is considered the same by the enterprise, which is fundamentally wrong. Because controls on the firewalls and proxies are restricted to be IP based, the only way to segment vendors and guests is to create a separate pool and a separate routing domain, which will result in operational overhead for operations teams.

In hindsight, one could argue that it cannot be that bad, could it? We are following good security practice by restricting access to specific users only. Externally, this might seem to be a good practice at first, but in reality, the access being provided is extremely restrictive and is not very sustainable to follow from a usability perspective. Humans are the weakest link in the chain of security, and unfortunately when you make life more difficult for someone, they figure out a way to circumvent the protection. This is where Zero Trust aims to give more freedom to any user, irrespective of vendor or employee, based on identity. With technology like TACACS+, you can restrict what commands a vendor can perform on the device and completely isolate guests from internal subnets.

Cannot Control Endpoint Posture

As discussed in the "VPN-Less Access of Enterprise Services" section, VPN is a common solution for third parties to access enterprise workload, as it is generally frowned upon to expose an internal server to the Internet. With VPN being the primary method of access provisioning, enterprises start adding posture checks to validate the endpoint security of devices connecting to the network. There is one drawback, however. The enterprise cannot enforce its endpoint security posture on a vendor or third-party laptop. An enterprise might have an agent that needs administrative access on the device, and the partner's information security policy might not allow privileged applications to run. The end user could just ignore posture checks, and if posture checks cannot be validated or ignored, alternate means of connections could lead to many more backdoors to the organization. Enterprises also lose visibility into the unmanaged devices connecting to the network, and without a strong visibility solution, policies cannot be enforced accurately. The only option for inspecting and blocking traffic is at the VPN headend, which isn't very effective because it does not have a lot of visibility into the endpoint except for IP address.

The Zero Trust Advantage

Consider the illustration in Figure 4-8, which depicts a third-party user accessing enterprise workload. A vendor could access a SaaS hosted by the enterprise or an internal server farm–hosted application. Depending on the architecture, access would be varied.

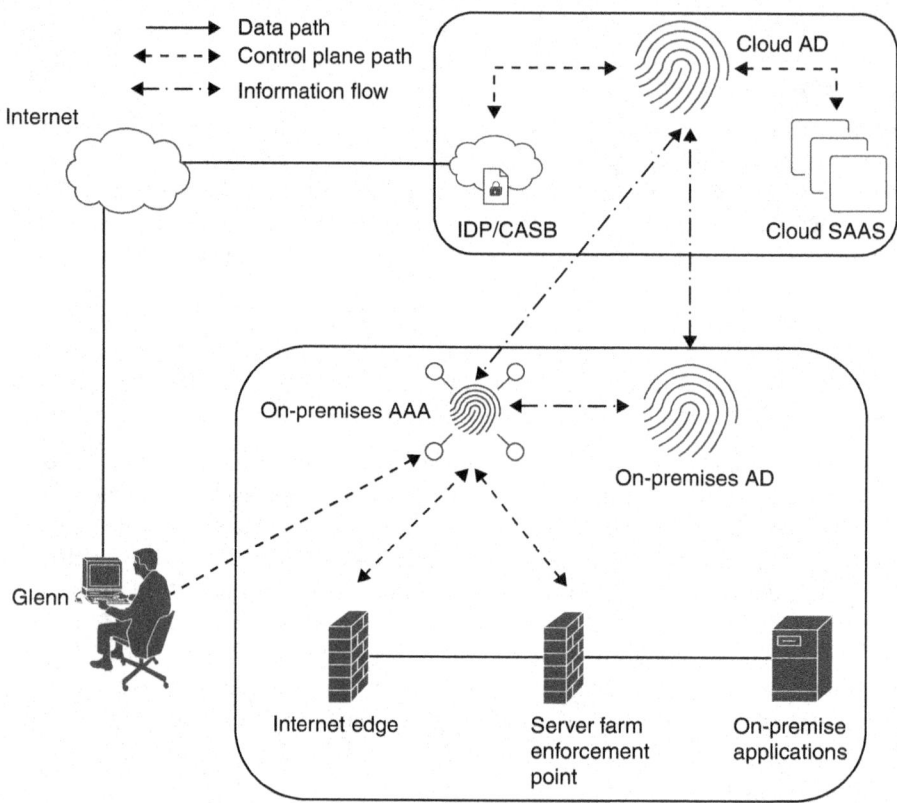

Figure 4-8 *Vendor Accessing Zero Trust Workload*

Enterprises will eventually have to make peace with the fact that some endpoints (especially for third-party and guest users) will not have postured endpoints, and control must be at the network and application access level rather than identifying users and blocking them at the network level. This is critical because when enterprises implement Zero Trust, they might choose to enforce a Zero Trust agent, which essentially works as a Zero Trust posture agent as well. Irrespective of being a combined agent or having no agent at all, not having control over the user endpoint doesn't mean that we give up on security posture.

Vendors might be given access to any workload as is deemed necessary as part of the project they are supporting. The enforcement points on the path will be existing on-premises devices as well as new cloud enforcement points (like a secure Internet gateway and so on). The decision points in the network (like Cisco ISE and Microsoft Active Directory) must be synced with each other as well as with the on-premises and cloud enforcement points to make sure that the policy enforced is uniform. Cloud workload as well as on-prem workload are easily managed without having to go through the additional operational overhead of rule creation. As long as access to the application is protected by multifactor authentication and application-specific enforcement protections, the access is evaluated based on the subject's context and risk profile. If the risk of an enterprise being accessed by a vendor is high (due to existing or past supply chain attack history), access can be denied after risk processing.

Large enterprises that have multiple projects running in parallel, with multiple vendors accessing the network at various times, will see Zero Trust as a secure access model to move toward without impacting the user experience. This is usually adopted by enterprises that wish to uplift their user identity and access management infrastructure as an initial step toward Zero Trust Architectures.

Keep this in mind and frame a mission statement for Zenith Trust Bank. Write down the statement to present to leadership.

Key Attribute

Mission Statements:

- *Aim to provide secure workload and network device access to vendors and third-party users with accurate segmentation and differentiated policies.*

- *Create simple, scalable, and timely configuration for Day 2 operations to manage.*

- *Make task completion simple for vendors and third-party users by enforcing uniform policies on all enforcement points without impacting the user experience by building risk-based profiles.*

Cloud Migration and Multicloud Architectures

As validated in previous chapters, cloud adoption and migration is a strong business as well as technology driver for Zero Trust adoption. Migration to the cloud is just a matter of time, and when the migration occurs, planets collide. Access control in the cloud is a completely different beast, and taming this beast is not easy. When workload moves to the cloud, it is imperative to understand the technical challenges of extending the perimeter or setting up VPNs to access these resources.

Challenges of Migration to the Cloud without Considering Zero Trust

The sections that follow detail some common challenges that most enterprises will face when they consider migration of workload to the cloud.

Unplanned Lift-shift or Re-platforming Migrations

The lift-shift method of migration involves moving existing applications and workloads from on-premises infrastructure to the cloud with minimal modifications. The application runs on the same operating system in the cloud as it did on the premises, but it is hosted on a cloud-based infrastructure. This by far brings the most diverse set of workloads together. You are moving applications that are used to being completely private and placing them in the cloud where access is ubiquitous. Compliance factors change, and security is stringent. Without the right security strategic road map, it is inevitable that there will be a security lapse when protecting assets in the cloud.

Re-platforming occurs when certain aspects of the application architecture are changed, and in most cases, it is likely that the security architecture is not modified to fit the cloud model. An enterprise cannot survive on the cloud by moving an entire flat network all the way to a single virtual private cloud (VPC). Concepts like segmentation and Security as a Service will start contributing to the new security regime. Application and security architects will try their best to make sure that the additional re-platforming aligns with the original business requirements. Yet, it is common that enterprises migrate workload first, expecting to make changes to the security architecture later, only to understand that without planning the overall security architecture before migration, movement to cloud is rarely successful.

In essence, unplanned security architecture is the challenge enterprises will face. Their goal is to get the application to work and not to make sure that the application is secure. This, however, will not align with the overall vision of making data easily available and maintaining security. This is where Zero Trust supports the overall business vision. Deploying a Zero Trust model on traditional networks makes the migration to the cloud easy to map because cloud security models are already based on Zero Trust. For example, if segmentation has already been implemented at the user and network level, translating that into the cloud architecture is relatively less work in comparison to rearchitecting it on the cloud.

Lack of Cloud-Native Security

This aspect deals with applications that are created on the cloud instead of the usual re-platforming or lift and shift. Generally, if enterprises do not have an option to lift and shift or re-platform, they just subscribe to another vendor's SaaS (such as Gmail) that provides the same function. The SaaS provider handles all the security aspects of the application, which has been tailor-made for cloud environments. Alternatively, there are customers who just wish to rewrite the applications by themselves for the cloud from scratch. A key takeaway from all migration discussions regarding the cloud, including scenarios involving a complete re-write, is that similar to the cloud-native code, we also have cloud-native security, which is very different from traditional on-premises security. When enterprises move to the cloud and visualize some of these changes in security control, they see the value-add of Zero Trust and how it can be useful when implemented on premises, but are unable to substantiate it in their code. Cloud-native applications, when deployed with the apt cloud security architecture, provide extremely well-planned Zero Trust–focused security controls. The challenge is to get application owners to incorporate the Zero Trust tenets into the new cloud-native code. Initiatives like DevSecOps greatly help accelerate security insertion into cloud-native applications and are an important aspect of Zero Trust strategies.

Perimeter Extension Mentality

The main tenet of cloud networks is ubiquitous access, especially over the Internet, and enterprises decelerate this setup by connecting it back on premises with a direct link or using a VPN. Most enterprises are not ready to let go of the access control that an on-premises device provides. There are other reasons why an enterprise may not entirely move workload to the cloud, data governance being one of them, however, the general norm when it comes to cloud or hybrid cloud environments is that enterprises wish to *extend* the on-premises perimeter rather than separating them and providing access to the workload based on identity-context. Figure 4-9 illustrates one such type of cloud access.

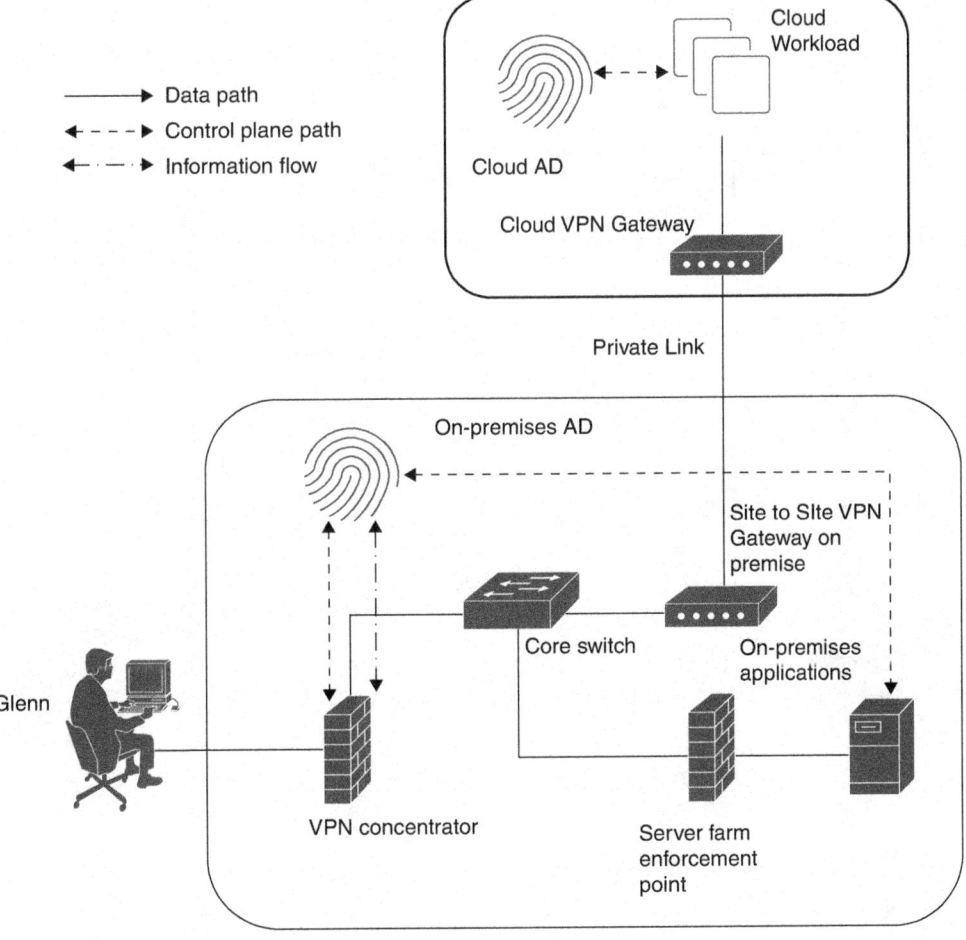

Figure 4-9 *Cloud Perimeter Extension with a Private Link (Type 1)*

In the first type of workload access, the enterprise buys a direct link to connect directly to a centralized routing or switching entity like a core switch or a server farm distribution switch. Alternatively, this direct connection might connect to the Internet firewall as well, depending on the risk appetite and security control needed by the enterprise. The catch here is that if an employee needs to access this cloud workload, the method of accessing it is via a remote access VPN to the server farm and then accessing the cloud workload via the direct link because the enterprise will not assign a public IP to this workload.

A second reason this method of access might be adopted is if the enterprise wants to restrict the IPs that can be used to access the workload in the cloud. This is generally to avoid public access to the cloud portals or workload and control the communications to the cloud itself. Along the same lines of perimeter extension, Figure 4-10 illustrates another example of cloud connectivity.

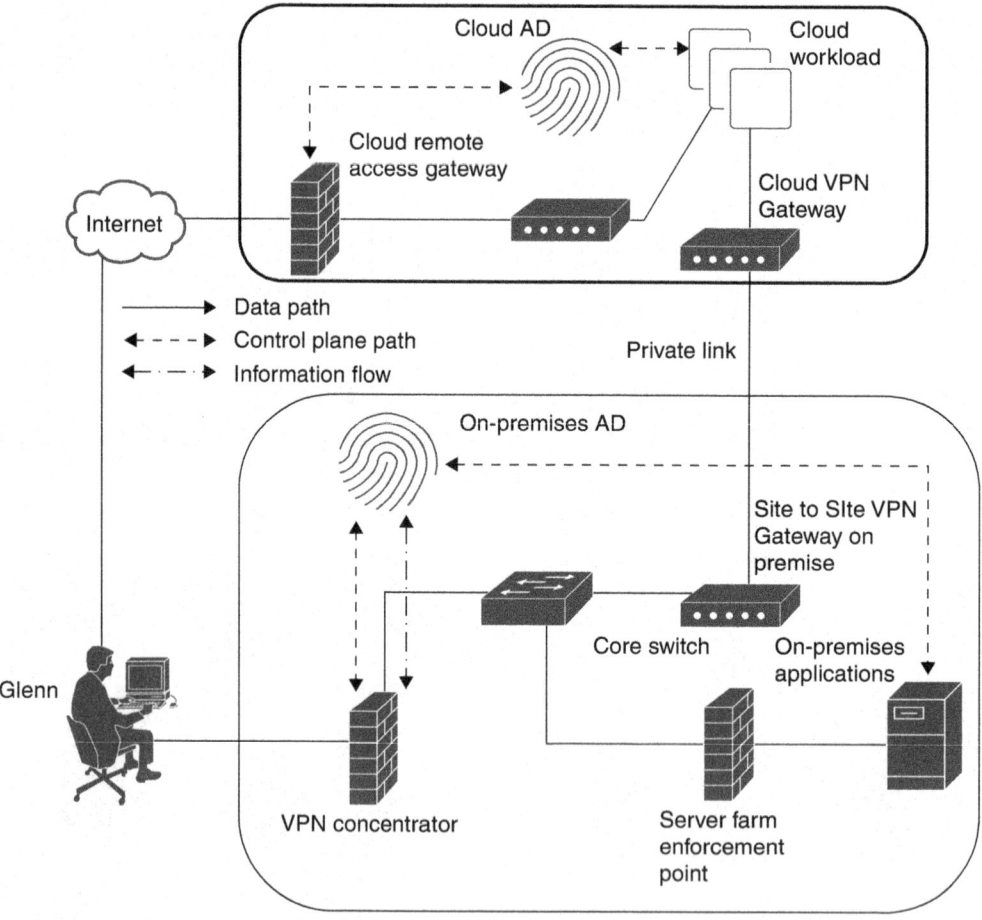

Figure 4-10 *Cloud Access to Workload via a Cloud-Based Remote Access VPN Solution (Type 2)*

The second type of access to cloud workloads is similar to the first one. Both involve using VPN to access the workload. The difference here is that users access the cloud workload via a separate remote access VPN to the cloud. The enterprises are still not ready to expose the workload IPs to the Internet, and hence server farm applications will use the direct link to communicate to the cloud workload, and all end users will access the cloud workload via a separate remote access VPN directly to the cloud provider. A key difference here is that with Type 1, the server farm firewall and the VPN firewall use the on-premises Active Directory for authentication, whereas with Type 2, the remote access firewall is on the cloud and uses the cloud Active Directory for authentication, which consistently breaks the uniform identity policy paradigm.

The second concern is that even by moving to a high-performing cloud, you are restricting the performance by the limitations of scale on the VPN concentrator, which gains you nothing except perceived additional security.

The Zero Trust Advantage

The core tenet to explore in this type of multicloud environment is to achieve a common decision point to unify policy access across the various networks. When specific workload is moved to the cloud, there are substantial changes in the way they can be accessed and authenticated. The enterprise usually wishes to move to a common access model when it has had workload on the cloud for a long time. A common access model can be achieved with Zero Trust–enabling technology like the Security Service Edge (SSE) or multicloud security solutions like Cisco Multicloud Defense. The SSE has a decision point that integrates with the cloud identity provider, which in turn synchronizes with all the existing information points like on-premises Active Directory and cloud Active Directory. The user can connect directly to the Internet via the SSE and connect to the public IP of the workload. Multicloud Defense allows for orchestrating and abstracting policies across multiple cloud vendors.

With a change in overall mindset, workload will need to be exposed to the Internet with a public IP to keep the core tenet of ubiquitous access, but with the right security controls of MFA authentication and application security controls, the flow is still secure. An on-prem application would connect to this workload via the Security Service Edge (server-to-server communication, which we will cover in the next use case). All enforcement points will communicate to the cloud decision-maker, which is the SSE. Note that the flows we are focusing on here are related to workload, which can have a public IP. All subjects will access the public IP of the workload. All communication is encrypted, and irrespective of how the user is accessing this application, as long as the access model is the same, movement to the cloud will be simple because all that needs to be done is to validate access to the newly migrated workload on the cloud. The enforcement points might vary based on the extent of adoption of the Security Service Edge, but the decision points must be in sync and function as one logical decision-maker. For example, the Active Directory would move to the cloud but will still be in sync with an on-prem Active Directory server, and this logical information point in turn stays in sync with the logical decision-maker (SSE in the cloud or the on-prem firewall). Figure 4-11 illustrates the concept where a subject (user) gets controlled access to a cloud workload via the CASB and Firewall as a Service (FWaaS), which is part of the SSE capability and is still controlled locally at each enforcement point when accessing the on-premises version of the workload.

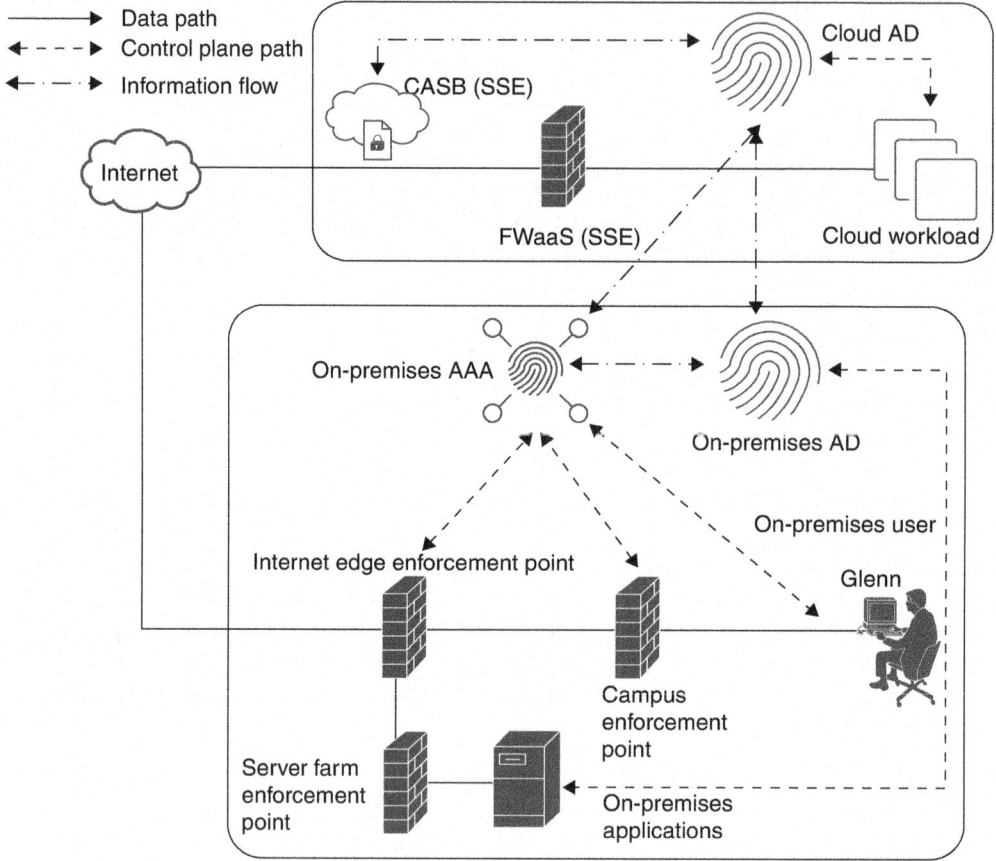

Figure 4-11   *Zero Trust Model for Cloud Migration*

Remember that movement to the cloud might appear daunting, especially when considering all the security aspects and changes required, but in the end the cloud follows the simple subject/object/context formula. If this is built into the enterprise security architecture, migration to the cloud becomes simple and easy to manage, and hence many enterprises foresee this early in the migration strategy discussion and plan to move the on-premises enterprise architecture to a Zero Trust access model before they shift workload to the cloud.

Keep this in mind and frame a mission statement for Zenith Trust Bank. Write down the statement to present to leadership.

Key Attribute

Mission Statements:

- *Aim to provide state-of-the-art security for services both on premises as well as the cloud workload.*

- *Empower businesses to embrace the cloud with confidence. Provide seamless migration of on-premises workload to the cloud, prioritizing continuous monitoring, authentication, segmentation, and access control.*

- *Make policy design and operations on the cloud simple with a unified security policy.*

- *Abstract the access policy and remove the dependency of product or platform specific rules to achieve simpler, manageable and unified access policy.*

Inter-Workload Visibility and Control

This use case is a more unconventional approach to Zero Trust security, but it does not mean that the approach is impossible. Whereas this is not a common reason, there are enterprises that wish to control server-to-server access via a Zero Trust policy. The reason this is not commonly done is because servers are generally assumed to be trusted, and the focus for most enterprises has always been to secure the user-to-service control, but enterprises who have tasted the success of Zero Trust and implemented it for their user-to-server access segments will see a value-add in extending Zero Trust to the server segment as well.

Alternatively, enterprises that have completed risk profiling and high-level micro-segmentation in their networks might choose to embark on the Zero Trust journey by focusing on workloads. There are multiple methods of achieving this, and based on the Zero Trust system that has been implemented, the deployment model will be either user-to-user based or server-to-server based. It is assumed that the enterprise has achieved a certain level of advanced maturity with Zero Trust to consider this use case. There are challenges, however, that enterprises will face when they embark on this workload-to-workload communication journey.

Challenges to Implement Effective Inter-Workload Zero Trust Flows

Controlling service-to-service communication has multiple challenges—the main one being that the thought of controlling this specific type of flow is an afterthought in most enterprises. Some key reasons are mentioned in the subsequent subsecions.

General Lack of Flow Visibility and Micro-Segmentation

One of the key detractors for workload-to-workload Zero Trust implementation is the overall lack of flow visibility. As touched upon earlier, enterprises consider the workload as trusted segments and do not spend a lot of time analyzing the types of flows relevant to workload in the server farms or in the cloud. Products such as Cisco Secure Workload and Cisco Application Centric Infrastructure can provide much needed insight into the workload flows, but without a strong flow mapping tool, many organizations depend on original application documentation to identify flow information, and some of these are outdated by several years. At this point, enterprises just accept that the applications are too vast and the beast cannot be tamed anymore. Hence, many organizations just move workload Zero Trust security to a later date when the application flows are more visible.

Without flow visibility, micro-segmentation also becomes difficult to implement, as there is no factor to classify specific flows. This makes it impossible to create context-based access control rules.

Siloed Decision and Enforcement Points

Today, security is separately handled at each enforcement point. Flows through the Internet firewall are governed by separate policies, whereas the flows in the server farm firewall have another set of policies. Even operations of the firewalls and decision points are isolated, and hence each flow is treated differently at different points in the network.

It is possible that the actual enforcement and decision for a flow may be taken in the cloud, based on whether endpoint Zero Trust agents are being installed. The business driver here is to achieve end-to-end Zero Trust with common enforcement policies, which means with the right micro-segmentation strategy and enforcement point placement, accurate access control will be enforced for both user-to-server and server-to-server workload. The server-to-server flow gets treated the same way across all enforcement points because the decision-maker is a logical constant. To be more specific, rather than having the Security Service Edge solution handle the user-sourced Internet-bound (outbound) flows and then control the micro-segmented inter-segment flows on the premises via NGFW or NSX, the enterprise would want a common enforcement point on the cloud, where all policies can be configured and managed. Enterprises wish to move away from multiple decision-makers due to siloed security policies and management. Fine-grain policies at the process level may be retained at the application itself, but flow-based control across applications is desired to be on a common enforcement point. SASE architectures will follow the same principle and are a common road map for enterprises embarking on Zero Trust journeys. The on-premises will connect to the public cloud using SD-WAN technology and the SD-WAN will integrate with SSE to provide the unified SASE architecture when the on-premises workload connects to the cloud workload. This is easier said than done, however, and without the right visibility and segmentation, it will remain a road map item and will see delays in implementation. The recommendation is to retain the intra-segment access control locally within each on-premises segment(billing, identity, monitoring) controlled with features like Trustsec. Use NGFW for inter-segment traffic (billing to identity) withing a location and utilize SDWAN to propagate segmentation and provide data in motion encryption for inter location(branch, campus) traffic. The SD-WAN will be enhanced by integrating with SSE to provide unified SASE. Edge inbound and outbound traffic to the internet will be controlled via SSE.

It is also important to highlight that server-to-server Zero Trust policy does not mandate SSE capability. Most server-to-server control is usually viewed as east-west traffic control, and as long as the enterprise-wide application access policy can be achieved across the subject service and the object service, Zero Trust policy can still be enforced. Based on the cloud adoption, latency requirements as well as existing Internet bandwidth, SSE might be considered as an option for uniform control of east-west communication. In all practicality, the east-west service-to-service communication will be enforced independently of the edge north-south traffic. For enterprises that consider it feasible to move service traffic to the cloud, policies will be unified.

The core detractor or challenge being highlighted here is the lack of a common enforcement point, specifically when the traffic is more lateral in nature. It is not a common use case to transfer traffic and inspection outside the enterprise's perceived trust boundary, however with enhanced SASE architectures, this use case is becoming more common with enterprises expanding aggressively to the cloud.

General Resistance to Expose Workload to Internet

What humans don't understand, they fear—and the Internet is no bed of roses. Business owners are not comfortable exposing the workload to the Internet for inspection to be performed on the SSE. In spite of end-to-end encryption, the argument is the same, which is that the Internet is untrusted and that the server farm is more trusted. This mentality of perimeter extension and implicit trust is a challenge when it comes to taking the next step in protecting workload.

The Zero Trust Advantage

A web server and a database generally have no security control between them in monolithic application environments because the traffic is considered trusted, but Zero Trust mandates the need-to-know concept. As with service-to-service communication, if a subject and object have flows that are to be allowed implicitly, they can be segmented into one enclave, and other enclaves will need to pass through an enforcement point to access the respective resources.

Consider the illustration in Figure 4-12. If the use case is considered where the application-service-to-database-service flow is implemented, one will observe some unique characteristics.

The next aspect to consider is the enforcement on the cloud. The workload is moving from an on-premises application server to the cloud and then back to an on-prem database server. This can only be done via agents installed on the applications, which build tunnels to the inspection cloud, or via a gateway sitting in front of the application itself, typically called an "application connector". Applications will reach the cloud segment with full encryption without having to be translated at the network level. For an application flow, it appears to communicate business as usual, but this traffic is actually being sent to the cloud via an encrypted channel, inspected, and then sent back to the data center via another tunnel. The cloud is the data processor. It seems counterintuitive to take traffic out of the data center and then back in, but the speed at which inspections can now be performed on the cloud justifies the detour. It promotes a hybrid workplace and workload enforcement model and a full micro-segmented network per user, per destination, per session; however, Internet links will need to be sized accordingly to cater to the increase in traffic to the cloud.

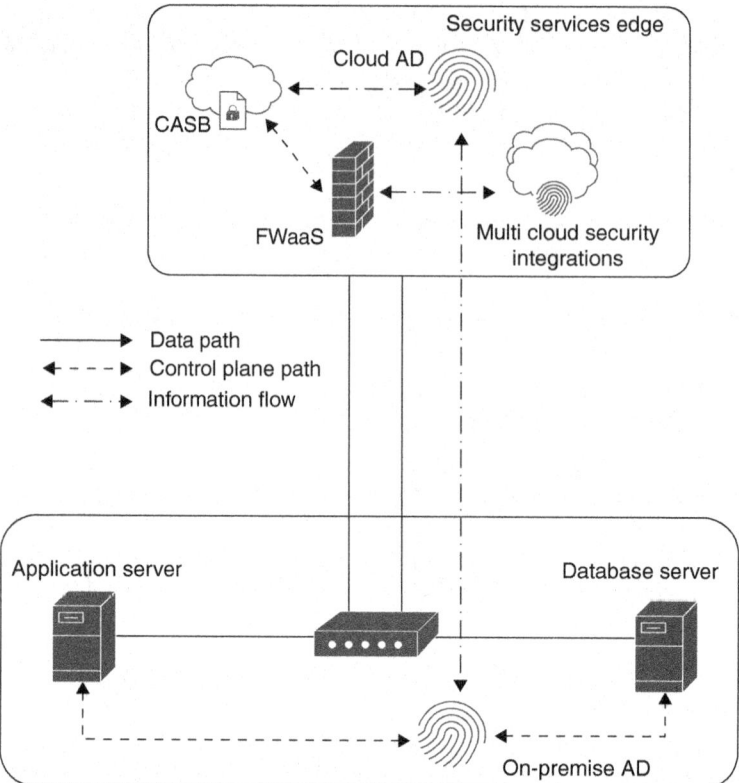

Figure 4-12   *Inter-Server Visibility and Control*

This use case is a common driver when enterprises have already done their micro-segmentation tasks and are moving toward identifying the service-to-service flows. If the enterprise chooses to keep inter-segment inspection within the enterprise boundary, SD-WAN should be utilized to achieve a hybrid workplace with data-in-motion encryption and micro-segmentation support. The core idea is that it is possible to achieve uniform policy by sending both user and server traffic to the SSE as well as though hybrid SASE models where the user traffic traverses the SSE and inter-workload communication stays within the enterprise. If SSE has been implemented, the next optional step for enterprises would be to consider service-to-service enforcement via SSE, and this becomes a use case driver for such enterprises to achieve the uniform policy principle.

Key Attribute

Mission Statements:

- *Aim to provide users access to private workloads securely with the same protection as public workloads.*

- *Allow secure and controlled communication between workloads with micro-segmentation.*

What Matters Is Why

At the end of this discussion, it is important to consider that even before you select a strategy or product, you need to know why you are embarking on this journey. Identify key business drivers and align the technical drivers to the business use cases. Once the drivers are identified, all your metrics must map back to the key drivers. Spend some time with the representatives of the enterprise you are supporting and understand their drivers.

Zero Trust sets itself apart from other security initiatives in its innate alignment to business drivers. As an information security model, Zero Trust does not start with selection of a product. Without aligning a product selection activity or even policy creation initiative to business drivers, the risk of a failed initiative increases manifold. The core advantage of Zero Trust is the importance that is given to aligning the technical and business drivers with what Zero Trust can provide. Every driver gets mapped to a metric and essentially a set of atomic projects that, when completed, accelerate the enterprise's Zero Trust journey. There is end-to-end traceability all the way from business drivers to project completion.

The Interview

[Glenn pauses to make sure that Mr. Smith, Ms. Lee, Mr. Chen and Mr. Eaton have clearly understand why it is important that he, Glenn, understands why Zenith Trust Bank wants to embark on the Zero Trust journey.]

Glenn: So essentially, I have gathered what your enterprise does, and as a bank there are some important use cases that are strong drivers for you to move toward Zero Trust. What I want to understand now is what are your crown jewels? What are your key assets you feel need to be protected with this new strategy?

Mr. Smith: So essentially being a bank, the financial assets are critical to our survival. Some of the key aspects to protect include our customers' deposits and essentially their personally identifiable information, including credit card details. I feel it is my personal responsibility that the right protection is provided to this information so that it is not lost or, worst, stolen. We also have to protect physical assets in various branches like gold, cash, etc., which demands the right process to be in place to protect the assets.

Mr. Chen: We also have intellectual properties of how we support our customers with algorithms to predict investment opportunities along with other disruptive technology. Of course, I cannot imagine achieving anything without our employees, and their data is as important to us as is the customers'. So, yes, a lot of physical assets and a lot of data assets.

Glenn: Great! That's very useful information. From what I understand, moving away from VPN and segmentation are key technical drivers to achieve your business drivers of customer and employee data protection. Just to understand, has Zenith Trust Bank been breached before?

Mr. Smith: We have had breach attempts but no breach that has led to stolen data. That's what makes me a bit antsy. Looking at the Nexus Bank situation and considering the entire audit report that was performed, I want to proactively avoid an attack and be prepared in case there is one.

Glenn: In all likelihood, it is impossible to avoid an attack, but it is definitely possible to reduce the blast radius. I believe you identified many reasons to move to a Zero Trust access model, and I will help you get support from various stakeholders in your enterprise. In essence, I will summarize what I understand as your primary business and technical drivers, and once we agree, we will be identifying key stakeholders to communicate this message to, but before that we need to identify some metrics:

Zenith Trust Bank's Strategic Business Vision:

To be the customer's trusted banking partner and their one-stop shop for all banking needs.

Zenith Trust Bank's Strategic Security Vision That Aligns with Zero Trust:

Zenith Trust Bank's vision is to provide ubiquitous, secure, and convenient access to enterprise workloads for both customers as well as employees.

Zenith Trust Bank's Tactical Mission That Aligns with Zero Trust:

- **Provide convenient and easy data access for customers and employees.**

  - Make access to network and workload simple and streamlined.

  - Empower our users with efficient, easy, and intuitive access to the information and resources they need, while ensuring their privacy and security.

  - Aim to provide secure workload access to vendors and third-party users with accurate segmentation and differentiated policies.

  - Aim to provide users access to private workloads securely with the same protection as public workloads.

- **Protect customers and employee data from fraud.**

  - Aim to provide high-quality services while maintaining the privacy and data protection of our clients.

  - Make sure that personal data is collected, processed, and stored securely and in compliance with all relevant data protection laws and regulations.

  - Protect all client and employment data at rest and when it's in motion, and inspect the traffic for anomalies efficiently without impacting performance.

- **Prioritize performance and scale supported by rapid expansion to multicloud architectures.**

  - Build comprehensive multicloud security solutions that enable the enterprise to operate securely in a diverse hybrid environment while maintaining strict security compliance standards.

- Utilize technology that scales with an increasing user base and expanding workload with minimal risk of downtime.

- Empower businesses to embrace the cloud with confidence. Provide seamless migration of on-premises workload to the cloud, prioritizing continuous monitoring, authentication, segmentation, and access control.

- **Proactive approaches to managing breach risk.**

  - Implement risk-based profiles to identify and mitigate potential security threats.

  - Continuously improve our protection practices by regularly reviewing the policies and procedures, identifying areas for improvement, and implementing new measures to safeguard against potential risks.

  - Aim to provide the apt security to a user's flow with the right context and risk awareness rather than provide siloed security controls.

  - Provide the same level of contextual security control to users irrespective of their location.

  - Allow secure and controlled communication between workloads with effective micro-segmentation.

- **Create simple and efficient application, network, and user architectures that support daily operations.**

- Implement robust security measures and ensure simple and secure access to network and workload through streamlined authentication and authorization processes and with the least operational complexity.

Does this sound accurate Mr. Smith, Mr. Chen, Ms. Lee, and Mr. Eaton?

[Mr. Smith, Mr. Chen, Ms. Lee, and Mr. Eaton agree.]

Glenn: Alright, before we move to the next activity, I want to highlight some key takeaways from our conversation as well:

- Since the enterprise has not been exposed to Zero Trust, we spent some time understanding what exactly Zero Trust is and why the enterprise would want to move towards Zero Trust.

- There are business drivers that highlight how Zero Trust drives the business objectives, and there are technology drivers that highlight how Zero Trust benefits and supports the technological challenges.

- Apart from generic business and technology drivers, there are use cases that are niche drivers for most enterprises and cover close to 90% of drivers toward Zero Trust adoption. They are:

  - VPN-less access of enterprise services

  - Uniform policy enforcement across locations

- Secure third-party or vendor access

- Cloud migration and multicloud architectures

- Inter-workload visibility and control

Mr. Smith: Alright. I think this is a lot of information to digest. Let's reconvene after five weeks and discuss the next steps. In the meantime, is there some information we can provide you to further help us begin this journey?

Glenn: Yes, could you let me know who your operations and security leads are? I can speak to them and understand some of your pain points to feed into our next phase.

Ms. Lee: Sure, please reach out to Mariam, our infrastructure lead, and William, our security lead. They are led by Jed, who should be able to provide any strategy-related information and insights. If you need other teams to get involved, let me know as well. I want to make sure we are fully equipped with all information I need to be able to present this to the board.

Glenn: Thank you Mr. Smith, Mr. Eaton, Ms. Lee, and Mr. Chen. See you in five weeks.

[Five weeks have passed, and Glenn has arrived at the Zenith Trust Bank premises to speak to Mr. Jonathan Smith, Mr. Christopher Eaton, Ms. Samantha Lee, and Mr. David Chen about how he has spent the three weeks working with the infrastructure and security leads understanding key pain points and analyzing their network infrastructure. The last time they had met, Mr. Smith had asked for some time to digest the core concepts of Zero Trust. After brief small talk, Mr. Smith gets right to the point.]

Mr. Smith: Alright, Glenn, Zero Trust does look like something we would like to explore. Where do we begin ?

Glenn: Let's now move to the next activity, which is building key metrics. You need to sell the vision, the approach, and mission and how you are going to measure their success to the board. I have spent quite some time with your infrastructure and security leads and have been able to capture some key aspects of your infrastructure and I would like to thank them for all the support and transparency. Before we go into the details, let me explain why metrics are important in the context of beginning a Zero Trust journey.

Endnotes

1. "CISO Perspectives and Progress in Deploying Zero Trust," https://cloudsecurity-alliance.org/artifacts/ciso-perspectives-and-progress-in-deploying-zero-trust

2. "Encrypted Traffic Analytics (ETA)," https://www.cisco.com/c/en/us/solutions/enterprise-networks/enterprise-network-security/eta.html

3. "Encrypted Visibility Engine (EVE)," https://secure.cisco.com/secure-firewall/docs/encrypted-visibility-engine

Chapter 5

Measuring Zero Trust Success

Initially, Glenn the consultant had the CIO's (Mr. Jonathan Smith's) attention. Now Glenn has Mr. Smith's curiosity. With more detailed discussions, leadership teams appreciate the value of Zero Trust as a concept; however, they want to know how to apply the concept to their own enterprise. Hence, the logical next step is to map specific business drivers to measurable outcomes that are aligned to the larger Zero Trust vision. Effective metric creation will help enable strategic discussions on identifying Zero Trust mission statements to drive adoption across diverse teams within the organization. A common challenge enterprises face is to convince other leaders (ops, finance, and so on) within the organization that Zero Trust has a larger impact, not only to the security architecture but to overall enterprise risk, strategy, and cost. For example, business operations teams may not see the benefit of adopting Zero Trust unless there is a tangible metric that aligns with their strategy. For a vendor-neutral consultant, it appears obvious to move to a secure access model; however, many intricate dependencies such as cost implications, politics, and overall organization position in the market need to be considered when proposing a metric, as adverse conditions might deter enterprises from implementing Zero Trust. Some of the common deterrents to adoption have been discussed in Chapter 1, "When It All Begins."

As translators of strategy and operational requirements, if consultants are unable to craft the right metric that is acceptable to leadership, they will not be able to showcase the value that Zero Trust brings to the enterprise infrastructure, processes, and people. Similarly, if the daily operational problems are not considered when crafting metrics, end users and employees will not fully appreciate the value of Zero Trust and its impact to their workflow. It is common knowledge that leadership speaks in the language of performance and risk, and the focus of this chapter will be to help craft metrics based on these key constructs, such as risk and performance. Risk and performance metrics are standard measurements that can be consumed by all business units within the enterprise.

Once leaders and adopters see the value of Zero Trust as a concept for their enterprises, they will be keen to understand from vendors how they incorporate Zero Trust not only into their products but also within the vendor's enterprise itself. For example, Cisco has been on the Zero Trust journey for quite some time and hence is a good reference point to showcase to other enterprises how they can begin their own journey. Cisco's journey also helps enterprises understand how to craft customized Zero Trust metrics to validate the efficacy of the Zero Trust initiative.

By driving Zero Trust in the enterprise, you are essentially committing to improve the security posture of the enterprise. Since you are looking at the adoption process of Zero Trust holistically, you must acknowledge that there are very few people who really see the entire Zero Trust picture. The key observants and enablers are leadership stakeholders. Unfortunately, without metrics, the value Zero Trust provides is conjecture at best for most stakeholders. Budgeting is another touchy subject. The board members must buy into your vision, and you must be able to showcase to them that the initiative will bring back quantifiable success in terms of performance improvements or risk reduction and eventually monetary gains and organizational stability. You cannot achieve these broader strategic goals without taking all precautions to protect the data of the customers and employees. At the same time, you do not want to let customers or employees create backdoors due to the extreme lengths the organization goes to secure data. That is where intelligent metrics come in.

There are some common metrics you identify to set a baseline that can be utilized to craft enterprise-specific metrics. Before any discussion about metrics begins, it is important to understand what a good metric is and why it is important to create tangible metrics.

Importance of Measurement

As a consultant, you need to help the enterprise identify a measurable metric. Consider a common example of showing the current status of a movie download in movie download software. It is a common strategy to see status messages on the software user that show "Almost done" rather than "99%" completion. There could be end users comfortable just knowing that the download will finish soon, and there might be other users who want to see the exact download percentage. Another example is the traffic lights showing a countdown to the next light change. Some people consider this a good feature on traffic lights because they prefer to switch off their vehicles when the traffic light is red and turn on the engine seconds before the light turns green. However, the number of such vehicle owners is lesser than the majority population that do not care about the time frame and just keep the engine running. What the architecture and design team must do is run a survey to understand the percentage of each of these users and decide which option to lean toward, which can bring in more utilization and value. Leadership is typically interested in maximizing recurring subscription to services or products along with increase in customer promoters. It is a strategic decision to decide what metric to consider when measuring a specific strategy. A metric like "reduce impact of an attack" is qualitative at best and in reality is very vague and broad scoped. When creating metrics, you must consider that each metric is a means to convince the listener that the scope can be measured and that the strategy is working from each stakeholder's perspective. Crafting metrics is an adoption strategy by itself and therefore requires the knack of understanding what your target audience wants. To an operations lead, the metrics should resonate with availability and ease of operations. To an enterprise architect, it would resonate with providing the right architecture and design following all best practices and compliance. A CxO would be more concerned with support to the business, recurring revenue, and risk reduction. Hence, Zero Trust shouldn't be restricted to one type of metric. It is usually an amalgamation of many metrics targeting all the stakeholders.

Another decision is the final state that has been envisioned for the enterprise. Once the metric has been crafted, the enterprise needs to decide where it would like to be from a Zero Trust access perspective, which *aligns* with their business vision. Should the enterprise target the highest maturity level or should it consider the asset value and context and decide which is the right state to be in. Enterprises need to build a meaningful and achievable metric to be able to show immediate value with tactical and operational returns and, in turn, propose more details about the strategic goals. By possessing some of the characteristics mentioned in subsequent sections, a metric can help an organization to identify, prioritize, and mitigate security risks and maintain a strong security posture to reach the desired performance and alignment to business. This will in turn drive the security budget requirement.

Deciding final state will hence depend on what the enterprise feels critical. A banking enterprise might want to consider any financial activity–related applications as important and applications handling personal identifiable information (PII) data as critical. Data classification, flow mapping, asset inventory, and segmentation will help identify the maturity vision of the enterprise based on the critical infrastructure present and identified. Once the vision is clear, the next step is to build observable, simple metrics to ensure that the capabilities around protecting these critical assets are in place.

The Metrics Lifecycle

The metrics lifecycle is usually part of the overall Zero Trust lifecycle but can still be independently showcased to understand its position and importance in the overall strategy.

There are four key steps in the lifecycle of a metric, as illustrated in Figure 5-1.

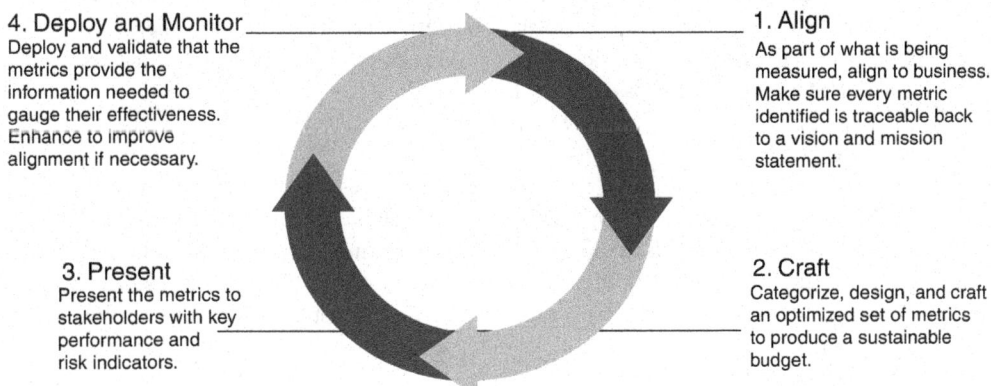

4. Deploy and Monitor
Deploy and validate that the metrics provide the information needed to gauge their effectiveness. Enhance to improve alignment if necessary.

1. Align
As part of what is being measured, align to business. Make sure every metric identified is traceable back to a vision and mission statement.

3. Present
Present the metrics to stakeholders with key performance and risk indicators.

2. Craft
Categorize, design, and craft an optimized set of metrics to produce a sustainable budget.

Figure 5-1 *Zero Trust Metrics Lifecycle*

The first step is to align any metric to a business driver and, in effect, a security driver. This is a primary reason why metrics are not crafted early on in the discussion with the CIO. One cannot meet a CIO and commit to provide 100% availability and uptime without understanding their business drivers and organizational dynamics. Once the metrics are aligned to the Zero Trust vision and mission, the next step is to craft intelligent metrics that are achievable and quantifiable. In this chapter, the focus is going to be on aligning and crafting metrics. Once metrics are crafted, they need to be presented to the respective stakeholders for approval. This usually happens when the overall strategy is being presented along with the Zero Trust team. Once the Zero Trust initiative has been deployed and metrics are being actively measured, feedback from the implementation and operation teams must be incorporated to the overall metric definition to make the metrics more robust.

Step 1: Align Metrics

Creating metrics must always begin with alignment to the vision and mission. The final goal of metric creation is to measure an activity, process, or capability that can produce an actionable and measurable outcome to support the overall vision or initiatives. Classification of metrics can be varied, depending on whether they are goal-oriented or based on how they are derived.

Types of Metrics Based on Target Goals

The most common taxonomy of metrics includes strategic, operational, and tactical. These are goal-oriented and are crafted based on the type of goal or activity that is being measured.

- Strategic Metrics

  Strategic metrics are extremely high-level metrics that measure the overall success of an organization in achieving its long-term goals. Strategic metrics typically focus on outcomes, such as revenue growth, market share, or customer satisfaction. They are often used by senior executives and stakeholders to evaluate the performance of the organization as a whole. When overlapped with Zero Trust, strategic metrics measure how well Zero Trust has been adopted in the enterprise and how it has reduced the risk to overall business.

- Operational Metrics

  Operational metrics measure the day-to-day activities of an organization and the efficiency of its processes. Operational metrics typically focus on inputs, such as the number of sales calls made or the amount of time it takes to complete a task. They are often used by middle managers to monitor performance and identify areas for improvement in day-to-day activities. In a Zero Trust context, an example would be to measure the number of attacks detected in a day or the number of automated incidents handled.

■ Tactical Metrics

Tactical metrics are used to measure the performance of specific projects or initiatives within an organization. Tactical metrics typically focus on outputs, such as the number of products shipped or the percentage of customers who renew their contracts. They are often used by project managers to track progress of specific organizational initiatives and make adjustments as needed. In a Zero Trust context, an example of a tactical metric would be the measurement of how much infrastructure has been segmented as part of the segmentation initiative, where the extent of segmentation is the initiative being measured.

Strategic metrics measure overall efficacy of the strategy. For example, augmenting existing identity and access management would be a strategic goal. When you speak to a CxO, you need to show revenue growth and business alignment. These are strategic in nature and look far into the future. The alternate aspect of security is "operations," implying that you are communicating to personnel who handle uptime of infrastructure. Another aspect is building skillset for the management of products, staffing and so on, which can be a major metric for enterprises in locations where there is a dearth of skilled workforce. An operational goal would be reducing the number of incidents by implementing better visibility into the network. Operational metrics will measure how well the enterprise is performing at the grassroots level. This will take into considerations risks that enterprises see every day.

Tactical metrics are somewhere in between strategic and operational metrics. They measure a specific program. For example, a Zero Trust transformation project could be considered a tactical metric to achieve the overall strategic metric of protecting customers' data. Tactical metrics are focused metrics that most security engineers have not been inherently crafting. A tactical metric would be achieving 99% security awareness for the entire workforce since security awareness itself is an initiative. Attack vectors, threat actors, and many other threat hunting tactics will be deployed to measure the effectiveness of the security initiative.

Types of Metrics Based on Method of Data Analysis

The following classification of metrics is based on how the metrics are calculated and conveyed. There are two major classifications in this type of taxonomy:

■ **Quantitative metrics:** Quantitative metrics involve numerical measurements, and they are typically used to measure objective data such as the number of sales, the amount of revenue generated, or the percentage of website visitors who make a purchase. In a Zero Trust context, measurable metrics like failed attacks, reduced risk percentage, and so on are quantitative metrics. Quantitative metrics are often used to track progress toward a specific goal or to make data-driven decisions. These metrics are used for activities that can be measured with numbers, charts, and so on.

■ **Qualitative metrics:** Qualitative metrics, on the other hand, are based on subjective assessments and are typically used to measure more intangible factors such as customer satisfaction, brand perception, and employee morale. Qualitative metrics often involve gathering data through methods such as surveys, interviews, and focus groups. These metrics measure over a range of high, medium, or low. The scales can be as granular as needed.

Both qualitative and quantitative metrics are important in data analysis and research, and they often work together to provide a more complete picture of a particular phenomenon—in this case, the overall Zero Trust strategy. While quantitative metrics can provide hard numbers and measurable results, qualitative metrics can offer deeper insights into the reasons behind the data and provide a more nuanced understanding of complex issues.

Be a Translator

As a consultant, you will always find yourselves wearing various hats. Sometimes operations teams need an explanation on how the product or solution works, and on the other hand senior leadership needs another completely different explanation about metrics and strategy. Being a translator is an important aspect of consulting. Metrics are the language that the consultant needs to know as a translator. Once the right metrics are identified, it is important to translate one set of metrics to the other. An operational lead will understand performance metrics, but a translator needs to explain to leadership why a specific strategy is in place and how it drives multiple operational metrics to enhance performance and reduce risk. There must essentially be a translation between a performance metric and a risk metric, as is illustrated in Figure 5-2.

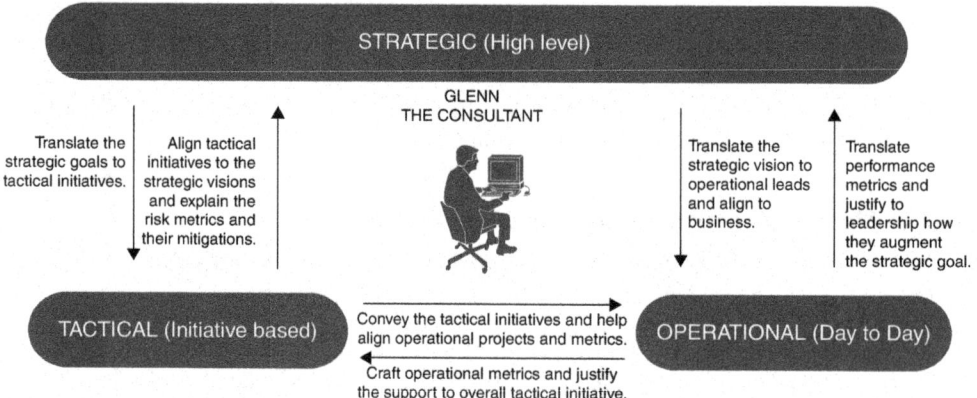

Figure 5-2 *A Translator in Action*

Operational leaders need to build performance metrics, which architects translate as a future state that the enterprise wants to be in. For example, when explaining the need for segmentation in the network, architects will explain the current state of segmentation in the network and showcase how increasing the number of micro-perimeters will help make granular policies and facilitate only need-to-know access, thus boosting productivity. This metric therefore shows a key performance indicator (KPI). Tactical leaders need to know the risk to their assets as well and how the current strategy can reduce or mitigate the risk. When architects translate the same metric to leadership, they explain the current attack surface and then showcase how the segmentation initiative augments the enterprise by protecting it from lateral movement and reducing the blast radius of an attack. These are called key risk indicators (KRIs) and are a list of threats to an enterprise and their mitigations. To recap the same segmentation example, we used the performance indicator of better granular rule creation to boost productivity. This is then translated to a risk indicator, stating that segmentation reduces the blast radius and thus lowers the risk of data loss, which is a strategic metric. A combination of this strategic metric with an operational metric adds context to what the enterprise wants to achieve. The highlight in this case is an effective combination of the key performance metrics (provide only need-to-know access, increase overall security posture) as well as risk metrics and their mitigation (reduce blast radius) to achieve alignment to overall strategic vision.

Being a translator is fascinating because, to be a translator, one not only needs to know multiple languages, but also needs to be able to understand the context of each statement and deliver it with the same emotion and context.

As a translator, you need to get feedback from tactical, strategic, and operational leads and to craft metrics that appeal to all stakeholders. Often metrics are considered as just information or as targets to achieve, where you tick a box on a huge list of security controls. A security audit would list a large set of control gaps and their mitigations, and most enterprises craft metrics based on those gaps. In a Zero Trust context, metrics are not primarily driven by technology or compliance gaps. They are driven by the business and security drivers. Without metrics, it becomes extremely difficult to quantify the effectiveness of a solution or a program. Metrics are key language definitions for leadership and leads alike to gauge how well the Zero Trust initiative is faring. This will effectively influence leaders to provide more support and budget for the tactical Zero Trust mission, which in this case is to provide an accurate access model for your assets.

When you're considering metrics, the rule of thumb is to keep them simple and achievable. There are metrics that tell you where you want to be and metrics that tell you what your risks are. From the context of creating the right metric, security governance must also be considered to make sure the metrics align with security and the strategy. As a translator, you will not only be translating tactical and operational metrics to senior leadership but also to other non-security-focused business units of the enterprise so that they understand your metrics and build their metrics while keeping their pain points in mind.

Remember the rule of thumb:

- Strategic metrics are high level and generally map to measuring the efficacy of a vision or mission statement.

- Operational metrics deal with the people and process aspects of enterprise security and help with measuring overall performance and day-to-day activities.

- Tactical metrics are specific directional metrics that map to measuring larger initiatives and driving a certain strategic goal. These deal with risks and provide key mitigations controls to be implemented.

■ All strategic, tactical, and operational metrics can be qualitative, which means they are measured on broader scales and are not supported with numbers. They can also be quantitative, which means they are measured with numbers and graphs.

Step 2: Craft Metrics

Once the vision and mission are established and desired metrics are aligned, it is time to mold the metrics into measurable and convincing outcomes to help garner support from all stakeholders.

Crafting Effective Metrics: A Bicycle Case Study

Metrics are all around us. You measure the efficacy of almost everything in your day-to-day life before investing your money and time. IT infrastructure is no different, and most enterprises have metrics in all business units, as this helps to showcase their effectiveness. This stresses the need to identify what should be measured and how well it needs to be packaged to each stakeholder.

Let's look at an example of how important metrics are. Consider a high schooler named Gary asking his father for a bicycle and mentioning the following: "My goal is to get a bicycle. I am going to pass all my exams, which should be sufficient incentive for you to buy me one." To begin with, the first and most critical error on Gary's part was not having a conversation with his father about what his father wanted. He assumed that passing his exams was sufficient to get his father to buy him a bicycle. This is a common error that many consultants make. They assume what is good for the enterprise without aligning with what the enterprise wants. The next concern is the nature of the metric of "passing" an exam. This might have been unintentional from Gary's perspective, but in reality, the metric is very open and vague. There is no quantitative information on what score he should receive in each of his exams. Gary did not concern himself with measuring his passing score since he did not confer with his father about the required criteria for buying a bike, which ties in to the first error highlighted.

Let's dive deeper into some other characteristics. Passing an exam in itself requires valid proof that cannot be refuted. Gary could craft a fake report card or even say he lost the report card and have no tangible way to prove that he passed. The key aspect of a metric is to be able to reliably achieve the same result and measure the success or failure consistently. A report card showcases pass or fail consistently, whereas speaking to other parents or even the teacher may not be as consistent, as they may not have full information on who passed and failed or may be subject to biases. The metric also does not showcase a timeframe, which traces back again to lose alignment with the father. Gary claims he will pass the exam, but when? During which semester? By studying for how much time? These are all critical pieces of information that are missing.

If Gary had spent some time with his father to understand what his father really wants, the metric would be more accurate. For example, "My goal is to be able to ride a bicycle with my friends. For me to achieve this goal, I need a bicycle. I do not have money to buy it, but I know my father does. I spoke to him and he says he can provide me some money if I can prove to him that I am academically fluent (notice the strategic metric). For me to be academically fluent, I need to pass all my exams throughout this year, which covers five subject exams and one elective. I need to pass all four quarters by achieving a grade of 80% or above. My father will save some money as I pass my first two quarters and will buy me the bike once all quarters are complete and all passing criteria have been achieved. If I am unable to achieve my passing score, there is a chance that the money saved by my father might be spent elsewhere, and he might not trust me well enough to revisit or renegotiate my success criteria of 80%." (Notice the risk metric.)

Two important metrics can be visualized in this long statement. One metric measures the future (achieving 80% or above) and the second metric monitors the present with feedback from past experiences (distribution of bike funds and loss of trust). These are performance and risk metrics, respectively. The metric that Gary now has is much more specific and will be modified depending on his quarter progression, but is a fairly good start to lead to other discussions with his father.

The preceding example should shed some light on the importance of metrics and pave the way to the discussion on crafting the right metrics based on the right business driver. The following are some of the characteristics of a security metric aligned with the Zero Trust mission:

- **Relevant and aligned:** A security metric should be relevant to the organization's security goals and objectives. It should be able to provide meaningful insights and help the organization make informed decisions to address security risks of not adopting the overall strategy. It must also be aligned with how well the organization wishes to perform to achieve its overall Zero Trust vision. Some examples would be "support the bank to achieve its vision of being a trusted banking partner by protecting the customers' and employees' data."

- **Measurable:** A security metric should be measurable and provide a quantitative or qualitative value that can be compared over time. This helps to track the effectiveness of security measures and identify areas of improvement. An example would be "reduce the number of open showstopper incidents to less than two per year."

- **Actionable:** A metric should be actionable, meaning that it should provide information that can be used to take specific actions to mitigate or prevent security risks or enhance performance—for example, "provide more information about network traffic to enhance application visibility. This will in turn help SOC operations to make the right decisions either manually or through SOAR automation."

Additionally, a metric should not, for example, measure success based on something not happening because then it will not lead to an action. For example, consider the metric for the strategic goal "no breaches occur in the network." This is a fair vision and goal but unfortunately it leads to no action. What happens when there is a breach? How would you measure the risk? A more fruitful metric would be "augment the current segmentation and reduce the blast radius when a breach occurs." This leads to the action of completing and implementing segmentation, which is a security control that also augments the vision irrespective of whether or not a breach occurs. A metric must enable the vision and strategy of the enterprise.

- **Reliable:** A security metric should be based on reliable data sources and calculations to ensure its accuracy and usefulness in decision-making. For example, measuring the response times of engineers to an incident is a reliable method of measuring the efficacy of the SOC skillset. However, "reduced number of attacks" is not representative of an effective security control. It could be impacted by a lot of factors (like cost, malicious actor deterrence, interest in assets, and so on), which makes the measurement unreliable.

- **Comprehensive:** A security metric should be comprehensive in scope and cover all aspects of the organization's security posture, including physical, technical, and administrative controls. Zero Trust is not just about implementing technical controls. It must cover administrative, physical, and operational aspects of the processes and people involved in the enterprise. For example, adding biometric authentication at secure server farms can be a metric to fulfil a mission statement of using advanced multifactor authentication solutions to achieve secure data center access.

- **Understandable:** A security metric should be easily understandable by all stakeholders, including technical and non-technical personnel, to ensure effective communication and collaboration in addressing security risks. For example, "secure data at rest" is vague because the specifics of the metric are not available for a technical crowd. "Use AES 256 encryption on all storage devices" is also vague because leadership will not relate to the security mechanisms implemented. They will care about how it helps the business. Hence, a more valid metric would be "secure our customer data by storing only the information we need for the amount of time we need. Encrypt the data in storage to make sure that confidentiality of information is maintained. Use encryption technology with stronger keys like Elliptical Curve Cryptography."

This is where the initial conversation of being a translator becomes relevant. The metric highlighted here has a non-technical statement that appeals to the leadership and a tail-end technical statement appealing to the operational leads. As translators, architects need to point out the relevant aspect of the metric that each stakeholder cares about.

- **Timely:** A security metric should be timely and provide up-to-date information to ensure timely decision-making and response to security incidents. To make sure we understand this metric clearly, it is important to highlight that it does not relate to building a timeframe for the metric. This metric relates to providing feedback on the measured attribute constantly and creating multiplexed checkpoints to validate what is being measured so that more granular decisions can be made. Let's take the case study of buying a bicycle. A tactical metric Gary created was to pass his exams with 80% or more to achieve his strategic metric of riding a bike. The father, however, also created an operational metric, which he measures every week by validating the course studied over the week with a weekly quiz that has thirty questions. Gary must pass with 25 correct answers. He must consistently get 25 or above during all the relevant weeks to convince his father that he is studying and on the right direction to achieve his tactical goal of passing his exams.

■ **Ease of metric creation:** Finally, one must consider the ease with which the metric can be measured. This has dependency on the maturity stage the enterprise is in as well. For example, an enterprise at "measurable" maturity will have a SOC that is mature and have more visibility options; hence, metrics such as application flow visibility and dynamic context-based access monitoring are still viable. When enterprises are starting out with Zero Trust and are evaluating their existing security controls, metrics like SOC maturity may need more manpower and skill with limited automation. A balance must be maintained between the cost to set up a metric, the long-term benefits, as well as the business alignment. The metric must be easy to craft and implement, and the possibility of measurement must be validated at the crafting stage.

When metric characteristics are considered from an IT infrastructure, they can be visualized as shown in Table 5-1.

Table 5-1 *Examples of Various Hybrid Metrics*

Hybrid Metrics		Goal-Based Metrics		
		Strategic Metric	Tactical Metric	Operational Metric
Metrics based on data collection method	Qualitative Metric	Successfully deploy Zero Trust and reach a maturity level of "Quantitatively Managed" as part of CMMI.	User awareness of Zero Trust movement in the organization.	Implement multifactor password-less authentication, and measure adoption of strong authentication methods.
	Quantitative Metric	Number of applications that have been migrate to Zero Trust per month.	Achieve 100% endpoint compliance as part of larger Zero Trust compliance initiative.	Reduce mean time to detect (MTTD) to 15 minutes for suspicious endpoints.

The Capability Maturity Model Integration (CMMI) was created by the Software Engineering Institute (SEI) at Carnegie Mellon University. The SEI is a federally funded research and development center that focuses on advancing software engineering and cybersecurity practices. CMMI was developed by a team of researchers and experts at the SEI, and it has since become a widely recognized framework for assessing and improving organizational processes across various domains, including software development, systems engineering, and acquisition.

CMMI has gained widespread adoption in enterprise settings due to its systematic and structured approach to process improvement. It provides organizations with a clear roadmap for enhancing their operational, tactical, or strategic maturity by defining a series of maturity levels, from Initial to Optimizing. This framework fosters a culture of continuous improvement, ensuring that enterprises can adapt to evolving market demands and stay competitive. CMMI's global applicability and proven success across various industries make it an attractive choice for enterprises seeking to standardize and optimize their processes. Additionally, it often serves as a strategic advantage when bidding for contracts with government agencies or major customers that require demonstrated process maturity, reducing risks, and enhancing the overall quality of products and services.

By offering a common language for discussing and benchmarking processes, CMMI facilitates collaboration within large and diverse organizations. It also helps mitigate risks associated with project delays, budget overruns, and quality issues, resulting in cost savings and improved customer satisfaction. CMMI's structured methodology, adaptability, and proven track record make it a valuable tool for enterprise maturity, enabling organizations to consistently deliver high-quality products and services while maintaining a competitive edge in today's dynamic business landscape.

Consider the metrics in Table 5-1 and superimpose the principles that have been listed in the previous sections. When the overall maturity of the Zero Trust implementation is measured, it cannot be mapped to a numerical value. It is achieved after measuring the adoption rate of several initiatives like password-less authentication and user awareness. The overall maturity will be mapped against a broader spectrum inspired from the CMMI maturity model. Hence, this specific metric of Zero Trust deployment becomes a strategic metric measured qualitatively. At the same time, we want to measure how many applications and workloads have been moved to the Zero Trust model. This, however, can be measured and mapped over time, making it a quantitative metric. Migration of workload to a micro-segmented microservices architecture measures how well the Application strategy is being adopted and aligned with the broader Zero Trust Scope.

Consider the tactical metrics and how they align with initiatives, which when combined achieve a strategy or mission. For example, as with the previous metric, user awareness is an important metric when it comes to understanding how well Zero Trust is being accepted in the employee community. These are usually measured with surveys and are still measured as a range of high adoption, medium adoption, or low adoptions. On the contrary, endpoint compliance is an initiative that aligns with the overall Zero Trust deployment mission, and compliance can be measured with a number or percentage of devices that are compliant.

Finally, operational metrics are day-to-day metrics that provide details into specifics of implementation projects. For example, in an ongoing visibility or SOC deployment project, a key metric would be mean time to detect (MTTD) for unknown or suspicious computers. This can be measured in time units and is therefore quantitative. On the other hand, adoption of users authenticating with MFA and password-less methods are qualitative and are broader measurements like high, medium, or low.

Overall, it is imperative to understand that all metrics are tied into each other. A large number of operational metrics measure the efficacy of a specific tactical metric. Multiple tactical metrics will measure the effective adoption of a strategic metric. How the enterprise wishes to measure and showcase the metrics determines if the metrics are qualitative or quantitative. Another disclaimer to highlight here is that certain metrics are ambiguous as to whether they are qualitative or quantitative. Some enterprises might possess the means to measure a specific metric, which other enterprises might not have. Deciding on whether a metric should be qualitative or quantitative is an enterprise-specific decision and cannot be standardized. For example, some enterprises may choose to measure adoption of MFA with number of users as well. There is no wrong metric. There are good metrics and better metrics and they key factor that influences the crafting of these metrics is how it aligns to the enterprise's requirement.

Measurement Targets for Zero Trust

This section highlights the two main types of metrics: performance and risk metrics.

Performance Metrics

If executive leadership does not understand the risk to business of implementing (or not implementing) a technology, they will not be able to put in the right controls to protect critical assets, and that is almost always the most common reason why enterprises just buy technology to buff up the security initiative but get attacked anyway. An enterprise could have all the security controls, hardware, and endpoint protection in the world and still get attacked if users "approve" instead of "deny" when they get a push message on their MFA solution. There is a common notion that security hinders the business by making access more restrictive and complicated, but the point to remember is, like most mechanisms, security needs to be planned and baked into the solution. The base security strategy must encompass all the business drivers, and security must align with the general business direction. Good security governance goes a long way to promoting the enterprise's business. Showing key performance improvements by improving security is one of the first steps in measuring Zero Trust success.

A performance metric is a measure used to evaluate how well a particular strategy, process, initiative, or product is performing. It is usually both a quantitative and qualitative indicator that helps in assessing the effectiveness, efficiency, and quality of the measured activity or product. Performance metrics look into the future.

A key aspect of performance metrics are that they are based on the current lack of performance by a specific product or a specific initiative. The feedback on lack of performance may be from senior leadership but is usually from middle managers and operational leads who see day-to-day performance gaps and expect that the Zero Trust initiative will help improve the gaps. The subsections that follow cover some common Zero Trust performance metrics that can be mapped to almost all types of enterprises and are a good starting point to tailor enterprise-specific performance metrics for enterprises.

Adaptability of Security Governance and Business Agility

Organizational needs change over time, and enterprises pivot products and services to suit the general business. Most enterprises do not spend time creating a blueprint or template to fit all security needs of the enterprise. Those who have already got this blueprint do not realize that this blueprint was likely created *after* the enterprise started its basic functionality and did not take into account any future changes to the business. Unfortunately, security has usually been an afterthought and has never been able to adapt to the changing needs of the business. This has been a concern in the field for most security practitioners defining enterprise security architectures. Zero Trust aims to change that perception. The security vision must be able to adapt to any change in business and support it by making sure the infrastructure and its data gets the right level of protection based on its context. With the right support in place from security, a lot of tasks such as mergers or time to market will be completed much faster. When customers see you take security seriously, you automatically build trust with them, even before they buy your product or service. With a unified architecture framework for security, enterprise business can adapt to any specific organizational change in policies. New types of devices or entities will automatically fall into the right segment and get the appropriate security control.

Security needs to consider different aspects of the business and should not just be reactive. Security must be proactive at protecting the network and making sure incidents are validated, evidence collected, and incident response is being followed to a T. Security must also proactively strategize supporting the existing business and any future changes in the business model. Business must drive technology, not vice versa. A simple example is if an enterprise is selling perfumes and suddenly switches to selling toothbrushes; the overall security vision of protecting the customer's data should not change.

Adaptability is not quantitative and usually maps to governance and overall vision and mission. This is therefore a strategic qualitative metric. The metric statement would be as follows: "The Zero Trust strategy must allow the security controls implemented to be adaptable to changing business needs."

Revenue Generation and Cost Savings from Zero Trust Initiatives

In general, a strategy must always align with revenue generation and cost savings, which is the primary focus at any executive level. Without a revenue stake, security will risk being considered an add-on. Zero Trust saves cost for the enterprise by reducing capital expenditure and shifting to operational expenditures. With efficient operations, most processes can get streamlined and optimized fairly quickly in contrast to buying and fitting new hardware, which usually takes months or even years. With the right personnel, accurate information can be extracted from a security device and can be used for multiple purposes like incident response or health monitoring. As you drive a car slower, it becomes easy to control. Similarly, the less complex the business operations are, the easier it is to secure the larger enterprise and the easier it is to identify and isolate key infrastructure.

Along the same lines, automating simpler processes (like incident management, account provisioning, and so on) allows users to allocate their time for more important tasks like incident analysis or even security awareness, which in the long run is a measurable metric in the form of skilled labor. With simpler processes, it becomes easier for various teams to communicate their requirements to each other. Hence, if a specific product is to enter the market with a simpler and transparent development process, it becomes easy to break existing silos and incorporate security from the start.

Another aspect is incorporating metrics to improve the Zero Trust capabilities of a product. Metrics like a Zero Trust index should be allocated for a product to measure the extent to which the product can support a Zero Trust strategy. This would include capabilities like context-based policing, visibility capability and so on.

Revenue is almost never a qualitative metric. Revenue generation is a business requirement. The metric statement will read, "The annual target revenue to be generated by implementing the Zero Trust initiatives is $10 million or above." Another metric would read, "The measured Zero Trust index for products sold by the enterprise must be more than seven measured on a scale of one to ten." Observe that one metric is a metric for the enterprise itself and how implementing the Zero Trust initiative saves cost, the second metric is relating to creating products that support customers and generate revenue. Both aspects align with business bringing in more profit margins. The metric stays strategic but is quantitative in nature.

Technology Innovation and Improvements

Security must support any disrupting technology that changes the direction of the business. It must adapt to changing security control and provide a better control strategy for any business models. Any technology innovation must be easily absorbed by the security and access control strategy.

Do not be scared of innovation. Necessity is the mother of invention, and that is exactly why we shouldn't be breaking that cycle. Many enterprise departments consider innovation as a hindrance to the business and are very wary to take a risk, especially with security. Blocking innovation is almost always detrimental to business. The perceived risk is never worth the returns that innovation could bring, and this is what determines how effective a leadership board is. They need to be able to identify a good innovative initiative and support it with clear understanding (and a measurable metric) of how it will bring back revenue and support the business. Supporting security innovations is an effective way of baking in security rather than bolting it on later, as modern designs and technology mandate that security be considered in all early discussions.

Microservices, for example, greatly support implementation of Zero Trust. Microservices comprise a services-oriented architectural approach to designing applications. It involves breaking down large, monolithic software applications into smaller, independently deployable services, preferably as ephemeral instances that can be redeployed in a matter of minutes. Containers are typically an example of microservices implementation, where web, application, and database services reside on separate container instances. This is not a concept that can be implemented by an enterprise in a day and needs well-planned application migration or transformation strategies. These microservices communicate via APIs and can be developed, deployed, and scaled independently. Microservices, with their fine-grained control over access and communication, can play a pivotal role in implementing Zero Trust security by facilitating granular access controls and security policies. Addition of the API-based information exchange helps Zero Trust architects to craft micro-segmentation around these flows. Each microservice can be treated as an independent entity with its own security rules, ensuring that only authorized entities can access specific services or data. Additionally, microservices can provide detailed logs and telemetry data, making it easier to monitor and detect suspicious activity, which is another fundamental aspect of Zero Trust.

It is important to clarify that implementing microservices does not mean Zero Trust is in place. You could have all the segmentation in the world, and if you allow all services to talk to each other, you are essentially following the implicit trust model.

Consider that you are buying a product that allows an administrator to escalate to root to perform troubleshooting capabilities. This product comes "as is" from the vendor, and you cannot really disable the root access provided. If this device is compromised, it could essentially give full access to your network. You need the product to facilitate your business, but you do not see it at the right security level. In a traditional model, one would consider the product as a larger risk, even though it greatly augments business. With Zero Trust, as long as you can control device access to a restricted set of people with specific roles and attributes, the blast radius of compromise is restricted to a specific user segment. Isolation from other network devices via VLANs and VRFs also helps provide network segmentation. Only device administrators can access the device over management, and the device can communicate only to specific systems that need to consume its information. By selectively providing access and making sure only the right servers and subjects can access this product, you have reduced the blast radius, increased security posture, and still allowed the business to continue. Over time, the vulnerability can be patched, but this does not need to hamper the business.

Technology supports initiatives, and support of innovative technology is a tactical performance indicator for the enterprise. The metric statement will read as follows: "Adopt innovative technology into all enterprise-driven initiatives." Each initiative would get a detailed metric statement; for example, "Adopt innovative inspection methods when monitoring encrypted traffic." This is tactical because it is supporting a specific initiative and is quantitative because we are measuring the adoption as a percentage of traffic inspected (90% of traffic inspected with Encrypted Traffic Analysis).

Efficacy of Customer Experience

Security controls must augment and support the business, user, or customer experience. If you are a security vendor, your products must not hamper business but must support and improve the way users access the devices and implement security policy. Operational support from the vendor as well as services rendered to support the product constitute customer experience, and from a Zero Trust maturity perspective, restrictive policies must not hinder business as usual (BAU). All policy creation must be backed and substantiated by continuously updated asset management, flow analysis, and segmentation.

Customer experience front desk agents are an example of how Zero Trust can help build trust with customers. Front desk agents now have access to more data than was previously considered relevant. In all practicality, they potentially have access to sensitive information as well to make the customer experience more customized. In a traditional setup, the agents would not be provided access to sensitive data and their communication to customers would be fairly dull and routine. With Zero Trust and secure API access, customer experience agents can be given just enough access to data so that the experience with each customer is unique. This will increase customer retention with a more effective subscription model.

When viewed from a different lens, security can be proactive or reactive. Generally, security is perceived as a reactive control. Risk analysis considers the threats to a specific asset, the possible attacks that could have occurred and provides recommendations for the controls to be implemented. This, however, measures success based on not being attacked. Zero Trust, on the other hand, assumes a breach and measures success based on how well you restrict that breach from exfiltrating your critical infrastructure. This way, you need to make sure you identify critical infrastructure and protect it well with the right access control, thus augmenting business and user experience. The likelihood of a threat might be extremely low, yet if the impact of a breach is high, the right controls and protection must be implemented.

Another example is migration of applications to the cloud. The cloud allows you to make minor changes and still move applications in a lift-and-shift model; however, when you consider the cloud, enterprise boundaries are not the same, the actors are not the same, and under no circumstance are the networks the same. Then how would an enterprise design its on-premises security around a completely different infrastructure? The simpler approach would be to rearchitect the security model and fit the cloud access model. Hence, a strategic quantitative metric here will be "support the customers and improve customer experience by reducing the downtime caused by enterprise migrations. The maximum tolerable downtime for a migration is 1 hour." This should be backed by the following tactical metric: "Adapt all workload to a multicloud architecture seamlessly with least impact to customer applications and least need for modifications. Achieve workload agility across various platforms and networks." This is measured with the number of workloads migrated (quantitative) and customer satisfaction (qualitative).

Evaluating the Preparedness of the Enterprise

The Information Technology Infrastructure Library (ITIL) service catalog plays a valuable role in the operation and management of the overall Zero Trust security strategy, architecture, and implementation. It also helps build a framework for measuring how ready the enterprise is to manage and improve on the Zero Trust solution.

The ITIL service catalog is a centralized repository that contains detailed information about the IT services offered by an organization. It provides a structured and standardized view of available services, including their descriptions, service levels, dependencies, and associated costs. In the context of Zero Trust, the service catalog serves as a critical tool for defining and managing various services. Critical services include the following:

- Policy Management services (including access controls and permissions for various IT services and resources)

- Logging and event correlation services

- Incident response services

- Security Operations Center Analyst services

- Identity and Access Management services (user identity lifecycle management)

- Data Management services (data lifecycle management)

- System and application management services (application lifecycle management)

- Digital Risk Management services

- Compliance Management services

The service catalog can facilitate access request and approval workflows. When users need access to specific resources or services, they can use the catalog to request access. These requests can trigger approval processes, ensuring that access is granted only to authorized individuals. This allows an enterprise to embark on the automation path with the right service definition and outcome.

ITIL also emphasizes the service lifecycle, which includes stages like service design, transition, operation, and continual service improvement. To showcase the level of preparedness as a performance indicator, you need to define what preparedness means for your organization. Depending on your industry and business operations, preparedness can have different meanings. For example, preparedness could mean being ready to respond to a crisis or an incident, having the necessary resources to meet customer demand, or having robust cybersecurity measures in place. You will need to craft metrics on how to measure the readiness of critical services needed to make sure the Zero Trust architecture and its necessary services is being implemented according to the original business vision and mission. You need to establish measurable objectives, which involves identifying the specific outcomes that will indicate the level of preparedness. For instance, if preparedness means having the resources to meet customer demand, you could set objectives around inventory levels, delivery times, and customer satisfaction rates. Some other examples are business continuity plans (BCPs), supply chain management, and crisis management. Once measurable objectives are created, you need to determine the specific targets that are to be achieved to demonstrate the desired level of preparedness based on the identified outcomes. These targets should be achievable, realistic, and aligned with your overall business strategy.

An example of a metric statement would be, "Measure the current zero trust maturity, identify key initiatives to invest in, and measure how well they have been integrated and adopted." Adoption of initiatives is a tactical qualitative metric. Note that enterprise preparedness is different from business agility. Business agility is strategic, and enterprise preparedness is tactical and maps to multiple technical and service initiatives.

Protection from Unauthorized Access Attempts

An important performance metric relating to access control is protection from unauthorized external access. This aims to measure how well the enterprise prevents external agents from attacking or entering their environment. Consider this as an operational task by any SOC in the enterprise. As an operational metric, it measures how well an enterprise can block out external attacks. This can be measured as a percentage of blocks determined by the number of blocks across a number of attempts over a timeframe. This make is a quantitative operational metric. Considering that Zero Trust is identity-centric, another important metric is to validate how many failed authentications are seen in the network. For some time during the learning phase, the failed authentication may increase but then overall the number of failed authentications must decrease from the current value showcasing the effective implementation of user authentication with MFA. This is also an operational quantitative metric, and the statement will be "percentage of unauthorized access attempts blocked must not be lesser than 98%."

Efficacy of Network and Endpoint Visibility

This metric is a measure of how well the enterprise monitors for network- and endpoint-based incidents and events. It also includes similar metrics like the following:

- Percentage visibility of managed and unmanaged devices
- Number of security incidents it has recorded successfully
- Number of incidents remediated
- Incidents identified and recorded fast (lower mean time to detection)
- Incidents mitigated fast (lower mean time to resolution).

Incident response capabilities are quantitative performance metrics, especially metrics like time to contain an incident (mean time to contain) and mean time to resolve an incident (mean time to resolution). These metrics could be qualitative or quantitative but are usually operationally motivated and hence considered measurable and quantitative. For example, a metric statement will read, "Mean time to incident resolution must not be more than one day."

Effective and Optimized Policy Creation

An enterprise should already have started mapping assets and flows or at least put asset inventory into its road map. After a certain stage of asset and flow mapping, the enterprise will be mature enough to perform trust modeling. Context of trust here is not just limited to flows but also to general security and governance policy. A business unit might have a different risk appetite in response to a certain threat. For example, an incident response team might consider failed authentications as a larger threat, but the HR department might not see it as a threat but rather as an operational concern. The trust factor and the promise to protect customer data in both cases should be the same, not only for the two business units but also enterprise-wide. Therefore, protecting all forms of customer data handled by various business units must also be considered. Trust conversations and modeling are key, as are risk conversations.

There are both operational and tactical metrics in this aspect of performance measurement. An operational qualitative metric will read, "Reduce the complexity of the rule creation by making rules contextual in nature, thus making operations simpler." This metric can be measured as Easy, Medium, or Hard. A tactical qualitative metric, however, would read, "Move the enterprise to create context-based rules to improve the effectiveness of the rules and make them more operationally simple and contextually relevant." This measures the effectiveness of the rule base from most simple and effective to complex and ineffective and relates to an enterprise-wide initiative of creating context-based policies.

Risk Metrics

The second type of metric measures the risk of threats exploiting vulnerabilities in the enterprise. With dynamic software-defined perimeters and a changing threat landscape, perimeter-based security is being perceived as less effective. With the right social engineering tactics, a malicious entity doesn't need to traverse your Internet and DMZ zone but could potentially be placed right into the heart of the enterprise's server farm without having to bat an eye. As senior leadership, how would one judge the controls needed without knowing the true nature of the asset and the risk to that asset? Once the context of an asset is understood, one will realize that placing an asset within a fixed boundary is moot.

Risk metrics also showcase a different picture to leadership. Rather than focusing on the future with performance and cost, risk metrics showcase the current risk profile and measure how the enterprise can reduce the risk to a more acceptable state. This includes reducing blast radius, assuming a breach, decreasing overall risk exposure and so on. Continuing with the front desk operator example discussed earlier, a front desk operator handles communication to customers and doesn't need to have access to server farm servers or other DMZ segments. As part of their communication, they must access their application, which in turn needs to extract customer PII or critical information from the server application via an API. If this situation is observed in more detail, the safe moat for the information is gone, and you have created a zipline from the untrusted segments to the trusted server farm, which completely depends on the security awareness of the front desk operator. The point being made is that data flow is no longer in a definite direction. It flows everywhere, and depending on how important the information or asset is, the risk is higher. If it is handled at different parts of the network, it must be protected with the same context. PII needs to be secured at the edge as well as in the server farm. Enterprises deploy defense-in-depth concepts by deploying security controls from various vendors; however, if there is a need for uniform policy, the security capabilities must match for all vendors. If a Cisco firewall detects Facebook chat but a Check Point firewall does not, uniform policy is lost. Now the server application itself might need to provide API access to public cloud applications, and you've basically allowed an enterprise-owned asset in an IaaS to access your application on your premises without the right security control. Boundaries are changing and perimeters are no longer static. Defense in depth is no longer as effective as it has been many years ago, and measuring the risk is an important aspect of creating metrics because our final goal is to lower the risk. Remember, risk cannot be entirely eliminated.

Another viewpoint is at the CxO level, where the risk to an enterprise is large scale and less technical. Risk at that level needs to show quantifiable outcomes and still cover a larger scale like vulnerable devices, risk to reputation, and risk of revenue loss. Solutions like RiskLens or Cisco Kenna provide a much needed alignment of enterprise-specific risk metrics to industry-standard solutions to make sure that the risk appetite is quantified and clearly measured for an organization. Maintaining risk metrics is not only optimal but also critical to measure the exact security posture of an enterprise.

Protection of assets must be focused and security controls must be closer to the asset. This is what drives risk discussions. Access to our resources is no longer restricted to specific defined subjects. A server is not just managed by a server administrator. Subjects from cloud networks, virtual machine admins, and so on need access to various aspects of the network, and a compromise of any of these accounts can compromise the entire network without the right access model. Service accounts facilitate services to log in and begin communication across the network without the intervention of a human user, which makes these accounts common targets for account compromise and privilege escalation. This is especially true if workload is on the cloud. The best example to explain the risk of cloud workload is that of a house. The premise of the cloud is to basically be ubiquitous and accessible to everyone. That's like building a house and telling everyone that they can access it, which of course is not true. In reality, your house is *already* in the public domain, and everyone who is motivated can *find* your house but they cannot enter it. That is exactly what Zero Trust architectures help enterprises achieve. Controlling access based on context and risk profile is the final goal.

Most risks are measured quantitatively, and this comes directly based on certain common aspects like impact, vulnerability, annual loss expectancy, and so on; however, there are risks measured qualitatively as well such as threat event frequency and the like. Generally, when qualitative analysis is considered, multiple dependent teams need to get involved.

Applications, processes, systems, and network users are all assets that bring with them their own inherent risks. For enterprises to be able to perform qualitative analysis, risk must always be considered. In Zero Trust, risk analysis is even stricter because it needs to assume a breach has happened rather than the impact when a breach happens. Risk analysis must be performed for all the assets, along with threat models to make sure that the right risks are prioritized and the right metric can be crafted and achieved. The sections that follow describe some common risk metrics.

Asset-Focused Risk Management

An asset-focused risk management approach places the asset at the core and attempts to understand the risk of loss. The quantitative metrics that can be used to identify the impact of loss per year are as follows:

- **Asset value (AV)** represents the estimated monetary value of the asset that is at risk. This could include assets like data, intellectual property, equipment, and other tangible or intangible assets. People are considered assets as well. AV is measured in currency to represent monetary value.

- **Exposure factor (EF)** is the percentage of the asset's value that is expected to be lost in the event of a successful attack from an external threat. It is usually measured as a percentage of the total asset value.

- **Single loss expectancy (SLE)** is a term used in risk management to describe the expected financial loss from a single security incident or event. It is a metric that helps organizations to quantify the potential impact of a security breach, which in turn can help them to prioritize their security efforts.

The SLE is calculated by multiplying the asset value (AV) by the exposure factor (EF):

SLE = AV × EF

Essentially, if you have $10 worth of candy and the chance that your brother will take it is high, and if he does, he'll take three quarters of your candy, your SLE is 75% of $10, which is $7.50.

- **Annualized rate of occurrence (ARO)** represents the estimated frequency at which the particular security incident or event is expected to occur within a year.

- **Annual loss expectancy (ALE)** is a term used in risk management to describe the expected financial loss per year from a particular security incident or event. It is a metric that helps organizations to quantify the potential impact of a security breach on an annual basis.

The annual loss expectancy is calculated by multiplying the single loss expectancy (SLE) by the annualized rate of occurrence (ARO).

ALE = SLE × ARO

Following the previous example, if your brother takes 75% of your candy every day, then in a year your annual loss expectancy is 0.75×10×365, which is $2,737.50.

An annual loss expectancy is a clear indication of the impact of loss. For example, if your PII is exfiltrated, the impact loss is $1M. This provides a very useful metric to leadership on how important an asset is and to prioritize security controls for the asset.

Context-Based Risk Management: Open FAIR Risk Analysis

The second less-utilized but more relevant option is context-based risk management and its derivative metric. Here, the focus is not only on the asset but its entire environment, including its threats. The center shifts to how much of a threat is a specific activity and is not just restricted to external threats. Context-based metrics are risk indicators or measurements that are tailored to a specific situation or context. These metrics take into account the unique characteristics of the situation, such as the goals of the organization, the industry, the market, or the audience. These are important because they provide more relevant and accurate information than generic metrics that apply to all situations. By focusing on the specific context, organizations can better evaluate their performance and make more informed decisions. Solutions like RiskLens utilize a methodology to consider the entire enterprise as the scope and provide a clear measurable risk. This allows the enterprise to evaluate its own risk appetite in alignment with industry-standard solutions.

A well-known contextual risk analysis framework is Open FAIR, which is a method of risk analysis well aligned with the Zero Trust narrative because Open FAIR looks at a failure use case rather than an asset specifically. It looks at threats and impact after a breach to validate specific metrics, which aligns with the overall context of the asset and data flow rather than just the asset. It is similar to the asset-focused risk analysis, except for the scope of the metrics, which covers more context. Open FAIR was built on the basis of the original FAIR method of analysis created in 2007. Over time, with collaboration with the Open Group, the Open FAIR Risk Analysis method was created in 2009.[1]

Note that asset-focused risk metrics focus only on the impact of loss of an asset. Context-based metrics focus not only on impact of the loss but on impact of loss under various conditions when exposed to different threats. Having multiple threats leads to multiple loss impacts. Loss of PII in general has a dollar value attached to it, but loss of PII to a belligerent country is worse and has catastrophic repercussions. Impact of loss is not only to tangible assets but also to abstract assets such as reputation of the company. Thus, context-based metrics showcase the entire end-to-end impact of the loss.

To understand why Open FAIR is relevant to the Zero Trust conversation and to understand how different the methodology is from asset-based risk analysis, the following section will cover the overall phases of the Open FAIR risk analysis. There are five major phases of Open FAIR risk analysis. In this section, a baseline of some of these metrics will be created, which will subsequently be utilized in the interview with the CIO, CISO, COO, and CTO. Figure 5-3 illustrates the five phases of the Open FAIR methodology.

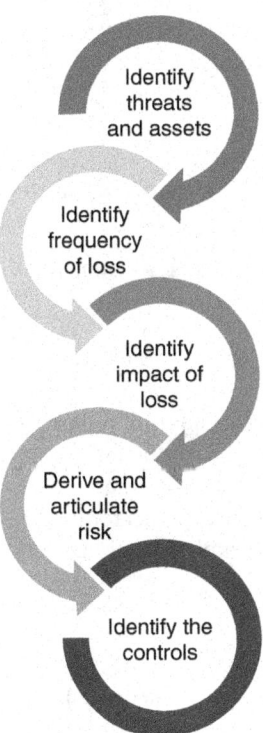

Identify threats and assets

Identify frequency of loss

Identify impact of loss

Derive and articulate risk

Identify the controls

Figure 5-3 *The Open FAIR Methodology*

Phase 1: Identify the Threat and Asset

Similar to Exposure factor, which is derived from threats, the first phase of Open FAIR is to identify a *Loss scenario*, which is derived when an *asset* is compromised by a *threat* by exploiting a *vulnerability*, which leads to an *incident* and subsequent loss of money or reputation under *specific conditions*. These are key distinguishing factors when assigning a loss value and are not as straightforward as assigning a simple dollar value to an asset. In this case, we are looking at a loss scenario and not the loss value based on intrinsic asset value. Essentially this is an entire kill chain, starting with identifying the target and going to the means of attack. The Cyber Kill Chain is a concept developed by Lockheed Martin that describes the stages of a cyberattack. It is intended to provide a framework for understanding and defending against sophisticated cyberattacks.

The Cyber Kill Chain consists of the following stages:

- **Reconnaissance:** The attacker collects information about the target system and its vulnerabilities.

- **Weaponization:** The attacker creates a weapon, such as a virus or a Trojan horse, that can be used to exploit a vulnerability in the target system.

- **Delivery:** The attacker delivers the weapon to the target system, often through phishing emails or other social engineering tactics.

- **Exploitation:** The weapon is used to exploit the vulnerability in the target system, allowing the attacker to gain access to sensitive data or take control of the system.

- **Installation:** The attacker installs malware or other tools on the compromised system to maintain access and control.

- **Command and control:** The attacker establishes a command and control (C2) channel to communicate with the compromised system and issue commands.

- **Actions on objectives:** The attacker carries out their intended actions, which may include stealing data, disrupting operations, or causing damage to the system.

If we consider the kill chain, we are looking holistically at a loss scenario, which is a useful metric to explain to leadership because it will drive the need for Zero Trust better. This also aligns with context-based policies, which help propagate uniform policies across the enterprise. Remember that nothing drives initiatives better than the fear of an attack. When leadership sees how easy it is to extract data and how well the strategy can be measured, they will be more open to accept the strategy and provide funds for implementation. A Zero Trust strategy presentation usually has a relevant kill chain scenario to showcase how Zero Trust reduces some of the risk.

Phase 2: Identify Frequency of a Loss

In this phase, the overall frequency of a specific loss event is calculated, which is called *loss event frequency*. This metric is analogous to the annual rate of occurrence (ARO). There are certain metrics that help identify the loss event frequency for each loss event.

Threat Event Frequency

Threat event frequency refers to the rate at which a particular type of threat occurs within a given time period. For example, floods in Singapore happen once in 25 years. It is often used as a metric for assessing the likelihood or probability of a threat occurring and is a key component of risk management.

Threat event frequency can be measured in different ways, depending on the specific threat being assessed and the available data. For example, it might be measured as the number of attempted cyberattacks per day, the number of incidents of employee theft per year, or the number of natural disasters per decade. To quantify this in more detail, it is measured by calculating the *contact frequency*, which is the probability that a threat agent will come in contact with an asset in a given timeframe (an example would be the number of times an external agent fails to authenticate to the network) as well as the *probability of action*, which basically measures what the chances are that the threat will take an action when in contact with an asset in a given timeframe (the probability of action on a DMZ web server is far more than the probability of action on an internal router).

To calculate threat event frequency, data is typically collected and analyzed over a specific time period to determine the number of instances in which the threat occurred. This data can be used to develop statistical models to predict the likelihood of future occurrences. Note that frequency of loss is usually qualitative. Table 5-2 showcases the threat event frequency for a scenario of customer data being exfiltrated to a malicious entity. The quantity of 50 times and time frame of 5 years in Table 5-2 will be unique for each enterprise and must be statistically derived with relevant enterprise threat research.

Table 5-2 *Threat Event Frequency Rating*

Rating	Frequency of the event of occurring
VERY HIGH	Greater than 50 times a year
HIGH	10 times a year to 50 times a year
MODERATE	Once a year to 10 times a year
LOW	Once in 5 years to once a year (ARO = 1)
VERY LOW	Less than once in 5 years (0.2)

Threat Capability

Threat capability refers to the overall capability of a threat to take an action. It impacts the probability of action when evaluating threat event frequency because when the capability of a threat is large, the chances of it taking an action when in contact with an asset is higher. A DDoS attack is a threat. The capability of state-sponsored actors executing a DDoS attack is high because they have infinite money and time and a fixed target set (which is identified in frequency of loss). If a state-sponsored actor comes in contact with customer data, the probability of action is very high and the threat capability is very high, which means the risk is much higher than just measuring annual loss expectancy.

Table 5-3 showcases an example of measuring the threat capability of a malicious actor to exfiltrate customer data. This is based on multiple factors such as motivation, technical skills, and availability of time and money. Threat capability is also measured qualitatively. The number 2% and 15% in Table 5-3 will be unique for each enterprise and must be statistically derived with relevant enterprise threat research.

Table 5-3   *Threat Capability of an External Actor*

Rating	Capability scale
VERY HIGH	Top 2%
HIGH	Top 15% of attackers
MODERATE	Average between 15% and 85%
LOW	Last 15%
VERY LOW	Last 2%

Control Strength or Resistance Strength

Control strength is the capability of a security control to resist the strength that a threat can apply on an asset. If segmentation as a control can be used to prevent a state-sponsored actor from accessing customer data, control strength measures how effective segmentation is and would be HIGH or VERY HIGH. In turn, it could reduce the loss frequency from VERY HIGH to MEDIUM based on its strength. Control strength also ties in with the threat capability of an actor. For example, effectiveness of a control's strength is much higher if it can deter threats with VERY HIGH capability. Overall, the control strength is also a qualitative metric that measures how effective it is to implement a control. Table 5-4, for example, showcases a rating system for control strength. The number 2% and 15% in Table 5-4 will be unique for each enterprise and must be statistically derived with relevant enterprise threat research.

Table 5-4   *Control Strength Capability Rating*

Rating	Capability scale
VERY HIGH	Protect against top 2% threat actors
HIGH	Protect against top 15% threat actors
MODERATE	Protect against average threat actors
LOW	Protect only against lower 15% threat actors
VERY LOW	Protect against lower 2% threat actors

Calculate Vulnerability

Vulnerability is the probability that a threat event will materialize into a loss event. This also means that the strength of the threat is greater than the controls in place. Your customer data is vulnerable to attack if segmentation doesn't deter or prevent a state-sponsored attacker from exfiltrating data. Remember, a vulnerability is always created when the threat capability is greater than existing controls. In simple terms, if the existing controls are VERY LOW and the threat capability is VERY HIGH, then the vulnerability is also VERY HIGH. The larger the control gap, the larger the vulnerability. Table 5-5 can be used as a mapping between threat capability and control strength. As is clear, even with strong control strength, the higher the capability of the threat, the higher the chance a vulnerability will be exploited.

Table 5-5 *Vulnerability Derivation*

Vulnerability		Control Strength				
Threat Capability		VL	L	M	H	VH
	VH	VH	VH	VH	H	M
	H	VH	VH	H	M	L
	M	VH	H	M	L	VL
	L	H	M	L	VL	VL
	VL	M	L	VL	VL	VL

VH=Very High

H=High

M=Medium

L=Low

VL=Very Low

Loss Event Frequency

The Loss Event frequency is the number of times a threat can impact an asset and lead to a potential loss within a specific timeframe. In context of specific threats mentioned before, the number of attack attempts doesn't showcase the loss event frequency because a loss has not occurred. The number of successful data exfiltration attempts maps to loss event frequency. The number of times a vulnerability is exploited within a specific time-frame is the key metric. If the vulnerability is VERY HIGH and the threat event frequency is VERY HIGH, the resulting loss event frequency is going to be VERY HIGH. A VERY HIGH loss event frequency signifies that a breach is almost certain to occur and steps must be taken to mitigate the threat. Table 5-6 showcases a loss even frequency derivation.

Table 5-6 *Loss Event Frequency Derivation*

Loss Event Frequency		Vulnerability				
Threat Event Frequency		VL	L	M	H	VH
	VH	M	H	VH	VH	VH
	H	M	M	H	H	VH
	M	L	L	M	H	H
	L	VL	VL	L	M	H
	VL	VL	VL	VL	L	M

VH=Very High

H=High

M=Medium

L=Low

VL=Very Low

Phase 3: Impact of Loss

The impact of loss is a measure used by enterprises to identify and evaluate the various environmental factors that can contribute to a loss of asset when there is a breach. The following constructs are critical to understanding the measurement and how the impact is evaluated.

Evaluate Probable Loss Magnitude (PLM)

Probable loss magnitude (PLM) is the probable loss that a loss event can incur for the enterprise. Since this is a probable loss, it need not be quantitative in nature and can be qualitative. For example, loss of customer data can lead to high loss of reputation.

Estimate Worst-Case Loss

The *worst-case loss* refers to the estimated maximum possible loss that an individual or organization can incur from a particular investment or decision. In other words, it represents the estimated largest amount of money that can be lost under the most unfavorable conditions or scenarios. This is the loss for the worst-case scenario. For example, if an investor is considering investing in a particular stock, the worst-case loss would be the maximum amount they could lose if the stock price were to plummet to zero. Similarly, if a business is considering a new project, the worst-case loss would be the largest possible financial loss they could incur if the project were to fail completely.

Estimate Probable Loss

Probable loss refers to a potential financial loss that is likely to occur in the future based on past experience, trends, or other available data. It represents the estimated amount of money that an individual or organization may lose as a result of a specific event or risk materializing. This is a more realistic value and what most enterprises will be prepared for.

For example, an insurance company may estimate the probable loss associated with a particular type of insurance policy by analyzing historical data on claims and losses. Similarly, a business might estimate the probable loss associated with a new project by analyzing market trends, competition, and other relevant factors. When and enterprise is considering measurement of probable loss, the dollar value associated with each qualitative range is based on the general impact of the loss in the specific enterprise stream. Loss of PII is a large impact for all verticals in the market, but loss of availability impacts banks more than a research institute. Table 5-7 showcases a possible magnitude range and the associated dollar values.

Table 5-7 *Magnitude of Impact: A Qualitative Mapping*

Magnitude of Impact	Range in $
SEVERE	Greater than 10 million
HIGH	1 million to 10 million

Magnitude of Impact	Range in $
MODERATE	100K to 1 million
LOW	10K to 100K
VERY LOW	Less than 10K

It is important to note that the magnitude of impact can be dependent on various external factors and not just monetary. Reputation loss, losing competitive edge, fines and compliance issues, loss of operations, loss of customer data, and so on are some other dimensions of how magnitude of impact is calculated. Enterprises ideally must use more than just one of these aspects to determine magnitude of impact.

Phase 4: Derive and Articulate Risk

In this phase you map all the findings and qualitative ratings to the actual risk of a threat scenario or risk of not addressing a specific security gap. The story of the risk metric must begin from the threat and asset loss use case followed by the strength of the threat and how frequently the threat will materialize. Then, based on existing controls, the impact of loss is derived. Finally, the frequency of the loss event is mapped to the impact of loss using the probable loss magnitude (PLM) to derive the contextual risk. Table 5-8 illustrates mapping the risk, and Table 5-9 illustrates the risk calculation.

Table 5-8   *Risk Severity Key*

Risk Key	Description
Critical	Catastrophic risk to overall IT infrastructure and enterprise reputation. Large-scale impact due to vulnerability being exploited.
High	Huge impact to overall business. Might lead to downtimes and there is high chance that there will be loss of data.
Medium	Impact to business as usual. Several processes may be halted. Chance that a vulnerability is exploited is not large.
Low	Very low chance that a vulnerability is exploited.

Phase 5: Identify the Controls

The final phase of any risk analysis framework is to have a tangible outcome to the risk derivation. In this case, it involves identifying the right security controls for the identified risks. The control, however, is not asset focused. It is not a response to a gap in asset protection. The controls identified are based on real threats and contextual environmental responses. It adds more contextual value to the controls identified. In this way, Open FAIR sets itself apart from most risk analysis frameworks and is fully aligned with the Zero Trust paradigm.

Table 5-9   *Risk Derivation*

Risk		Loss Event Frequency				
Probable Loss Magnitude		VL	L	M	H	VH
	S	H	H	C	C	C
	H	M	M	H	C	C
	M	L	L	M	H	H
	L	L	L	M	M	H
	VL	L	L	M	M	M

Loss Event Frequency:

VH=Very High

H=High

M=Medium

L=Low

VL=Very Low

Probable Loss Magnitude:

S=Severe

H=High

M=Moderate

L=Low

VL=Very Low

Step 3: Present the Metrics

After crafting metrics and making sure there are clear objectives and targets to measure, the next step in the metrics lifecycle involves accurate representation of metrics in the overall Zero Trust strategy and architecture presentation to all the relevant stakeholders. A more detailed representation of this phase is covered in Chapter 10, "Presenting the Zero Trust Strategy."

Step 4: Monitor Metrics

The final step in the metrics lifecycle is to implement the crafted metrics. In this phase customized metrics are deployed and implemented along with the overall Zero Trust architecture. Once in production, the metrics are monitored and then any changes, feedback, or improvements are incorporated into the overall metrics design and subsequently into the Zero Trust architecture. Details of implementation and monitoring are covered in Chapter 11, "Implementation and Continuous Monitoring."

A Hybrid Approach

Usually, most enterprises do not pick and choose a specific risk management approach since asset-based risk management and threat-based risk management both have their pros and cons. The asset-based risk management is a more traditional method of risk management, and crafting metrics from these methods would need a strong asset inventory setup. This is usually a long process and has multiple recurring cycles to enrich. Depending on asset inventory alone would greatly impact the crafting of metrics and its timelines. Threat-based risk management is faster and more effective when it comes to context-based evaluations. An enterprise would usually begin asset inventory and start identifying key assets. As each asset is identified, threat flow scenario–based risk management is performed, and over time the metric gets crafted or influenced by the threat flows identified for each type of asset. At this stage, during strategic and architectural discussions, the main goal is to spend time and effort to align and craft critical metrics that satisfy and resonate with all stakeholders.

The Follow-Up

[Glenn pauses and ends with a quick summary.]

Glenn: In summary, here are some highlights of what we want to achieve:

- Metrics are important to be able to drive adoption of the Zero Trust vision to senior leadership.

- The metric lifecycle consists of aligning metrics to business, crafting intelligent metrics, presenting the metrics to all stakeholders, and monitoring the metrics to make them more robust and relevant.

- In this discussion, we spent time only on aligning and crafting. We will present the metrics with the overall strategy and modify metrics if needed once they are deployed and monitored. This will happen post-implementation of the architecture.

- Alignment of metrics can be the following:

  - Goal based

    - Strategic

    - Tactical

    - Operational

  - Measurement based

    - Qualitative

    - Quantitative

- Metrics are crafted as either performance metrics or risk metrics.

- Performance metrics are future-looking; risk metrics measure the current state and gaps.

- Risk metrics can be asset focused or context focused. Our goal is to be able to showcase context-focused metrics because they consider the overall asset and threat and not just the intrinsic asset value.

Mr. Smith: Alright. Metrics are not new to me; however, it has always been a hassle to align the metrics with our security initiatives. I want you to spend some time with Mariam, Jed, and William and craft the relevant metrics for us relevant to the Zero Trust initiative.

Glenn: Yes, I have, and before I begin, I would like to start by doing a recap to make sure we have correctly understood your vision and mission. In short, the following graphic represents the vision and mission of Zenith Trust Bank (see Figure 5-4).

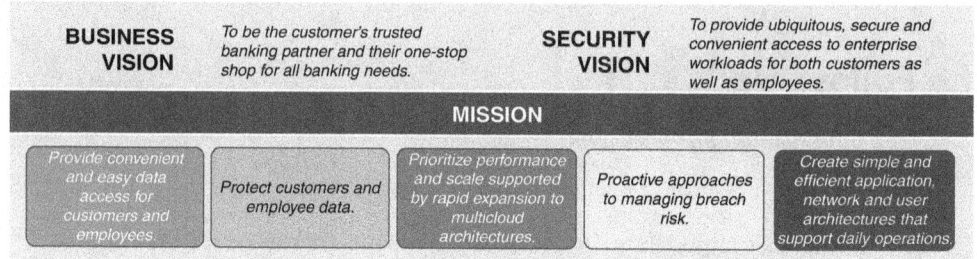

Figure 5-4   *Zenith Trust Bank Vision and Mission*

With the vision and mission in mind, let us start with what we want to measure to take the Zero Trust strategy forward.

Strategic Performance Metrics

- **Ease of access of user data measured qualitatively.** This metric relates to how easily data owners can access their data. Data owners usually include end-users like customers as well as employees, and the measurement is based on factors like availability from anywhere, strong authentication, and so on. A questionnaire will be sent to a sample of groups to understand the overall ease of access.

  - **High:** A scope of High means that the data is accessible easily.

  - **Medium:** A Medium scope points to possible issues with access or delays in access.

  - **Low:** A scope of Low means difficult availability or possibly unavailability of data when needed due to restrictive security measures.

    This metric aligns with the mission statement "Provide convenient and easy data access for customers and employees."

- **Strength of controls for critical data measured qualitatively.** This metric relates to control strength for critical data. This measures the security controls that have been considered when subjects access the data. This metric is asset focused and is measured as follows:

  - **High:** Strong security controls like multifactor authentication, with data encryption implemented. Endpoints are postured.

■ **Medium:** Security control is limited to data access only and is implemented for certain users. Employees can access data with fewer security controls.

■ **Low:** Protection for critical data is limited. Encryption at rest is not implemented.

This metric aligns with the mission statement "Provide convenient and easy data access for customers and employees."

■ **Organizational agility measured quantitatively:** Organizational agility is the capability of the enterprise to pivot to a different type of workload or strategy when performance and scale are critical. This specifically points to cloud movement and can be measured as a percentage of on-premises workload that has successfully moved to the cloud with the right security controls in place.

This metric aligns with the mission statement "Rapid expansion to a multicloud architecture."

■ **Total cost of ownership (TCO) measured quantitatively:** TCO is a financial metric that measures the total cost of a technology investment over its entire lifecycle. In the context of cloud migration and Zero Trust adoption, TCO can be used to compare the cost of running applications and services in a traditional on-premises environment versus the cost of running them in the cloud. To measure TCO, companies can consider factors such as hardware and software costs, maintenance and support expenses, energy consumption, and personnel costs associated with managing the infrastructure. By quantifying these costs and comparing them to the cost of running the same applications and services in the cloud, companies can determine the potential cost savings of cloud migration. With the right Zero Trust strategy, movement to cloud is simpler and measured separately. Quantitative measures of TCO will include cost savings achieved through reduced hardware and software expenses, lower energy consumption, and more efficient use of IT staff. This metric can be tracked over time to measure the ongoing cost benefits of cloud migration and to identify opportunities for further optimization.

This metric aligns with the mission statement "Rapid expansion to a multicloud architecture."

Tactical Performance Metrics

■ **Reduce the existing blast radius by segmenting the network and applications measured quantitatively.** Segmentation is a critical tactical goal when it comes to Zero Trust to reduce the blast radius of an attack. Segmentation needs to be achieved at the user, workload, and network architecture and traffic levels and can be tracked as a percentage of the total. Workload segmentation can be tracked as a percentage of the total workload, and network segmentation can be tracked as a percentage of the total network setup.

This metric aligns with the mission statement "Proactive approaches to managing breach risk."

- **Achieve endpoint posturing to augment existing subject context measured quantitatively:** Endpoint posturing refers to the security posture of an endpoint device, such as a computer or a mobile device, in a network environment. It involves the measures taken to secure the device and its data, including the installation of security software, the implementation of security policies, and the application of patches and updates. As a tactical metric, endpoint posturing can be measured by assessing the security posture of each endpoint device in a network. This assessment typically involves evaluating the endpoint's compliance with security policies, the presence of security software and updates, and the vulnerable threat surface exposed to attack vectors, and it will be expressed as a percentage of the entire device asset inventory of the enterprise.

  This metric aligns with "Protect customers' and employees' data" as well as "Proactive approaches to manage breach risk."

- **Reduce incident response time measured quantitatively:** Incident response is important when it comes to Zero Trust, and a key metric to measure the incident response effectiveness is how soon an incident can be isolated, artifacts captured, and reports created. With a larger automation and orchestration (SOAR) initiative, the aim of this metric is to measure how fast the enterprise can isolate, identify, and take effective action either automatically or manually. This is measured in minutes, hours, or days, depending on the overall average.

  This aligns with the mission statement "Proactive approach to breach risk."

Operational Performance Metrics

- **Reduce troubleshooting time during incidents measured quantitatively:** With simple contextual policies and simple design by Zero Trust, troubleshooting configuration and flow issues should be simpler. With more detailed visibility provided by Zero Trust initiatives, troubleshooting and root cause analysis must take less time, measured in hours or days.

  This metric aligns with "Create simple and efficient application, network, and user architectures."

- **Reduce mean time to detection of security incidents measured quantitatively:** As an operational metric, this measures how fast the enterprise can detect anomalous incidents. This will include a combination of technology like user and endpoint behavior analysis (UEBA), behavior analytics, and other behavior-based solutions to provide accurate information to reduce detection time. The faster the solution can identify the type of flow, the faster it can detect whether it's an anomaly and is measured in minutes, hours or days.

  The metric aligns with "Proactive approach to managing breach risk" and "Simple application, user and network architecture."

- **Successfully block fraudulent activities measured quantitatively:** This metric measures how well the enterprise is able to detect and block fraudulent activities.

With the right visibility controls, the metric is measured as a percentage of total activities recorded. The goal is to achieve 98% or above block rate.

A tactical visibility and automation initiative is needed to align with the "Protect customers and employee data" mission statement.

■ **Reduce downtime of applications and services measured quantitatively:** This metric will measure how much availability the application can provide, along with security controls in place. The goal will be achieved if the network provides 99.99% or more uptime.

This metric aligns with the "Provide convenient and easy access to data" as well as "Prioritize performance and scale."

As demonstrated, all metrics align with your key mission statements and each metric gets mapped to tactical projects that need to be completed based on your feedback. With the overall cost of projects at hand, we can provide a security budget, but we would like to make sure your metrics are completely accurate. We have spent quite some time with Mariam and William and have understood pain points and what exactly you would like to measure based on your vision.

Mr. Smith: Honestly, I am not sure if we have ever aligned our metrics to our vision in such detail for other strategies. I am impressed. Sam, what do you think?

Ms. Lee: I think this is a good start for us to align. So, what exactly is our next step?

Glenn: Currently we have identified these key performance metrics that you can use to drive adoption for the Zero Trust ideas to other stakeholders. Here is how your tactical road map looks (see Figure 5-5).

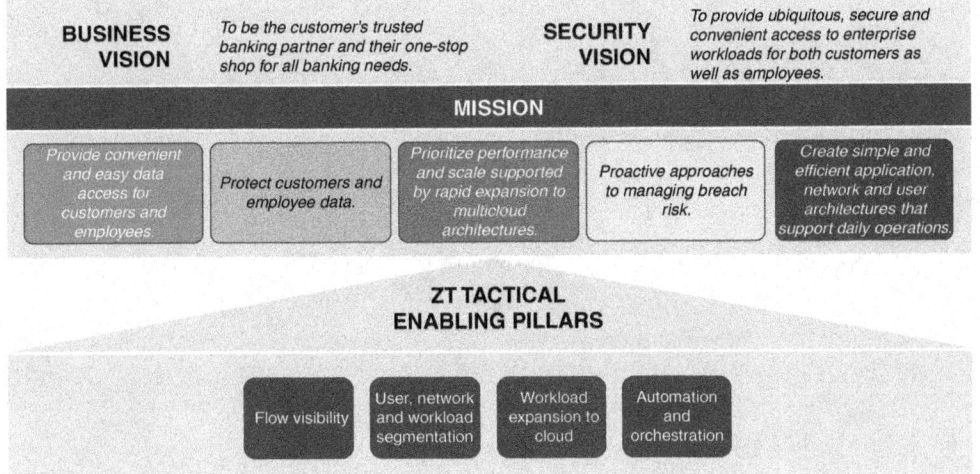

Figure 5-5   *Zenith Trust Bank's Vision Tactical Enablers Based on Performance Metrics*

Like I mentioned, based on your pain points and discussions with your infrastructure and security leads, we have also been able to run some key threat scenarios and have identified gaps in the infrastructure. With these gaps we have performed a maturity assessment, which has also helped us craft risk metrics and an implementation road map to help you showcase the true value of Zero Trust as a security framework for Zenith Trust Bank.

Mr. Chen: Gap analysis? What exactly have you been able to extract and what was your reference?

Glenn: We spent time discovering your network and identifying all the possible segments, assets and threat actors to produce a maturity assessment. The assessment is a reference for you as well as a baseline for us that helps create new metrics based on risk, which we added to your existing list of metrics. This will also reveal the additional security controls needed to reduce the risk in your infrastructure, which will eventually impact the overall security budget.

Mr. Smith: Alright, tell me more about this maturity assessment. Is it similar to the audit we had done?

Glenn: It is not, but there is some scope that overlap. Let me explain in detail.

Endnote

1. "What Is Open FAIR and Who Is the Open Group," https://www.fairinstitute.org/blog/what-is-open-fair-and-who-is-the-open-group

Chapter 6

Understanding Zero Trust Maturity

At this point it is expected that, as the consultant, Glenn has had the chance to speak to the key leads in the enterprise and identified some pain points and performance metrics that align with the overall mission statement. The next step is to gauge the existing maturity of the network and map out the threats, assets, and important cost considerations that can impact the environment and subsequently the business. This is an important step because it helps build metrics and tactical strategies for the enterprise to pursue as part of the overall Zero Trust vision.

The most common framework used to measure Zero Trust maturity is the second version of the Cybersecurity and Infrastructure Security Agency (CISA) maturity model (CISA v2). CISA Zero Trust Maturity Model v2 is a standard that has been created to aid in understanding the maturity of an enterprise's Zero Trust readiness. This model helps the enterprise assess how much expenditure (capital expenditure and operational expenditure) needs to be outlaid to be able to achieve it Zero Trust vision. It is important to understand that though the CISA provides a clear demarcation of maturity levels, every enterprise has its own requirement and will lay out its Zero Trust road map based to its vision rather than what CISA mentions, which is why the CISA v2 maturity model is just a guideline to measure the current state and the controls required to reach the desired maturity. As per the survey "CISO Perspectives and Progress in Deploying Zero Trust," 35% of enterprise CISOs believe that the CISA v2 model is closely aligned to a Zero Trust vision.[1]

When enterprises engage consultants for Zero Trust, it is a common observation that some organizations may have already begun their Zero Trust journey. It's true that it is rare to find a fully functional end-to-end Zero Trust strategy, yet it is also true that all enterprises have their own goals and visions and when a maturity analysis is performed, most enterprise find themselves benefitting from Zero Trust with scope for improvement.

The CISA v2 maturity model is a reference model to measure Zero Trust readiness and not a reflective indication of the enterprise risk. That is why this chapter will focus on measuring how ready the enterprise is to design and implement Zero Trust networks and solutions. This model must be enhanced with a broader information security measurement like NIST Cybersecurity Framework v2.0 or specific use-case-driven regulation guideline like Health Insurance Portability and Accountability (HIPAA) or the Payment Card Industry – Data Security Standard (PCI-DSS) framework. However, CISA Zero Trust maturity model v2 is a good starting point to understand the current capability readiness of the enterprise to adopt and deploy Zero Trust networks. As you will see in this chapter, CISA maturity model v2 and Capability Maturity Model Integration (CMMI) are complementary maturity models that, when combined, can provide a holistic view of the people, processes, and technology aspects of Zero Trust readiness. Each can also be evaluated as a standalone framework if the other is not acceptable to specific organizations. Forester's Zero Trust Extended frameworks also map to the CISA maturity model v2 and can be used as a reference to measure the maturity of a specific capability readiness as an alternative to CISA maturity model v2.[2]

The Five Pillars of Maturity for Zero Trust

CISA maturity model v2 has five pillars to support the Zero Trust strategy and three common concepts to strengthen the base.[3] Figure 6-1 provides a visualization of the five CISA pillars.

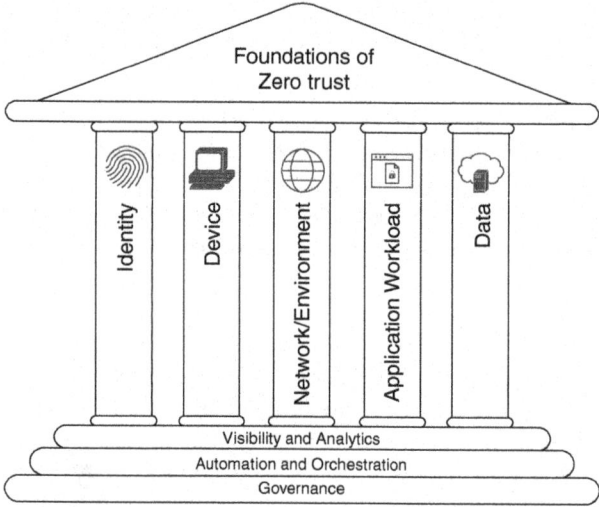

Figure 6-1 *Pillars of Zero Trust Maturity (Image Courtesy of Cisco's Guide to Zero Trust Security)*[4]

It is worthwhile to pause and compare our Zero Trust definition to the illustration in Figure 6-1. At its core, Zero Trust controls access to an object using the context of both the subject and the object over a secure communication channel. Here, both subjects and objects are unique identities. They are accessed across various networks/environments or "transports". The final goal is to be able to access the application workload and the data it processes. This access must be continuously monitored to make sure the access is still valid. Orchestrating simple responses requires making automation enhancements to specific business processes and the creation of more playbooks for timely responses. Finally, all the activities must align with the business needs.

Identity Pillar (Subjects and Objects)

The Identity pillar is the first pillar in the Zero Trust maturity model. It involves securing, limiting, and enforcing person, non-person, and federated entities' access to enterprise resources and encompasses the use of identity and access management (IAM) capabilities that include functions like multifactor authentication (MFA) and Single Sign On (SSO). Organizations need the ability to continuously authenticate, authorize, and monitor activity patterns to govern users' access and privileges while protecting and securing all interactions and transactions. Role-based access control (RBAC), attribute-based access control (ABAC), and context-based access control (CBAC) will apply to policies within this pillar to authorize users toward access of applications and data.

A federation is a mechanism that allows for the sharing of authentication and authorization information across multiple systems or organizations. It enables users to access resources in one security domain without having to create and manage separate credentials for each domain. In a Zero Trust environment, a federation can enhance security by ensuring that trust is established across different domains while adhering to the Zero Trust principle of continuous verification. This is particularly useful when considering vendors or third-party users accessing enterprise applications. The Identity pillar extends to making sure that not only the local user base but also the unmanaged user base are controlled and identified.

For example, if a user needs to access resources in multiple cloud services, a federation can provide a seamless and secure way for the user to authenticate once and access these resources without the need for multiple login credentials. Federation provides the ability to manage cross-organizational access decision-making. Single sign-on (SSO) is a similar concept, except that SSO is applicable to providing seamless access across applications within an enterprise.

An enterprise might be considering moving to a cloud Lightweight Directory Access Protocol (LDAP) server or might already be using cloud-based identity providers. Irrespective of which option the enterprise chooses to use, the recommendation will always be to utilize the mechanism that suits the business requirements. Bringing back the focus to business requirements, the final goal of Zero Trust is to dynamically evaluate identity to control apt resource access, which supports the basic function of the system. Maturity in this pillar would mean that the enterprise considers all aspects of the identity, including the risk profile. A VIP user is more prone to social engineering threats than an employee and hence providing full access to a VIP user is not best practice. The user's identity must not be authenticated just at the beginning but must be continuously validated to make sure the user's context has not changed.

This brings us to the second aspect of identity—dynamic profiling and automating the IAM lifecycle. The IAM lifecycle is usually automated in most enterprises, but dynamic profiling is not as commonly used. Dynamic profiling helps ensure that just-in-time access is maintained. Coupled with user and endpoint behavior analysis (UEBA), these information points help to add much needed endpoint and user context to be able to make dynamic decisions. These are the criteria an enterprise will be evaluated against. Products such as Cisco ISE, Cisco Secure Endpoint, and Cisco Secure Analytics, when combined as a solution, provide a very accurate representation of most users, endpoints, and their behavioral patterns.

To understand the current maturity of an enterprise in the space of identity, the following questions can be used to help initiate a discussion. This list is not complete and can be expanded based on engagement and the target enterprise. These questions are usually targeted to IAM administrators and security architects. The questions are aligned based on specific capabilities identified as part of the Identity pillar in CISA maturity model v2.

- **Identity and Access Management (IAM)**

    - Is there an identity management service or platform in the enterprise?

    - Is the identity and access management solution distributed?

    - Is role-based access in place for either resource access rules (traffic flow, object access) or network access?

    - Is access given based on subject or object risk profile?

    - Is identity proving performed (typically as part of hiring) before access to resources is provided?

    - Are new users given a default employee privilege? If yes, what is the extent of access an employee gets?

    - How are vendors and third parties given access to the network and applications? Is there a federation in place? Is there a VPN or virtual desktop infrastructure(VDI) solution?

    - Are users provided need-to-know access to resources?

    - How are access requests for various subjects to resources and workloads evaluated and implemented?

    - Are access requests subject to multiple approvals? Who are the approvers and what are their roles? (Manager approval or resource owner approval at a bare minimum must be included.)

    - Are different users dynamically or statically provided permissions(to network, applications or data) considering separation of duties?

    - Is there emergency ("break glass") access? How is it managed?

    - Are there any legacy solutions or operating systems integrated into the IAM system?

- Is admin management access segregated from the less-privileged user access toward applications and network?

- Is TACACS or RADIUS used for device management?

- Is there any solution that implements session-based access control for users?

- **Authentication**

  - Are multifactor authentication mechanisms used for user access across various transport media like LAN, WAN, VPN, and so on?

  - Is there location-based geographic restriction?

  - Is there time-of-day consideration?

  - What privileged access management (PAM) solutions are in place?

  - Is the management/admin/operator access to various applications regulated with the PAM solution?

  - Are applications SSO capable? What is the extent of SSO adoption in the enterprise?

  - Are container-based workloads integrated with external Active Directory–based access as part of the base image or is authentication local?

  - Are there subjects that utilize password-less authentication via certificates or other means?

  - How are certificates managed in the enterprise?

  - Are all entities (user and system) authenticated for access to the network and resources?

  - Is a posture assessment performed (Windows version, Windows patch level, antivirus) for a remote access user? If yes, what criteria are checked?

  - Do on-premised users have the same access policy as VPN users?

  - Are all entities (user and system) authorized for access to the network or network segments in addition to being authenticated?

  - Do wired and wireless access have different access policies?

  - Do end devices and servers perform mutual authentication (mTLS)?

- **Identity Stores**

  - Is the Active Directory self-managed or a managed service?

  - Does the identity management platform manage different forests?

  - What is the directory architecture and hierarchy?

  - Are there accounts in identity repositories that are not linked to an identity or user?

- Can users have more than one account in the same access repository? If so, under what circumstances?

- Does the enterprise have a use case to share user accounts?

- Are all subjects common across the entire enterprise and partners? Is there a master user record for identities?

- Is there segregation of administrative management access from the other user access to the Identity store?

- **Risk Assessment**

  - Are user risk profiles evaluated for access requests?

  - Is a risk assessment performed on creation of new roles?

  - Are third-party/vendor identities assessed for risk before being provided access to the enterprise?

  - Is the risk processing dynamic or static (for example, UEBA/UBA, user type, such as employee or contractor)?

  - Is the sensitivity/criticality of a resource being accessed taken into consideration when providing access to subjects?

- **Visibility and Analytics**

  - Is analysis and alerting in place for continued access failures for any user's account, even if across resources?

  - How soon can the SOC identify compromised accounts? Is sufficient logging enabled for the same?

  - How frequently are user logs analyzed? Are they automated?

  - Has user and entity behavior analytics (UEBA) been implemented for all user segments?

  - Are all access and user events correlated with contextual data to detect anomalous behaviors?

- **Automation and Orchestration**

  - Is identity creation initiated from the HR system or manually?

  - Are identities automatically removed when users are offboarded/end-dated in the HR system?

  - Is provisioning of users to identity repositories automated specifically when considering user authentication, authorization enablement, and per-user personalization?

  - How are permissions managed and controlled for users as part of the user lifecycle (provision/modify/deprovision)?

- Are risk calculations for resource access automated?

- Are access approvals automated based on known access requests and risk profiles?

- Is there a rapid containment mechanism of a compromised user in collaboration with an AAA server?

- Can accounts suspected of being compromised be suspended rapidly?

- Does suspension of accounts occur automatically in response to triggers such as continued authentication failures or UEBA analysis?

- **Governance**

  - Are periodic user access reviews performed?

  - Who attests that accounts as well as entitlements are still required?

  - Are accounts suspended if attestation is not made?

  - Are end-to-end identity-centric zero trust policies in place across various enforcement points?

  - Are User Access Matrices or Separation of Duty Documents in the enterprise that can be reviewed to understand allowed access intents?

Device Pillar (Endpoint Security)

Devices are also considered subjects in modern networks and an important context setter for a subject's identity. Devices add a different aspect to context. When checking maturity for Zero Trust, a device should match the enterprise's endpoint security policy, which measures the security posture for the device. The security posture of a device must be an addition to the context metadata so that unpostured devices are not provided full access. Of course, to be able to measure and implement security posturing, the device profile, communication, and access requirements must be identified, and this is made much simpler with automated asset management and user behavior analytics.

Having the ability to identify, authorize, list, isolate, secure, remediate, and control all devices is essential in a Zero Trust approach. Real-time attestation and patching of devices in an enterprise are critical functions with respect to maintaining endpoint security posture. Some solutions, such as "Mobile Device Managers" and "Comply to Connect" programs, provide data that can be useful for device confidence assessments. More validation should be conducted for every access request (some examples include examinations of compromise state, anomaly detection, software versions, protection status, and encryption enablement) to make sure that the entire security posture of a device is extracted and a subject identity context is rich with attributes.

Identifying posturing at the beginning of a connection or onboarding is not enough, and posture must be checked continuously. Generally, posture states are maintained for half an hour to an hour based on the access and UI requirements of the enterprise, but it is recommended to be validated at each subsequent and unique access request. Questions from the device pillar are usually targeted to the infrastructure teams that own corporate assets as well as security architecture teams. When you're trying to extract information regarding the device management practices in an enterprise, the following questions can be used as a base:

- **Policy Enforcement and Compliance Monitoring**

  - How is ongoing device compliance managed?

  - Do users have local admin privileges?

  - Do all devices have a standard hardened baseline golden image? Are all devices hardened using industry or vendor guidelines for a secure configuration?

  - Are all device images signed and validated?

- **Asset and Supply Chain management**

  - Are all end-user device types centrally managed with a secure configuration?

  - Are all workload systems and applications registered under asset management before go-live?

  - Are specific types of devices allocated to specific users? (For example, employees get phones and laptops, researchers get a PDA, and so on.)

  - How are headend devices like printers identified? Are they identified based on their MAC address or certificate?

  - Are non-enterprise assets checked for security posture and compliance?

  - Are third-party risk assessments performed when non-enterprise subjects or device request access to enterprise resources?

- **Resource Access**

  - What capabilities are included for posture measurement in each device?

  - Can specific users bring their own personal devices and access networks in a non-guest flow? (Is there a BYOD flow?)

  - What kind of mobile device management (MDM) is in place for mobile assets?

  - Are there operation technology environments that are tracked in inventory?

  - Is there consideration of end-user device risk (posture) for resource and network access?

  - Is the risk calculated dynamically or statically when providing access to resources?

- **Device Threat Protection**

  - Are there threat protection capabilities integrated with compliance initiatives specifically for certain types of devices (MAC, Linux, and so on)?

- Is threat intelligence available and updated such that it is current for all detective and preventative controls requiring it?

- What host-based threat protection capabilities are used on the devices? Examples include EDR, antivirus, file integrity monitoring, host-based firewall, process anomaly detection, application whitelisting, and intrusion detection.

- **Visibility and Analytics**

  - If the device is a valid asset, what happens if the device configuration is out of alignment with enterprise policy? Where is the monitoring for these situations?

  - Are end-user device compliance anomalies logged to SOC/SIEM?

  - Are device threat protection policies determined by threat/risk analysis derived from SIEM logs?

  - Is device health extracted and analyzed for all enterprise assets?

  - How are unauthorized devices on the network detected?

  - How are devices checked to determine whether they are enterprise assets? Are there specific logs sent to a central entity?

  - Are devices scanned to determine whether they align with policy? Where can the scanned results be viewed to correlate?

  - How is the security posture validated?

- **Automation and Orchestration**

  - How is ongoing device compliance managed?

  - Are there Infrastructure as Code (IaC) initiatives in the enterprise?

  - Are there devices that have their build process automated?

  - Is there an integrated continuous integration/continuous deployment (CI/CD) pipeline in place for IaC?

  - Is device patch management and vulnerability assessment automated?

  - Are unauthorized devices automatically discovered on the network and quarantined/isolated?

  - Are any security posture anomalies resolved using automation?

- **Governance**

  - Is there an endpoint policy document that can be provided?

  - How frequently are hardening baselines revaluated?

  - How frequently are devices updated?

Network and Multicloud Environment (Location and Access Media) Pillar

At a high level, Zero Trust aims to segment (both logically and physically), isolate, and control the network/environment (on-premises and off-premises) and create granular access and policy restrictions. Maturity in network or access media for traffic focuses on segmentation both at the micro and macro levels. The maturity level for the network would be an accurate atomic isolation of network segments into micro-parameters. Micro- and macro-segmentation are key drivers in Zero Trust strategy, and a move toward a Zero Trust network workload segmentation must be mature and network flows must be clear. This must be coupled with constant monitoring to make sure changes in flows reflect in policy. SSE deployments take this concept and provide per, user, per session, per application isolation thus reducing the dependency on network. However, network segmentation still remains an important aspect of security architecture.

As the perimeter becomes more granular through segmentation, there is greater protection and control over resource access. It is critical to control privileged access, manage internal and external data flows, and prevent lateral movement. Software-defined networking strategy adds the additional support of orchestration and automation that is now being adopted by many enterprises in addition to the SSE capabilities.

Another aspect to network segmentation is to make sure all the threat scenarios are considered when it comes to specific zones. Threats to the DMZ segment are different from threats to a server farm segment. These are not relevant to policy creation but can be criteria to build context. Network threat protection integration with security information and event management (SIEM) solutions provides additional context and visibility for access validation. Based on some of the capabilities in the network, it is expected that the network devices are sending traps and information of all types to a common log repository for processing and is subsequently sending this processed information to the policy decision point. If there are situations where there is a breach or a compliance mismatch for some endpoints, these situations must be automatically remediated. This pillar hence measures the maturity of the network elements in the infrastructure.

Some common queries targeted to the network infrastructure and security operations teams are as follows:

- **Network Segmentation**
  - What are the common connectivity media (web-app, RA VPN, wired, wireless, extranet, and so on) that subjects can access resources from?
  - Is the network segregated at the macro level (traditional network segmentation user/data center/DMZ) firewalls?
  - Is the network is segregated at the workload level—micro (application) segmentation?
  - Is traffic inside specific segments, segmented based on function? For example, are management and identity flows being isolated from each other?

- What is the general approach for network segmentation? Is there application segmentation/isolation/sandboxing? Is segmentation done by function, risk, data classification, or any other method? For example, is the data center in one security segment/zone? What is the strategy?

- **Network Traffic Management**

  - Do legacy networks depend on subnet-based control via enforcement points?

- **Traffic Encryption**

  - Is all network communication encrypted? If not, what is the criteria for encryption exemption?

  - Specifically, is there implicit trust for traffic from campus, branch or other WAN segments when entering the data center?

- **Network Resilience**

  - How is network resiliency deployed in the network? Is it designed at the beginning only or scaled automatically? (Cover various resiliency initiatives in the enterprise such as disaster recovery practices, failure recovery actions, and so on.)

  - Is the design and architecture of the enterprise workload and networks active/standby or active/active?

- **Visibility and Analytics**

  - How are network flows monitored? Are there visibility mechanisms to correlate flows end-to-end?

  - What are the key sources from where telemetry is extracted in the network?

  - Are advanced concepts like machine learning used to establish "normal" and therefore "anomalous" activity?

  - Are machine learning and AI techniques used to alert of potential policy breaches, including unauthorized access? (An example is network flow analytics.)

  - Are capabilities like network flow analytics used to enrich profiles with contextual data (workload, user, and so on)?

  - Is correlation performed across event sources and other types to alert for potential policy breaches, including unauthorized access?

- **Automation and Orchestration**

  - Are network devices suspected of being compromised quarantined/isolated rapidly? Is this process automated?

  - Are there plans to implement network automation initiatives such as software-defined access, software-defined wan, software-defined networking, and so on?

- **Governance**

  - Are there inbound and outbound network security policies? What are the deployed security controls based on compliance and regulation?

  - What tools and processes are in place to protect from and detect threats to the workplace? (For example, threat hunting, IPS, network malware protection, firewall traffic filtering, DNS.)

  - Are workplace threat protection capabilities determined by threat/risk analysis or through the implementation of a static policy?

  - What is the existing cloud adopted strategy?

- **Cloud Readiness**

  - Are there multiple cloud connections toward the enterprise?

  - How are multicloud networks connected to each other and on the premises?

  - What are the security features utilized when communicating across multiple cloud vendors?

  - Are applications' SaaS deployed with a public IP? Or are they private applications accessible over VPN or secure private access on the cloud?

  - Does the enterprise use a cloud access security broker (CASB) to protect cloud workloads?

  - What are the other capabilities present in the cloud to protect SaaS workload?

  - What are the other capabilities present in the cloud to protect IaaS and PaaS workload?

  - To achieve "access from anywhere" to "workloads anywhere," does the enterprise aim to subscribe to a Secure Access Service Edge (SASE) solution?

Application Pillar (Workload Profile)

As an extension to the network aspect, micro-segmentation of applications is also key to a mature Zero Trust enabled enterprise. Similar considerations are looked into, such as threat intelligence integration and dynamic and continuous access monitoring.

One aspect that really measures Zero Trust maturity for workload is the enterprise's confidence to expose its applications to the Internet. Cloud deployment of applications is based on this freedom of access without a NAT or an explicit web tier. Making the application available over the Internet shows confidence in not only securing access to the application but also in securing the transport with end-to-end encryption. This essentially implies that the enterprise has implemented all the right security controls to be confident enough to expose the application to the Internet. This also means that the application itself has been hardened and all security best practices have been considered when creating the application. This application can hence be categorized as "Zero Trust native." Security must be built into software creation as part of a broader DevSecOps strategy. Policies must be updated dynamically based on change in the application flow, which ties back to asset inventory. The bottom line is that one can understand application flows only when the assets in scope are discovered and their flows are mapped.

Applications and workloads include tasks on systems or services on-premises, as well as applications or services running in a cloud environment. Zero Trust workloads span the complete application stack, from application layer to hypervisor. Securing and properly managing the application layer as well as compute containers and virtual machines is central to Zero Trust adoption. Application delivery methods such as proxy technologies enable additional protections, including Zero Trust decision and enforcement points. Developed source code and common libraries are vetted through application security and secure development practices to secure applications from their inception. CI/CD toolchain integration, along with software-defined compute, supports orchestration and automation at time of deployment and ongoing change management. Similar to network segments, integration of application logs with the SIEM solutions provides additional context and visibility into inter- and intra-workload flows. Authorization of access to workloads is dynamic and considers security posture of accessing devices, environment, and users. When you are extracting the current state of the workload in an enterprise, the questions in this section are a good reference point to craft queries that extract more information about the overall application development and deployment architecture. These are usually targeted to application owners and custodians, along with overall application security architecture teams.

- **Application Access**

    - Is all access control policy enforcement local to the application? If yes, is authentication and authorization performed local to the application or externally by an AAA server (local, external, hybrid)?

    - Does the enterprise use access gateways such as reverse-authenticating proxies to control inbound access to applications?

    - Is any dynamic analysis of risk considered to authorize access to an application ?

- **Application Threat Protection**

    - Are there protocol restrictions present within workloads? Are there enterprise-specific custom protocols? Has there been a threat event or cyberattack against these applications?

    - Is the threat protection application technology aware? (For example, web access firewall (WAF), app security, and email security solutions.)

    - Is workload communication enforced based on required flows only? Are there segments that have free flow of information?

    - What is the general approach for workload segmentation? Is there application segmentation/isolation/sandboxing? Is segmentation done by function, risk, data classification, or any other method? Does it follow the network segmentation strategy?

- East-west flows are categorized as flows that do not traverse a gateway. Microservices incorporate a lot of east-west flows. How are east-west flows controlled?

- Is workload security posture available for any system at any time?

- Does a new workload get full access to other workloads, or are flows controlled from the first day of deployment?

- What host-based threat protection is used for workloads? (Examples include EDR, antivirus, file integrity monitoring, host-based firewall, process anomaly detection, application whitelisting, and intrusion detection.)

- **Accessible Applications**

  - Are there applications that can be exposed to the Internet for secure private access?

- **Secure Development**

  - How are container builds managed?

  - Are applications designed in a microservices format?

  - What kind of controls are in place for traffic enforcement across container-based services?

  - Is there a formal assurance process to determine if workload security is production ready?

  - Are container images and existing applications signed and is the authenticity/integrity validated?

  - Is there a standard hardened build for each application OS/version?

  - Is hardening performed at the application, middleware, and database levels?

  - What is the current DevSecOps strategy, if any?

  - Are there any implementations of the CI/CD pipeline for workload?

  - How are local code/libraries managed?

  - Is configuration compliance scanning performed?

- Are periodic penetration testing activities performed on applications?

- **Secure Testing**

  - Does the solution delivery process mandate security testing?

  - Is static application security testing performed?

  - Is dynamic application security testing performed?

  - Is production data used for testing new solutions?

- Is vulnerability and compliance scanning and penetration testing performed once the workload is deployed? Or is it performed in staging?

- Is vulnerability scanning or penetration testing performed periodically on all applications in production?

- Is production data used for testing new solutions?

- **Visibility and Analytics**

  - Are all application flows known? If not, what percentage of flows are visible to the enterprise and monitored?

  - Are application security events integrated into the SIEM solution?

  - Does the threat protection utilize behavior analysis of flows with understanding of workload? Do you alert/block behavioral anomalies?

  - How is rapid threat control performed for traditional or container-based workloads?

  - Does SOC have visibility of application and operation system access events as well? For example, does the SOC receive relevant logs when user accesses the application's user interface ? If an IT admin account accesses the operating system hosting an application, does the relevant access log get registered on the SOC ?

- **Orchestration and Automation**

  - Are system dependencies such as network and security tool configuration automated?

  - Is patch management automated?

- **Governance**

  - Are end-to-end application lifecycle policies created enterprise-wide?

  - Are there application access policy rules identified for all subjects and workloads?

  - Do all application enforcement points incorporate risk and compliance criteria before providing access to subjects?

  - Are there broader vulnerability management, patch management, change management, static and dynamic testing policies and compliance management documents that can be reviewed?

Data Pillar (The Core of the Maturity Model)

Zero Trust protects critical data, applications, and services. Though it is already part of the main protection cycle, data was added to the Zero Trust equation with Forrester's Extended Zero Trust model. Data must be considered part of the asset inventory, and hence automated asset inventory and management is always critical to improve Zero Trust Maturity at handling data. The main key maturity factor in this tenet is the **just in time** and **just enough** access to the data based on the context.

Encryption is an important maturity factor. The decision point must take into account the data classification at each level of access request. Top-secret data must be handled separately with proper security controls. A clear understanding of an organization's data is critical for a successful implementation of a Zero Trust architecture. Organizations need to categorize their information assets in terms of mission criticality and use this information to develop a comprehensive data management strategy as part of their overall Zero Trust approach. The data classification must always be reevaluated to make sure the controls are adequate and the access is moderated accordingly. This can be achieved through the categorization of data, developing data schemas, and encrypting data at rest and in transit. Solutions such as data loss prevention (DLP), software-defined storage, and granular data tagging and labelling are relevant in protecting critical data and enabling automation and orchestration.

It is impossible to identify and provide an accurate data access policy without monitoring the data assets dynamically for changes in context. Access should be constantly monitored and suspicious activity must be captured and logged. The information captured must also be fed into the decision point to be able to build a context dynamically which subsequently allows accurate change in policy as needed. When extracting data classification and other data-specific processes from data owners and custodians, you can use the following questions as a reference:

- **Data Inventory**

  - Does all data have an owner? Are owner and custodian responsibilities defined?

  - Are all data sources registered and tracked in a central inventory?

  - Is technology in place to discover data of value not listed in the inventory?

- **Data Classification**

  - Is all data classified?

  - Is all data labeled based on context?

  - Does data classification change dynamically or statically?

- **Data Availability**

  - Are there mechanisms to make sure that data is reliably served?

  - What are the data backup procedures in the enterprise? Is all high-value data backed up?

- **Data Access**

  - What is the current data access model being used?

  - Are data access requirements determined by classification?

  - Are data access requirements determined by risk (static or dynamic)? User, device, location, and so on?

  - Are there role-based controls in place to control data access?

- **Data Encryption**

    - Is data encrypted at rest?

    - How is key management performed in the enterprise?

    - How is data confidentiality and privacy protected? (Elaborate on data confidentiality practices.)

- **Visibility and Analytics**

    - Are data flow, data access, and visibility logs sent to a SOC/SIEM?

    - Is suspicious behavior associated with data access detected/analyzed and correlated with other security logs such as user logs, NetFlow logs, and application access logs?

    - Is analysis performed to determine whether access controls are commensurate with data value? Is this automatic or a manual process?

- **Automation and Orchestration**

    - Is data inventory updated automatically and dynamically?

    - How are new data segments developed through data aggregation classified? Are they automatically identified or do they need manual intervention?

    - Is software-defined storage being used/evaluated in the enterprise?

- **Governance**

    - What are the data loss prevention (DLP) tools in place?

    - Is data loss prevention utilized for data in storage, data in use, and data in transit?

    - How does the DLP solution identify or understand whether data loss is occurring and prevent it from happening? Are static lists used, or is it heuristic with machine learning?

    - Are DLP solutions integrated with other access control points to provide context-based blocking?

    - Is there a data governance initiative or document in the enterprise that details how data is managed (within its data lifecycle)?

Visibility and Analytics, Governance, Automation, and Orchestration

Note that visibility, context creation, governance, and automation are usually incorporated into each maturity requirement. Vital, contextual details provide greater understanding of performance, behavior, and activity baselines across other Zero Trust pillars. This visibility improves detection of anomalous behavior and provides the ability for decision points to make dynamic changes to security policy and real-time access decisions. Additionally, other monitoring systems, such as sensor data in addition to telemetry, will help paint the overall picture of what is happening with the environment and will aid in the triggering of alerts. A Zero Trust enterprise SOC will capture and inspect traffic, looking beyond network telemetry and into the packets themselves to accurately discover traffic on the network and observe the threats that are present. Armed with this information, the enterprise can orient defenses more intelligently.

Zero Trust maturity would also include automation of multiple manual security processes to make policy-based decision-making across the enterprise fast and scalable. Continuous diagnostics and remediation-based solutions, such as security orchestration, automation, and response (SOAR) improves security and decreases response times. Security orchestration integrates the security information and event management (SIEM) and other automated security tools, including the security operations center (SOC), and assists in managing disparate security systems as one cohesive unit. Automated security response requires defined processes and consistent security policy enforcement across all environments in a Zero Trust enterprise to provide proactive Zero Trust control.

Zero Trust policy is unified across pillars, ensuring consistent enforcement and holistic protection. Consistent policy enables runtime policy representation to be orchestrated and automated for building and configuration. It also enables discovery of runtime policy anomalies and deployment of automated remediation required for alignment. The target audience for the discovery activity for this segment is usually incident response professionals, SOC analysts, and governance professionals. When discovering the details about existing processes and automation in an environment, you can use the following questions as a base:

- Is there central monitoring of all access control and security events more akin to a unified SOC?

- Are there multiple logging destinations for different type of logs (user, application, NetFlow)?

- Are all the logs processed at the same centralized SOC or is the SOC distributed across various solutions?

- Are all logs integrated with the SIEM/SOC? If not, which logs are set to SIEM/SOC?

- Is analysis and alerting in place for continued access failures for any user's account, even if across resources?

- Are any device security configuration and policy changes automated as a result of a predefined playbook?

Once these questions are answered and there is clear visibility of users, devices, network, data, and application segments, a maturity assessment is performed to identify the current gaps. Each capability is evaluated using a scale of 0–3, where 0 is Traditional, 1 is Initial, 2 is Advanced, and 3 is Optimized. These levels are aligned with version 2 of the

CISA maturity model for Zero Trust security controls. Each vertical in CISA maturity model v2 consists of a set of capabilities. For example, the Identity pillar consists of seven verticals:

- Authentication
- Identity stores
- Risk management
- Access management
- Visibility and analytics
- Automation and orchestration
- Governance

Each vertical is evaluated individually and given a score from 0 to 3, and the average is calculated to position the overarching pillar. Once this is done, a prioritized road map is created based on the desired maturity and then an implementation map is developed where low-hanging fruit can be deployed to help get leadership buy-in for more activities over the next years. This feeds into the security budgeting as well to showcase critical projects that need to be initiated. The next section will explain the process of mapping the maturity of controls as well as processes.

Note that the scale here may vary and start with 1 instead of 0. The decision is purely at the consultant's discretion. The only aspect to make sure is that if the CISA v2 model that measures Zero Trust capability starts at 1, then it is imperative that the CMMI model that measures process and people skillset, also starts at 1 to maintain scale consistency.

Zero Trust Maturity Levels

Once the existing gaps are identified, the enterprise is placed at a specific maturity stage so that the right tactical initiatives can begin. When you are measuring maturity, it is important to consider both process maturity and implementation maturity. For example, implementing flow visibility with Cisco Secure Analytics to understand important flows evaluates the implementation of Zero Trust visibility, whereas evaluating user lifecycle and authentication strength is considered a process evaluation. Remember that technical capabilities need people and processes to maintain.

Zero Trust Process Maturity

The Capability Maturity Model Integration (CMMI) has defined five maturity levels that, when overlaid with implementation of Zero Trust architecture, are used to define how efficiently the enterprise deploys, manages, and monitors its security controls and mechanisms that align with Zero Trust. CMMI usually measures the effectiveness of the personnel managing the overall architecture and the processes in place to make sure the architecture is sustainable. They measurement levels are as follows:

- **Initial:** Organizations have a basic understanding of the Zero Trust concept and have implemented some initial Zero Trust processes.

- **Managed:** Enterprises have established policies and procedures for Zero Trust implementation and are actively monitoring and managing access requests.

- **Defined:** Enterprise have a well-defined and documented Zero Trust architecture, including clearly defined roles and responsibilities for implementation and ongoing management. Services needed to manage the Zero Trust architecture are defined.

- **Quantitatively Managed:** Enterprises are continuously measuring and analyzing Zero Trust implementation to identify areas for improvement and adjusting their approach accordingly. Critical services are providing measurable feedback to metrics and architecture.

- **Optimized:** Organizations have fully integrated Zero Trust principles into their overall security strategy and are continuously refining their Zero Trust architecture to address emerging threats and technologies. All operational services are defined and the outcomes are continuously monitored.

Zero Trust Security Control Maturity

CISA has also defined a maturity model that closely aligns with Zero Trust security controls. CISA has four major maturity levels, as illustrated in Figure 6-2.

The Zero Trust maturity model developed by the Cybersecurity and Infrastructure Security Agency (CISA) is based on four levels of maturity:

- **Traditional:** This level represents the basic approach to security that most organizations have been using for years. It is focused on perimeter-based security, which assumes that everything inside the organization's network is trusted and everything outside the network is not trusted. This approach is no longer effective, as cyber threats can easily penetrate an organization's perimeter.

- **Initial:** The initial level is where most enterprises find themselves when they embark on the Zero Trust journey. Most initiatives are just beginning, and most processes are manual with minimal automation. There is usually some level of macro-segmentation, but most security decisions are governed by the network architecture. Usually, data classification is either in progress or not begun.

- **Advanced:** This level represents a more proactive and dynamic approach to security. It is focused on implementing Zero Trust principles, such as identity and access management, network segmentation, and continuous monitoring processes. This approach assumes that no one and nothing is trusted by default, and every access request must be verified and authorized before access is granted. There is some level of automation expected at this maturity level.

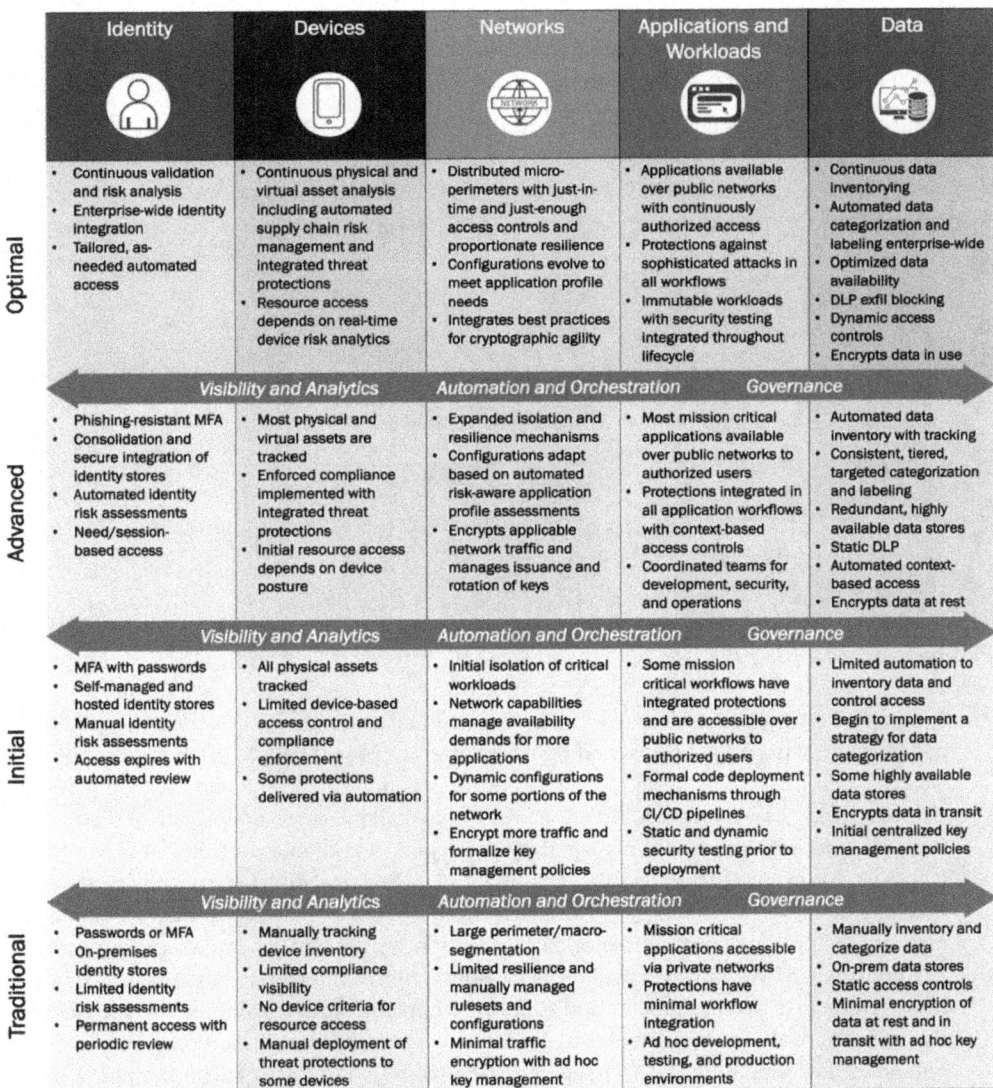

Figure 6-2 *CISA Maturity Levels (Image Courtesy of CISA Maturity Model[3])*

- **Optimal:** This level represents the most advanced and mature approach to security. It is focused on implementing an integrated and automated Zero Trust architecture that provides end-to-end security across the entire enterprise, including cloud environments, IoT devices, and third-party systems. This approach is data-centric, meaning that security controls are applied based on the sensitivity and value of the data being accessed as well as the context of the intent of access.

Note that achieving a higher level of maturity in Zero Trust does not guarantee complete security. Zero Trust is a continuous process that requires ongoing evaluation, monitoring, and improvement to keep up with evolving cyber threats. Organizations must also consider their unique business requirements, regulatory requirements, and risk tolerance when implementing Zero Trust.

Figure 6-3 showcases a mapping between the CISA and the CMMI.

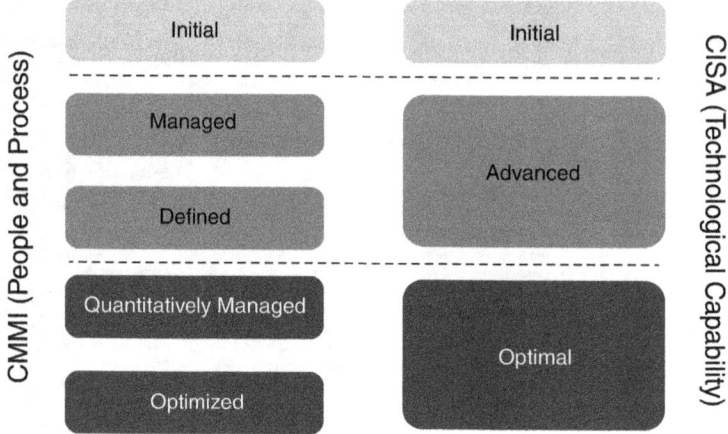

Figure 6-3   *Mapping Between CISA and CMMI*

It is important to draw a clear distinction between measuring security control capability and measuring personnel skillset and process effectiveness. The CMMI measures how well the enterprises puts in personnel and process controls. These could be administrative or specific to a process. CISA measures the overall presence or absence of a control itself. A process is an activity managed by personnel. A control is a technical capability. Consider, for example, the Identity pillar in CISA, which has multiple sub-pillars. If we take one of these sub-pillars (say, Authentication), the CISA maturity model v2 would measure the presence of multifactor authentication, single sign-on, and so on as a security control. CMMI, however, will measure how well the enterprise onboards users and how well the user base segmentation strategy is implemented. It will measure the capability of the Active Directory owners to automate certain activities based on clearly defined outcomes. Implementing the user segmentation strategy is another such example of process measured by CMMI. Though there are some overlaps in the measurement, the goal is to highlight that it is not enough to just put in the security controls. These controls must be managed by the right personnel and processes in place. The measured outcomes must also feed into the overall architecture in order to complete the feedback loop.

Zero Trust Maturity Goals

A maturity goal is a future state that the enterprise wants to be at when it performs a gap analysis and understands its current state. The sections that follow describe the states that most enterprises find themselves in when evaluating Zero Trust readiness. Note that Traditional level is not considered a goal and is usually the state for enterprises that have not adopted Zero Trust.

The Initial Level

Most enterprises find themselves at the Initial level. At this level, the strategic vision is to reach optimized maturity, but the tactical goal is to reach advanced Zero Trust processes and a more managed or defined implementation. Therefore, the enterprise must essentially begin to implement basic Zero Trust processes, manage identities, encrypt data, and protect apps that access sensitive data. Technically, it needs to architect and enforce segmentation control around sensitive data and apps. Identity and access management must be used profusely to limit and strictly enforce access control. Finally, the enterprise must aim to secure all resources and establish security operations fundamentals.

The Initial level is not generally a goal for enterprises; however, the path to Zero Trust is unique to each enterprise, and it is possible for enterprises to approach the journey tactically per year rather than strategically over a larger span of time. Either way, an enterprise that is at the Traditional level cannot jump directly to Advanced or Optimized and will therefore aim to reach the Initial level before aiming for its target goal.

The Advanced Level

When the enterprise is at the Advanced level, the next phase is to introduce more automation to incident response as well as to start threat detection capabilities. Technically this is a strong phase to begin workload segmentation and introduce processes such as flow visibility and asset inventory since the identity of users and devices should already be implemented. Achieving SSO is an important milestone, along with security analytics, for network and application flow visibility as well as AI-based anomaly detection. Data security also becomes critical at this stage, and DLP projects need to be given priority. Finally, SOC automation is ideal to have so that repetitive low-risk tasks can be automated. Usually at the Advanced level, the enterprise spends more time getting feedback from the deployed Zero Trust architecture to enhance the controls before moving on to automation and orchestration activities, which would take it to the next (Optimized) level.

The Optimized Level

At the Optimized level, macro-segmentation of broader routing domains should already be in place. Processes must be already set up to dynamically monitor changes in traffic profile. Micro-segmentation initiatives must be reaching completion and most visibility and security analytics must be mature enough to perform simple automation tasks and advanced incident response tasks as well. The primary targets in this phase will be to extend Zero Trust process orientation to all segments such as branches and to incorporate threat hunting strategies using multiple security intelligence sources. Finally, security orchestration is an important target to optimize existing processes. Technically, the aim is to consolidate, analyze, and adapt multiple threat intelligence sources and use threat hunting to reduce attacker dwell time. SOAR must be prioritized to automate and orchestrate complex human and machine tasks.

Measurement of Maturity

This section explores the key security mechanisms that are required in a Zero Trust strategy and map out how the maturity of processes and technology involved around these mechanisms can be measured, as this will be important to gauge what the overall Zero Trust readiness is. The CISA model of Zero Trust maturity is used to measure the maturity of the technological capability. The CMMI usually measures process maturity in an organization and is used to measure how well the strategy can be implemented.

Zero Trust Maturity for Users and Devices

A typical mapping of processes and controls for the Identity pillar is illustrated in Figure 6-4, and the Device pillar is illustrated in Figure 6-5.

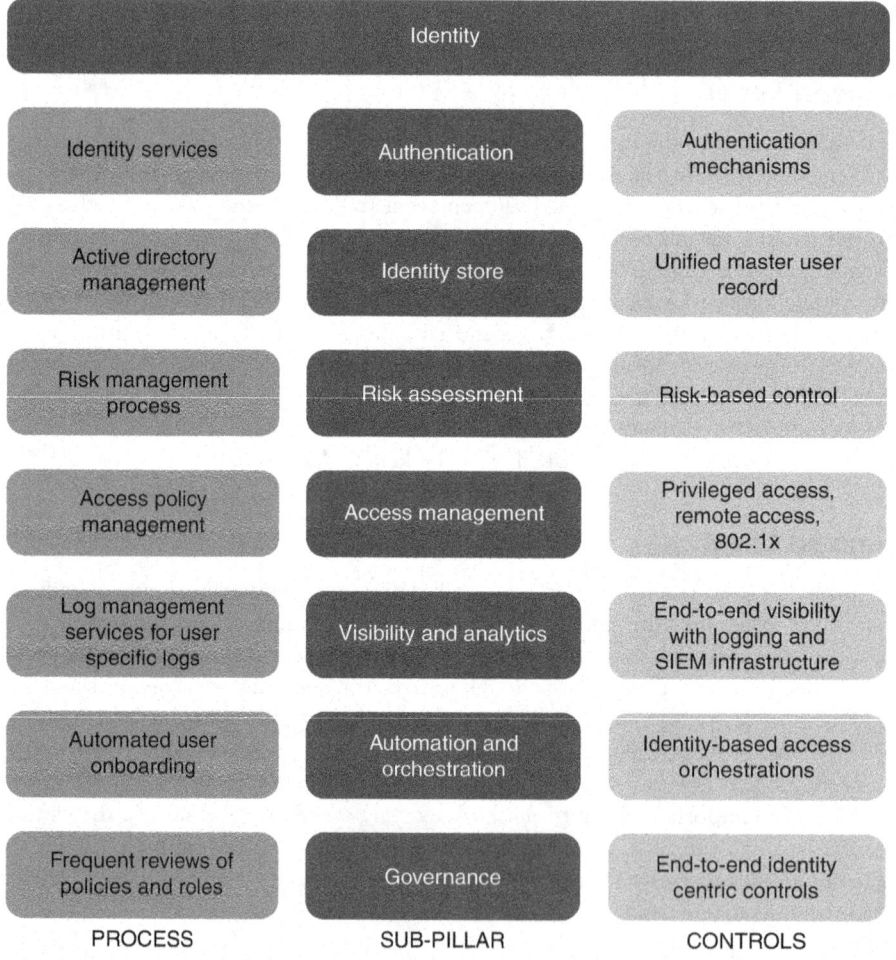

Figure 6-4   *An Example of Processes and Controls Critical for the Identity Pillar*

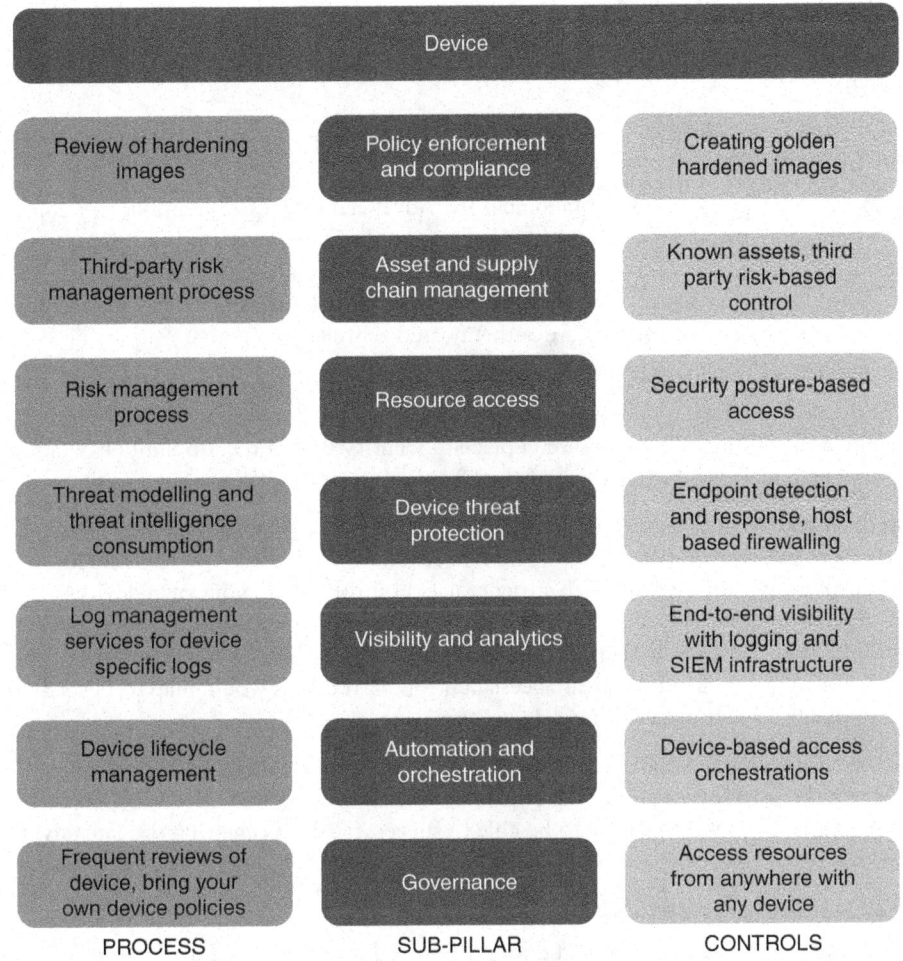

Device

PROCESS	SUB-PILLAR	CONTROLS
Review of hardening images	Policy enforcement and compliance	Creating golden hardened images
Third-party risk management process	Asset and supply chain management	Known assets, third party risk-based control
Risk management process	Resource access	Security posture-based access
Threat modelling and threat intelligence consumption	Device threat protection	Endpoint detection and response, host based firewalling
Log management services for device specific logs	Visibility and analytics	End-to-end visibility with logging and SIEM infrastructure
Device lifecycle management	Automation and orchestration	Device-based access orchestrations
Frequent reviews of device, bring your own device policies	Governance	Access resources from anywhere with any device

Figure 6-5 *An Example of Processes and Controls Critical for the Device Pillar*

Initial User and Device Controls

Initial security for workforce controls would include controls like IAM, profiling, device posture, BYOD, PAM, and asset management. Authentication is usually with passwords and, at best, with multifactor authentication (MFA). If MFA is no longer used, there is a larger security issue in the enterprise strategy. Risk management is fairly minimal and the Active Directory instances are distributed and on the premises.

Device asset management is likely manual, and compliance checks for the devices are with limited visibility. Users will be authenticated and verified with basic protection from compromised credentials. This will map to Level 1 (Initial) of the CMMI maturity model of Zero Trust because most processes around maintaining these controls will be at the very early stages.

Usually, the threat to enterprises at this stage are password guessing, dictionary attacks, along with phishing and whaling. The targets are either executive leadership or exploitable workforce.

Advanced User and Device Controls

Advancements in the existing IAM fall under the Advanced level of user and device controls. Enterprises with advanced maturity in users and devices ensure that end users are authenticated with MFA. Users are segmented, and risks are allocated for all user segments. Active Directory instances are mutually synced and cloud identity providers are used to make sure cloud workloads are authenticated as well.

Devices are postured with basic enterprise requirements such as antivirus and disk encryption. Posture is a key factor for providing network or data access and is not just configured as an add-on. Processes are in place to identify user and endpoint behavior. This would map to Level 2 (Managed) of the CMMI maturity model for Zero Trust.

Device asset management is automatic, and patch and vulnerability management is performed at a fixed interval. Over time, with more advancements in enterprise capabilities, more control is given to the user with stronger user controls. Bring your own device (BYOD) is enabled for devices, and subjects are dynamically allocated to networks after validation of their trust level. There are defined processes in place for managing the user and device life cycle with a clear access control matrix. This would map to Level 3 (Defined) of the CMMI maturity model for Zero Trust.

Threats are usually more sophisticated, with need to access a weaker system first and then move laterally. With a stable SOC and patch management system, the attack must exploit an uncommon vulnerability to be able to successfully get time to move laterally.

Optimized User and Device Controls

Identity is validated continuously, and risk is dynamically changed based on context. Workloads and users for the enterprise exist on the premises and the cloud, and the Active Directory instances are synced, making it one common decision-maker and logical unit. Risk is measured in real time using advanced machine-learning tactics. Password-less authentication is a common option explored in this phase. Device compliance checks and network and data access are dynamically calculated and constantly monitored.

All cloud and remote assets are tracked and integrated into a single asset management system for centralized tracking. Policies and rule creation are based on risk and context. Third-party risk is continuously and dynamically managed. There is automated processes in place for managing the user life cycle and access control. This maps to Level 4 (Quantitatively Managed) of the CMMI maturity model for Zero Trust.

As enterprise maturity increases, user-level segmentation is extended and continuous monitoring of user access is measured. Monitoring and response to high- or medium-risk user events are happening continuously. Users get a single sign-on experience, even though identity is being continuously monitored. Device posture is dynamically considered as part of access control. Orchestration has been incorporated for all business processes(not just critical ones) and the personnel are skilled enough to manage and maintain the user infrastructure. This would map to Level 5 (Optimized) of the CMMI maturity model for Zero Trust.

Threats during this phase are usually highly complicated attacks with advance persistence. Social engineering and push phishing are required to extract passwords and gain access. Lateral movement and complex exploits must be used to move across semi-trusted segments. Targets are not restricted to a specific set of people and are usually a set of multiple kill chains. The final goal is almost always to get access to intellectual property or personally identifiable information (PII).

Zero Trust Maturity for Network and Multicloud Infrastructure

This section deals with the overall network and cloud security maturity in the enterprise environment. Figure 6-6 illustrates a typical control and process requirement for network and cloud.

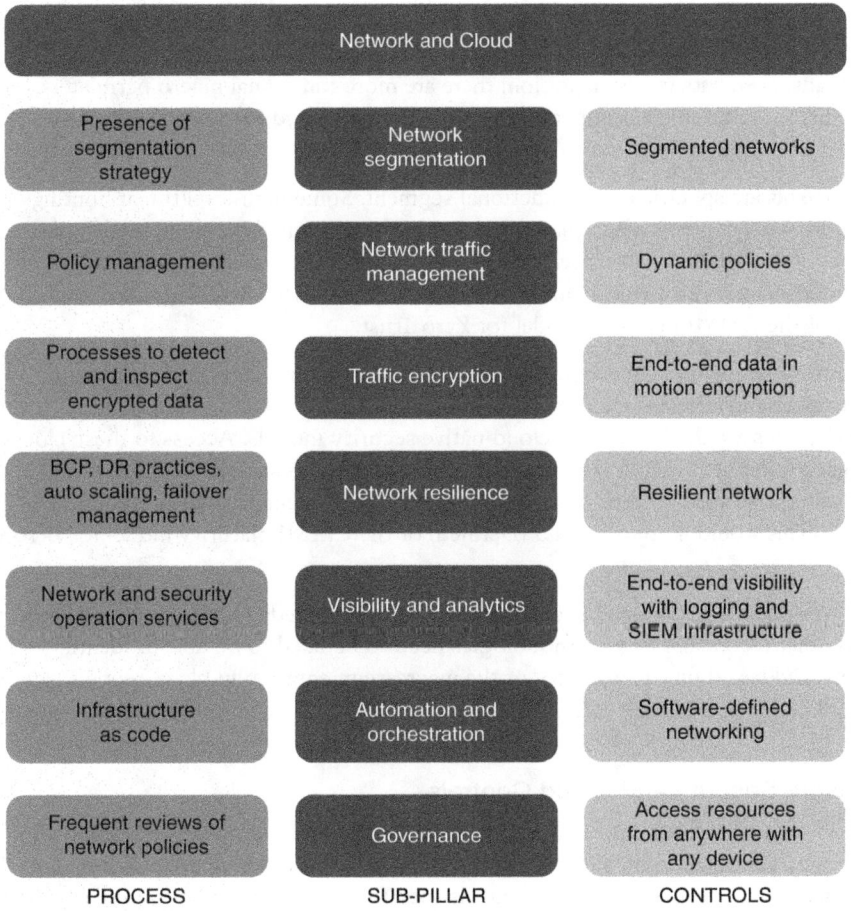

Figure 6-6 *An Example of Processes and Controls Critical for the Network and Cloud Pillar*

Initial Network and Cloud Security Controls

In an Initial-level network model, the segmentation is broad and covers simple segments like Internet, DMZ, and server farm. Within each segment, the networks are essentially flat. Not a lot of thought has been put into segmenting internal networks. There are fewer workloads on the cloud, and any workload on the cloud is not directly accessible from the Internet but is connected to the enterprise via a secure link.

Security controls are static and rule-based. Mostly, rules and deployments are network-centric and are less focused on the flow and more focused on providing per-packet access. The enforcement point itself doesn't encrypt or decrypt traffic; hence, enforcement is limited to the enforcement point's visibility of traffic. This will map to Level 1 (Initial) of the CMMI maturity model for Zero Trust.

Threats here are malicious insiders and persistent agents that reach out to command and control and move laterally due to the flat nature of each segment. Targets are usually not fixed, and the main concerns are network sniffing and lateral movement.

Advanced Network Security and Cloud Controls

With more advancements in segmentation, there are more functional macro perimeters, with some high level of micro-segmentation. Virtual routing and forwarding (VRF) is utilized as much as possible to isolate routing domains.

Security controls are specific to the functional segment. Some high-level threat hunting and behavior-based controls are in place. Encryption is performed on all internal applications. Network visibility is in place, with a functional security analytics solution. A single cloud exists where all workloads are deployed initially. This would map to Level 2 (Managed) of the CMMI maturity model for Zero Trust.

As an enterprise starts implementing advanced projects, network access control must become stronger and all users must be authenticated at network entry points. Cloud-native applications are deployed with cloud-native security models. Access to these cloud applications is still not fully over the Internet, and technologies such as VPN are still used. The enterprise starts exploring and deploying network-centric software-defined architecture. This would map to Level 3 (Defined) of the CMMI maturity model for Zero Trust.

Threats in this phase usually need sophisticated routing knowledge and lateral movement. Nmap and other reconnaissance methodologies need to be used to be able to identify key subnets, and based on the maturity of the macro-segmentation in place, recon scans might get all or some of the key critical subnets.

Optimal Network Security and Cloud Controls

Segmentation is a mature strategy in the enterprise. There are clear macro-segments with function-specific security controls both on the premises and on the cloud. There is also micro-segmentation of workloads with controls specific to flow rather than static rules.

Context-based, group-based policies are in place, and control is provided only to these groups. The default policy is to deny unless there is an explicit need to allow.

Access control uses technology like machine learning and business flow analysis to calculate risk and context-based rules.

All traffic in motion is encrypted. Deep packet inspection is performed for external applications to enhance access control preferably without decryption. Technologies such as encrypted flow analysis are implemented to make sure that encryption and decryption don't impact performance. Multicloud architecture has been adopted widely, where the security context of each workload is preserved across networks. Software-defined network access keeps networks secure and, at the same time, implements automation and orchestration to provide visibility and control over network and cloud access. This would map to Level 4 (Quantitatively Managed) of the CMMI maturity model for Zero Trust.

The final level of maturity (Level 5) would be obtained with more automated and dynamic changes to network segmentation in response to stimulus like detection of lateral movement. The network must be able to isolate itself from the affected segments dynamically with limited intervention of an incident response engineer.

Threats are usually reduced and sophisticated. Social engineering and possibly malicious insiders are required to know some aspects of the network before an attack is commenced. Targets are usually open, unsecure application servers.

Zero Trust Maturity for Applications and Data

The following section generally covers applications, servers, all service-providing entities, and the data they process. The typical examples of requirements for process and controls are listed for applications in Figure 6-7 and for data in Figure 6-8.

Initial Application and Data Security Controls

In Initial-level networks, access to internal servers and critical applications are via VPN only—the reason being that users are on a local Active Directory and access is controlled using static authorization and Active Directory attributes.

Application testing is mostly part of the overall waterfall model, and security testing is performed before deployment via static or manual testing. Application protection mechanisms are based on standard protection and do not consider the application-specific workflow, and they also do not integrate with the workflow itself. Not all workloads are known, and the enterprise is just embarking on data and application asset inventory creation and data classification. Classification itself is minimal and performed only for critical or sensitive data. This will map to Level 1 (Initial) of the CMMI maturity model for Zero Trust.

Threats are almost any attack that makes it past the DMZ or outside subnets. Because most data centers in this phase have a flat subnet with implicit trust, attack vectors are simple, and a vulnerable server can cause a breach to the entire server farm.

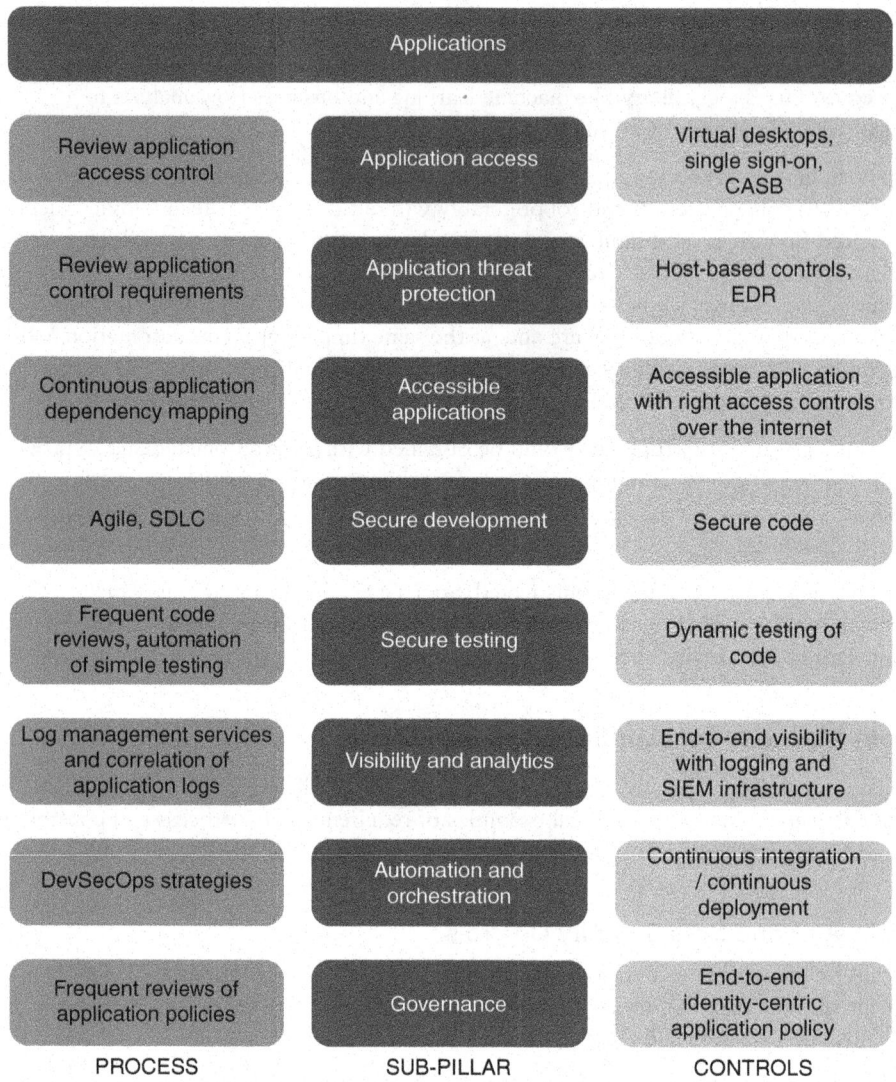

Figure 6-7   *An Example of Processes and Controls Critical for the Application Pillar*

Advanced Application and Data Security Controls

Similar to most modern networks, the authentication and authorization is handled centrally so that policy is driven above the application. Specific authorization control is retained at the application level. Threat protection is incorporated directly into workflows, which means that when there are specific types of anomalous flows that are seen in the network, threat protection can pick the anomaly and act accordingly. There is more visibility into flows within the data center.

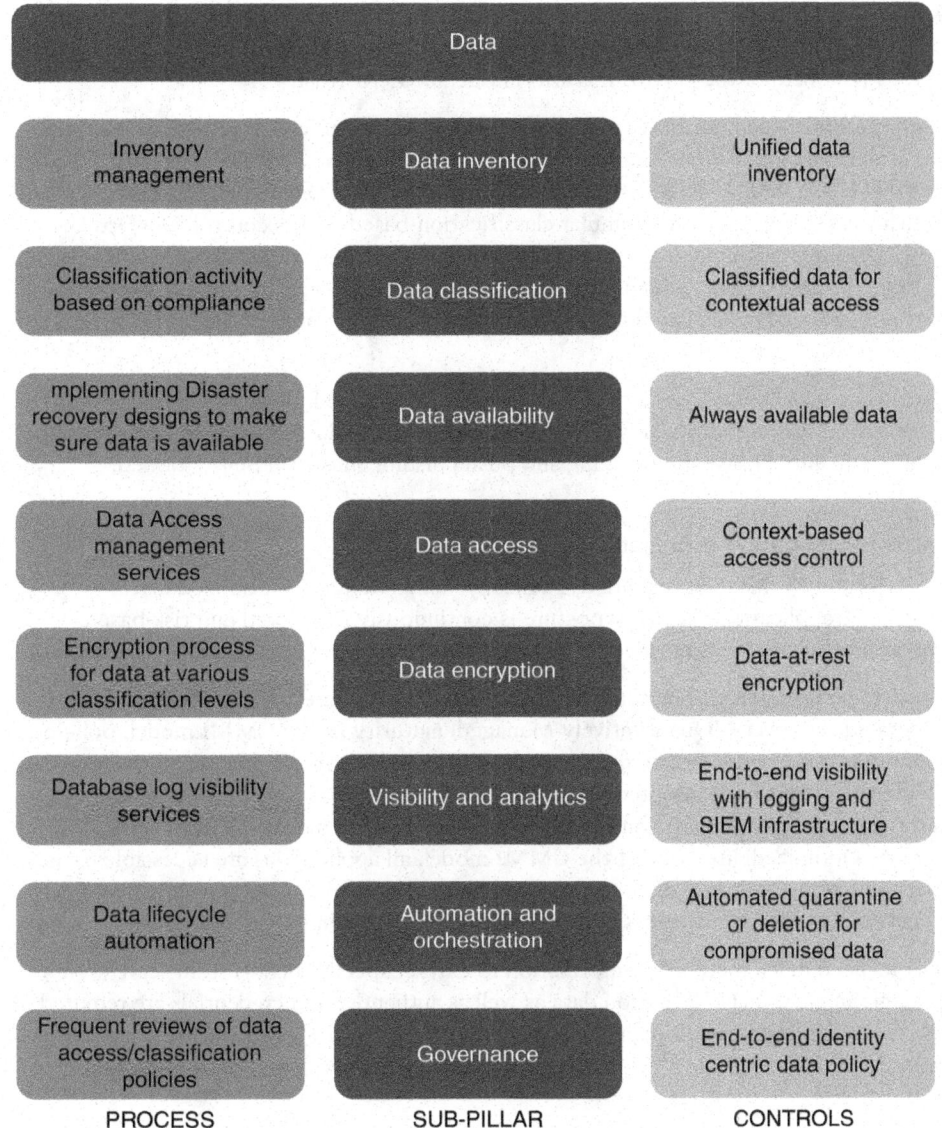

PROCESS	SUB-PILLAR	CONTROLS
Inventory management	Data inventory	Unified data inventory
Classification activity based on compliance	Data classification	Classified data for contextual access
mplementing Disaster recovery designs to make sure data is available	Data availability	Always available data
Data Access management services	Data access	Context-based access control
Encryption process for data at various classification levels	Data encryption	Data-at-rest encryption
Database log visibility services	Visibility and analytics	End-to-end visibility with logging and SIEM infrastructure
Data lifecycle automation	Automation and orchestration	Automated quarantine or deletion for compromised data
Frequent reviews of data access/classification policies	Governance	End-to-end identity centric data policy

Figure 6-8   *An Example of Processes and Controls Critical for the Data Pillar*

All cloud applications are accessible via the Internet with the right security controls. Some applications might still mandate a remote access VPN solution. No critical internal on-prem applications are available via the Internet, even with the appropriate security control. Site-to-site VPN is still present in the network for applications that cannot incorporate specific security controls. Data is centrally managed, and access is controlled based on roles.

Application testing is incorporated as part of the overall software lifecycle. Early adoptions of DevSecOps have begun. Testing includes dynamic methods as well. This maps to Level 2 (Defined) of the CMMI maturity model for Zero Trust.

With more advanced micro-segmentation projects, workload communications can be monitored and applications can be segmented to support the context-based policies that will be created. Application dependency mapping is automated to a large extent. Data access is managed with granular classification-based access control. DevOps and DevSecOps are beginning to be adopted in the software lifecycle with more automated testing. The application and data lifecycle are defined with clear processes, however they are managed manually. This would map to Level 3 (Defined) of the CMMI maturity model for Zero Trust.

Threats are usually challenged to access critical information infrastructure (CII) because these servers are accessible only via VPN; therefore, complicated advanced persistent attacks and initial password cracking and push phishing must still be performed.

Optimized Application and Data Security Controls

All users and workload access are dynamically validated in the Optimized maturity phase. The applications' security posture is continuously monitored and risk-based. Threat protections are flow-specific and behavior-based.

Application testing is throughout the lifecycle of the software, and DevSecOps is fully integrated. At Level 4 (Quantitatively Managed) maturity of the CMMI model, policies are uniform and centralized in nature, with very granular policies retained at the application level. Agile is more adopted than standard software development lifecycle (SDLC) and there is defined automation for all business application management processes. At Level 5 (Optimized) maturity of the CMMI model, all applications are accessible to users over the Internet with the right security controls and continuous monitoring. Data classification, inventory, and access control are dynamic and context-based.

Threats need much more prior information to extract specific details from servers. At a bare minimum, the HTTP session data as well as authentication credentials are required to be able to exfiltrate data, which is much tougher with HTTPS and authentication mechanisms in place.

Zero Trust Maturity for Visibility, Automation, and Orchestration

This section focuses on the overall visibility into the network and how well the network reacts to incidents.

Initial-Level Visibility, Automation, and Orchestration Controls

Realistically, in Initial-level networks, visibility and SOAR are not very mature. Monitoring of users is minimal. Most user-specific IAM activities are manual; for example, adding and removing users is performed by an HR administrator. Subsequently, all user credentials and access are audited manually. This leaves the administrator open to social engineering attacks.

Similarly, devices are inspected manually to check for compliance. Provisioning is manual with the help of IT helpdesk. The retirement and acquisition of devices are manual.

At a network level, visibility is restricted to perimeter only due to the overall perimeter-based security access. Rule changes and policy changes are manually recorded. Discovery of managed and unmanaged devices, users, or applications is manual.

Application monitoring is either manual or in silos. Policies for governance are not updated and are changed manually and much less frequently. Zero Trust policies do not exist. This will map to Level 1 (Initial) of the CMMI maturity model for Zero Trust.

Lack of visibility is an overall threat because there will be no audit trail of incidents or anomalous behavior. There is no information to feed decision points to make context-aware decisions.

Advanced-Level Visibility, Automation, and Orchestration Controls

There are common static playbooks for visibility that are then manually mapped to specific processes to automate. There is basic orchestration across Active Directory instances to permit administrative access. There are no shared accounts, and policy-based automated access is monitored for anomalies.

Devices are mapped against a static list, usually maintained manually in Level 2 maturity and automated with Level 3 maturity. Device lifecycle is still manual, with isolation of noncompliant devices being manual as well. Most devices support modern hardware and software security. Golden images are deployed manually. Legacy devices are reduced as much as possible, and, if not possible, visibility is enhanced for these segments.

From the network aspect, logs from a single perimeter device are not considered sufficient to build an end-to-end flow visibility. Multiple information sources are utilized from various macro-segments to provide the right visibility and subsequently more granular access control. Rule creation is automated with early implementations of automated workflows based on administrator feedback. Identification of legitimate subnets and segments is manual, so removal of unauthorized segments is also manual. This maps to Level 2 (Managed) of the CMMI model for Zero Trust.

As maturity increases, application monitoring is automated. The network can identify devices and users changing their context and control access accordingly. Policies are centrally managed and enforced. Information from SOC is fed into decision points to rapidly detect and isolate infected artifacts via rapid threat detection. This maps to Level 3 (Defined) of the CMMI model for Zero Trust.

Threats include overall lack of automation and playbooks for an effective SOC and incident response initiative, which could lead to attacks slipping through. One key differentiator between the Advanced and Optimal levels in visibility is the "correlation" factor. Most enterprises send all the necessary logs to a common collecting agent like a SIEM; however, there is no correlation between the respective logs to extract context.

Zero Trust policies are more defined. There is a clear access control matrix for subject to object access. Policies, standards, procedures are maintained with the right hierarchy with a minimum baseline procedure document for each business process in the enterprise (like BYOD, encryption, log retention and so on).

Optimal-Level Visibility, Automation, and Orchestration Controls

In the Optimal world, user visibility is centrally managed with UEBA to incorporate and monitor for anomalous behavior. Identity lifecycle is fully automated. User profiling and group membership are dynamic and risk-based for just enough access control. Enforcements for users are fully automated. This maps to Level 4 (Quantitatively Managed) of the CMMI model.

As automation and orchestration projects progress, DevSecOps and CI/CD are constantly used for scaling of software deployments. All posture checks are dynamically revalidated and continuously monitored. All security logs are stored centrally and correlated for more context.

Infrastructure as Code is used for CI/CD deployments. Various enforcement points trigger incidents, which are then used to perform automated incident analysis. Network controls are specific to the segment and data they are protecting. Because all devices are authenticated and authorized automatically, unregistered domains and actors are also identified automatically, and remediation is orchestrated. Asset discovery and management are automatic.

From an application perspective, all devices are monitored actively for health and downtime. Applications authorize specific subjects dynamically based on risk, and policy enforcement is automatic. Access control between applications changes and adapts based on environment. Data access is context-based. This maps to Level 5 (Optimized) of the CMMI model. Zero Trust policies are maintained at a high level subject to object access matrix which is dynamically updated on all enforcement points when there is a change in attributes of either subject or object.

Threats are more complicated because the biggest advantage of being Zero Trust mature is the continuous visibility; therefore, threats need to be able to exploit a credential and extract data in a very short period of time before the visibility agents pick up the anomaly.

Zero Trust Scoring Process

With the analysis performed based on target architecture and the gaps identified, a score is allocated to the respective pillars. In this section, we will use the Identity pillar as an example and calculate the maturity score of the Identity segment.

The CMMI for processes and people is scored from 0 to 4, where 0 is Initial, 1 is Managed, 2 is Defined, 3 is Quantitatively Managed, and 4 is Optimized. The CISA maturity model v2 for technology is scored from 0 to 3. where 0 is Traditional, 1 is Initial, 2 is Advanced, and 3 is Optimal. Figure 6-9 analyzes and presents the score for the Identity pillar.

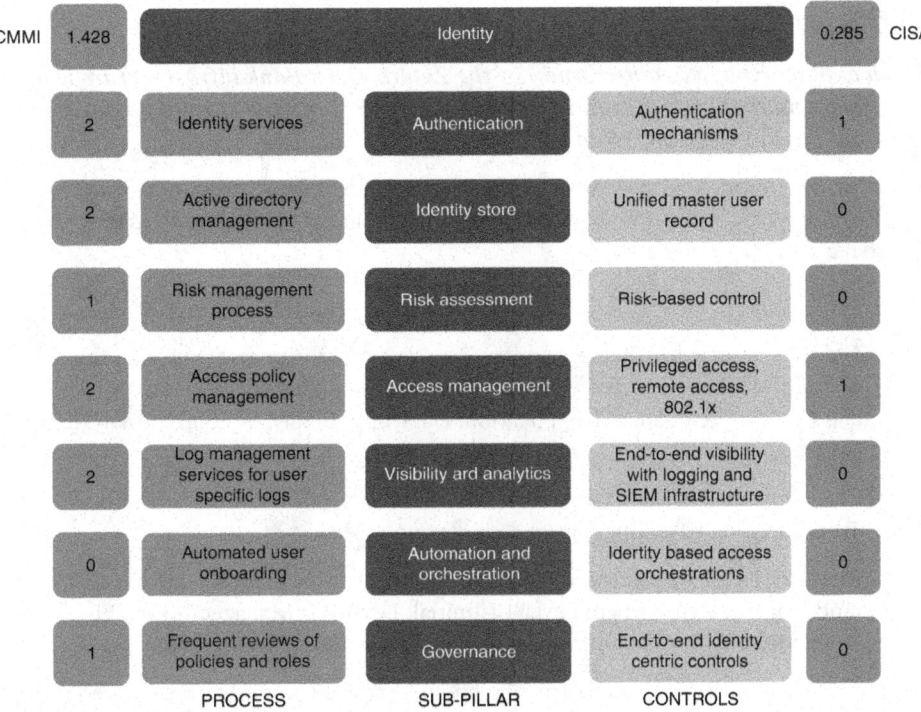

Figure 6-9   *A Sample Scoring Card for the Identity Pillar*

The maturity of the Identity services based on CMMI is 1.428(10 divided by 7) on a scale of 0 to 4, and the maturity of the identity controls is around 0.285(2 divided by 7) on a scale of 0 to 3. Because the main scale being referenced for maturity is the CISA scale, the CMMI value has to be normalized to the CISA scale. However, with the normalization formula, it is possible to normalize the CISA to the CMMI scale as well. Here is the standard normalization formula for the different scales:

$$\frac{(CISAMaxScore - CISAMinScore)}{(CMMIMaxScore - CMMIMinScore)} \times (CMMIMaxScore - Score(CMMI)) = CISAMaxScore - Score(CISA)$$

This becomes the following score:

$$\tfrac{3}{4} \times (4 - 1.428) = 3 - Score(CISA)$$

This scales the new normalized CMMI score to CISA as 1.071. This is the score for the process maturity. The average of the process maturity and the technological control maturity gives the overall maturity of the Identity pillar, which is the average of 1.071 and 0.285, or 0.678. Hence, the Identity pillar according to CISA maturity model v2 is Traditional after considering people, processes, and technologies. The same procedure is used for all other CISA maturity model v2 pillars. Remember that the number scale of the maturity model can be modified as long as the normalization equation is utilized to normalize the score to a specific scale and measurement model.

The Follow-Up

[Glenn explains that a detailed study of the Zenith Trust Bank infrastructure was performed and that he will present the findings along with the crafted performance metrics to the present leadership.]

Glenn: So, Mr. Smith, before we go into what we have identified in your network, I would like to provide a quick summary of what we really mean by a maturity assessment from a Zero Trust context.

To summarize:

- Contrary to existing security audits, Zero Trust maturity assessment focuses on the five pillars, which are Identity (subjects and objects), Devices, Network and Cloud, Workload and Application, Data, and, finally, Visibility, Analytics, Governance, Automation, and Orchestration.

- Zero Trust readiness measures how ready the enterprise is when it comes to Zero Trust adoption based on current security practices.

- Maturity levels are measured for Zero Trust process readiness using the CMMI model of five levels, where Level 1 is Initial, Level 2 is Managed, Level 3 is Defined, Level 4 is Quantitatively Managed, and Level 5 is Optimal.

- The maturity level for Zero Trust technology control readiness is measured using the CISA maturity model, which is Traditional, Initial, Advanced, and Optimal

We have completed the maturity assessment of your environment and, overall, we believe that overall, your enterprise is still at an **Initial** level of Zero Trust based on the CISA Maturity model v2 and normalized CMMI maturity model. Your Zero Trust control readiness is at **Initial** based on the CISA Maturity model v2 and the Zero Trust process incorporation is **Managed** based on the CMMI maturity model. To be specific:

Capabilities that exist with sustainable maturity:

- User segmentation

- Data-at-rest encryption

- Privileged access management integration

- Dynamic endpoint profiling

- Logging infrastructure

Capabilities that exist but where uplift is required:

- IAM improvements in process

- Macro-segmentation

- Network asset management

- Security operations center
- Network threat protection

Capabilities that do not exist or are extremely difficult to manage:

- Posturing
- Bring your own device (BYOD)
- Micro-segmentation
- Data and application asset management
- Threat and security intelligence
- Threat hunting
- Application threat protection
- Secure and data analytics
- Application security
- Data classification and data loss prevention
- Network flow visibility
- Automated asset and data discovery
- DevSecOps strategy
- Overall end-to-end Zero Trust access policy for network and application access

Mr. Chen: I am shocked, but at the same time I had expected a result like this when Jonathan reached out some weeks ago. So, basically, has our existing threat base and assets been covered as well?

Glenn: We have conducted threat hunting scenarios based on the MITRE Framework as well, and based on the current control gaps that we have identified, the following assets and threats have been gleaned:

Key Assets

- **Core banking system:** This is the primary IT system that manages all banking operations, such as customer accounts, deposits, loans, and transactions.
- **Customer relationship management (CRM) system:** This system manages the bank's interactions with customers, including customer data, preferences, and transaction history.
- **Online banking platform:** This is the web-based interface that allows customers to access their accounts, perform transactions, and manage their finances.
- **Mobile banking application:** This is a mobile app that allows customers to perform banking activities on their smartphones, including transactions, bill payments, and account management.

- **Data warehousing and business intelligence systems:** These systems provide the bank with tools to analyze customer behavior, identify trends, and make strategic decisions.

- **Security systems:** Banks need robust security systems to protect against cyber-attacks and data breaches. These systems include firewalls, intrusion detection/prevention systems, and access control systems.

- **Customer data and CII:** This is the critical information infrastructure that stores all of the customer data.

- **People (customers and employees):** All the human assets are included in this segment. Both customers and employees are critical actors and consumers of services and resources from the bank.

- **Disaster recovery and business continuity systems:** These systems ensure that the bank's IT operations can continue in the event of a disaster or disruption. This includes backup systems, redundant servers, and recovery procedures.

Threats

A threat hunting cyber exercise was performed across four sprints, where each sprint involved penetration testing of critical resources and validating the resiliency as well as incident response capability of the team managing the resources. The framework followed was MITRE, and tests included partially known, unknown, and known environment testing. This activity was correlated with a parallel threat modeling exercise and both the activities were conducted in parallel. At the end of the four sprints, which lasted four days, additional relevant threats were discovered. There are many potential threats to IT systems in a bank, but some of the most common ones we have identified that are relevant to Zenith Trust Bank are the following:

- **Phishing attacks:** Phishing attacks involve tricking bank employees into giving up sensitive information, such as usernames and passwords. This can be done through fake emails or websites that appear to be legitimate.

- **Malware and viruses:** Malware and viruses can be introduced into a bank's IT systems through email attachments, downloads from the Internet, or other means. Once installed, these malicious programs can cause significant damage to the bank's systems and compromise sensitive data.

- **DDoS attacks:** A distributed denial of service (DDoS) attack involves overwhelming a bank's IT systems with traffic, making it difficult or impossible for customers to access their accounts. DDoS attacks can be launched by cybercriminals or even by rival banks or political groups.

- **Insider threats:** Insider threats involve employees or contractors who have access to a bank's IT systems using their privileges to steal sensitive information or cause damage to the bank's systems.

- **Ransomware:** Ransomware is a type of malware that encrypts a bank's data, making it inaccessible until a ransom is paid. Ransomware attacks can be devastating to banks because they can disrupt operations and lead to significant financial losses.

Out of these, there are two loss scenarios that can prove largely impactful to the bank, and both relate to theft of customer data:

- **Scenario 1:** An insider gets access to customer data and exfiltrates it to a malicious actor.

- **Scenario 2:** An external entity gains access to a non-secure DMZ server and then moves laterally to get access to a server farm resource. Subsequently, the entity can move freely in the flat data center subnet and get access to customer data and exfiltrate it to a malicious actor.

The following is the risk analysis performed for both the scenarios:

Scenario 1:

The following qualitative values have been established after interviews with the team leads Jed, Mariam, and William and their respective teams members. With insider threat, **the threat event frequency** is **MODERATE**. The threat capability is also **MODERATE**. The desired control strength, which in this case is optimized user lifecycle process management, dynamic identity validation, and user segmentation combined, provides **VERY HIGH** protection but there is no dynamic identity verification as of today, and segmentation is minimal. Current control strength is **LOW**, which means the vulnerability is **HIGH**. With HIGH vulnerability and MODERATE frequency, the loss event frequency is **MODERATE**. The total valuation of Zenith Trust Bank is $3.5 billion, and the risk of data loss leading to financial loss is **SEVERE**, as it will not only impact customers and trust but also reputation and other business loss. With a MODERATE loss event frequency and SEVERE probable loss, the risk is **CRITICAL**, and immediate action must be taken to implement the following controls:

- Segmented micro-perimeters

- Dynamic identity validation and policy updates

- Automated incident response

- User lifecycle process enhancements

This increases control strength to **VERY HIGH** and reduces risk to **HIGH** since data loss is still a **HIGH** risk.

Scenario 2:

Similar to Scenario 1, the threat event frequency is **VERY HIGH** because there are constant access attempts that are successful within a day considering external-facing servers. The capabilities vary, but the worst case must be assumed and hence it is possible a state-sponsored attack with high talent and money can be attempting to access the bank's network. The threat capability is therefore **HIGH** because attacks are not always just state-sponsored. With the current controls, if a user gets access to the DMZ or any other server segment, lateral movement is easy, which means the control strength is **LOW** and the vulnerability is **VERY HIGH**. With a **VERY HIGH** vulnerability and a **VERY HIGH** threat event frequency, the loss event frequency is **VERY HIGH**. This means that due to the large number of state-sponsored attacks against the bank and the lack of overall segmentation and access control, an attack on the network with the critical information systems as the target will likely end in a loss event. With the importance of the critical data, the probable loss impact is **SEVERE**, making the risk **CRITICAL**. To mitigate this, the following controls must be considered:

- Software-defined macro-perimeters
- Patched software (to avoid software-based attacks)

This increases the control strength to **VERY HIGH**, which greatly reduces the risk, to **HIGH**. This still is a HIGH risk because it involves data loss for the enterprise as well as customers.

Addition of a New Performance Metric:

With the preceding critical risk path, we had already added metrics for incident response time and incident record times. An additional metric we will add is the **level of user awareness**:

Operational metric measured quantitatively or qualitatively : The metric will measure, as a percentage, how many end users and subjects accessing workloads have been educated on Zero Trust paradigms and how many of them understand and align with the organization's mission for ubiquitous secure access. The goal is achieved when 95% of the users have been educated and over 70% agree or align with the vision. This will measure how alignment with the vision reduces potential insider threats and/or human mistakes such as clicking on phishing links and so on. This metric could also be qualitatively measured with **High, Medium** and **Low** for both education and alignment depending on your preference. This aligns with **"Create simple application, network, and user architectures that support daily operations"** and **"Provide customers and employees convenient access to their data"** by making sure they are aware of the model and its nuances.

Tactical metric Software-defined network and policy adoption measured qualitatively or quantitatively: The metric will measure qualitatively as how much of the SDN paradigm is being adopted in the network. **High** showcases a higher percentage range of the infrastructure adopting SDN with more uniform software defined policies. **Medium** means moderate expansion to SDN networks, and **Low** represents negligible utilization of software-defined access and policies. These are qualitative metrics, however, if you feel you want a quantifiable value to showcase the adoption rate, the measure can be quantitative as well with customized thresholds. This aligns with **"Create simple and**

efficient network architecture" and "**Support rapid expansion to multicloud software-defined architecture.**" Note that we modified your mission statement to reflect the direction that Zenith Trust Bank wants to go based on feedback from your infrastructure and security teams.

Consider the following illustration for the overall strategic and tactical drivers identified for Zenith Trust Bank (see Figure 6-10).

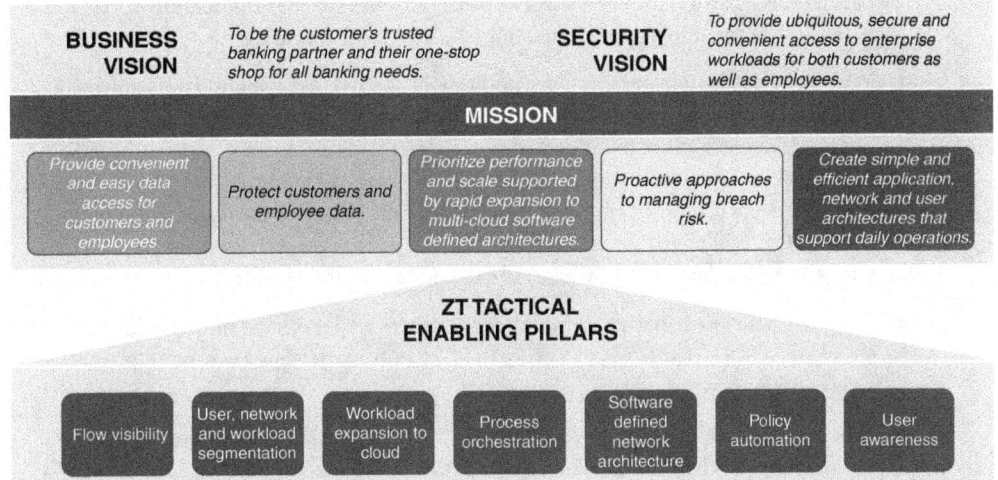

Figure 6-10   *Final Zenith Trust Bank Vision, Mission, and Tactical Initiatives*

A detailed tactical road map would look like this (see Figure 6-11):

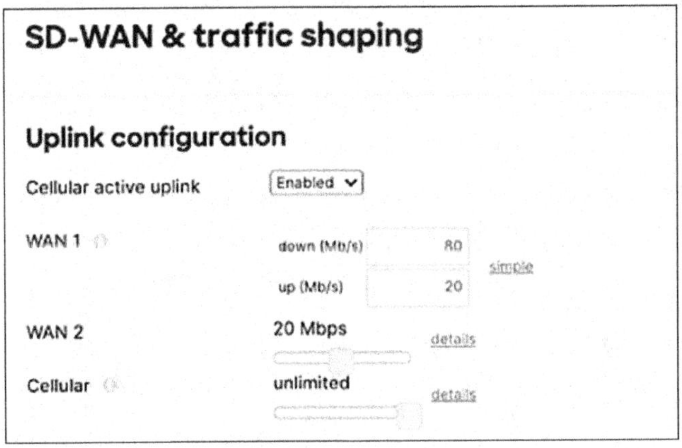

SD-WAN & traffic shaping

Uplink configuration

Cellular active uplink	Enabled ▾	
WAN 1	down (Mb/s) 80	simple
	up (Mb/s) 20	
WAN 2	20 Mbps	details
Cellular	unlimited	details

Figure 6-11   *Tactical Road Map*

The overall yearly breakdown is as follows:

- **Year 1:** Based on the priority explained by the CIO team, the first year will be dedicated to augmenting existing user based security controls along with building visibility capabilities for these users and their devices. Begin posture deployment for endpoints to support the visibility initiative.

- **Year 2:** Increase focus on enhancing user segmentation and building software-defined access for campus networks with network macro-segmentation. This will be a progressive movement to advanced Zero Trust maturity.

- **Year 3:** Focus on optimizing user and device Zero Trust capability by introducing BYOD policies and giving users flexibility to bring their personal devices. Network devices must be added to automated asset inventory, and workload assets will be manually added. Now that software-defined perimeters are identified and users are well segmented with the right behavior patterns being measured, SOC will get augmented to be able to process and trigger simple events with the right playbooks. Workload-based flow visibility will begin parallelly.

- **Year 4:** Focus on software-defined data centers and work on micro-segmentation strategies. Workload asset inventory will be automatic. The enterprise progressively moves toward advanced maturity.

- **Year 5:** The enterprise begins extending the Zero Trust paradigm to branches with SD-WAN implementations. Segmentation is extended across branches. Orchestration will be enabled for known processes to make sure incident response and other user and network access controls are automated.

The best course of action would be to attempt to get your workforce segment into a higher maturity level by improving IAM. The next step would be to improve on macro-segments, profiling, as well as automating network asset management. More people need to be incorporated into the SOC, and more automation and orchestration will be incorporated. Finally, when some of the user, device, and network initiatives are complete, focus will be directed to start mapping application flows, which will use our detailed report as a base reference.

Mr. Smith: Alright. We have the metrics and we have the road map on what we need to do. So, we should be good to get funding from the leadership, correct?

Glenn: Actually, no, Mr. Smith. What you have now is "why you want to embark on a Zero Trust journey" and "how you will measure the success of your adoption." What we need to plan out together is who will help achieve this vision from your enterprise. You cannot just shift the enterprise's security direction without the help of some key personnel. When we present this to your leadership, we also need to be able to showcase to them that a strong team has been formed to design and implement the Zero Trust network architecture. They will need to be aware of all Zero Trust architecture terminologies and then be able to articulate the design and complete the implementation.

Mr. Smith: Alright, I'm assuming you have this mapped out already?

Glenn: Yes, let me brief you in detail.

Endnotes

1. "CISO Perspectives and Progress in Deploying Zero Trust," https://cloudsecurity-alliance.org/artifacts/ciso-perspectives-and-progress-in-deploying-zero-trust

2. "OMB's Zero Trust Strategy: Government Gets Good," https://www.forrester.com/blogs/ombs-zero-trust-strategy-government-gets-good/

3. "Zero Trust Maturity Model, Version 2," https://www.cisa.gov/sites/default/files/2023-04/zero_trust_maturity_model_v2_508.pdf

4. "Cisco's Guide to Zero Trust Security," https://www.cisco.com/c/dam/en/us/products/collateral/security/zero-trust-field-guide.pdf

5. "Zero Trust Maturity Model, Version 2," https://www.cisa.gov/sites/default/files/2023-04/zero_trust_maturity_model_v2_508.pdf

Identify Key Stakeholders and Enable a Zero Trust Team

Once the tactical road map is in place and once the key performance and risk metrics are identified, the next step is to identify a team that will help build the strategy and implement the Zero Trust vision in the current infrastructure. This will involve personnel from not only the security and infrastructure teams but also from various other business units that will be impacted by the Zero Trust implementation. The following chapters are covered in this phase:

Chapter 7   Zero Trust Avengers, Assemble!

Zero Trust Avengers, Assemble!

Considering the story so far, Zero Trust has been introduced as an information security model and a paradigm shift from the current viewpoint of security. Once the key tenets of Zero Trust are understood, enterprises will spend time identifying the following:

- Key performance metrics

- Control gaps in the current architecture measured via CISA maturity model v2

- Maturity of the enterprise from a Zero Trust process perspective measured by the CMMI maturity model

- Threats to the key assets

- Key risk metrics

This chapter demonstrates that Zero Trust cannot be established as a single-vendor project running for a few months. Zero Trust is not just technology; it's also not just about changing the way we control access. It is an evolution, a journey, and, more importantly, a team effort. It is the progressive evolution of a traditional enterprise to a modern self-sufficient security ecosystem. Unlike security-specific initiatives, which are driven by security architects and other leaders, Zero Trust is driven by a wide variety of individuals. This chapter explores some key stakeholders as well as other team members who are critical in formulating and implementing a Zero Trust strategy. This chapter also emphasizes the importance of the broader working team in the successful implementation of Zero Trust, with greater detail into each role. CISOs and C-level executives are no longer focusing on large enterprise-grade hardware but are moving their focus to soft skills, security awareness, and effective and efficient security policy and policy makers.

Why Is the Team Critical?

A Zero Trust team is not a specific physical team focused on maintaining Zero Trust. Similar to how Zero Trust as an information security model is an overlay over the existing network, the Zero Trust team is also a virtual team overlaying multiple existing teams. With the exception of certain roles, almost all roles relevant to Zero Trust usually exist within the enterprise. What is unique when building a virtual team is the alignment to a common goal, which is to achieve the broader Zero Trust vision. For example, a Zero Trust team will have representatives from policy creators and policers (governance and audit), security operations (SOC), network operations (NOC), incident response (IR), and even legal teams. The best way to describe a Zero Trust team is as **a cohesive set of diverse personnel providing feedback on all aspects of the Zero Trust strategy and eventually playing a role in the successful design and implementation of the Zero Trust vision.** It is important to note that you will need to translate the feedback on the strategy from the larger Zero Trust team into valid security policies and procedures for the employee segment to utilize in a simple and efficient manner.

Multiple administrators and nontechnical users get directly or indirectly involved in shaping the Zero Trust strategy. IT stakeholders will receive a large chunk of the additional effort with identity and access management (IAM) because Zero Trust is primarily an identity-centric access model. Active Directory administrators get feedback from business owners on what kind of users can access specific applications. Finally, non-technical folks are included in the Zero Trust conversation (especially end users) in order to understand their feedback, as it will be them who will be performing business-as-usual (BAU) actions with the newly implemented strategy. Users will be more aligned to understand the enterprise-wide changes and consequently adhere to the policies when they understand why the enterprise is implementing Zero Trust. Senior leadership must communicate the Zero Trust vision and mission across the enterprise to make sure each and every user fully understands the motive for moving to Zero Trust. Some key aspects of why a Zero Trust team is important are described in the sections that follow.

Expertise to Architect, Design, and Implement

Although Zero Trust represents a very simple and versatile concept, it is not easy to adopt and definitely not easy to implement. There is a sea of vendors that provide security products that can implement Zero Trust networks, and you need to have security practitioners, leaders, and architects who can wade through this sea and identify the right solution for the enterprise. Leaders must understand that Zero Trust is not just about technology but about people and processes. There is a dearth of consultants and architects who can understand an existing organization's network and build a customized strategy based on the maturity of the organization since the overall tailoring of strategy gets lost in translation with conversations getting steered toward the products rather than approach and strategy.

In addition to strategizing Zero Trust, expertise is expected in the products that are selected to implement the Zero Trust strategy successfully. Once the strategy is in place, it is important to understand where each security control is placed, and engineers must be hired to support the implementation.

Collaboration and Feedback

Zero Trust cannot be implemented by just the security team. It cannot be a siloed initiative. A wide range of dependencies exist when deploying Zero Trust, especially getting information from business owners and building a viable proposal. There must be a common individual, usually the security architect, who understands all the languages—CxO; governance, risk, and compliance (GRC); incident response (IR); and legal—to be able to coordinate information and communication across various business units.

Maintenance and Operations

Zero Trust is not a one-time project but rather an ongoing process. The right team should be able to maintain the Zero Trust architecture and ensure it is updated and optimized over time. Feedback from the SOC and IR teams is critical here when it comes to processing logs and engaging automation. Security orchestration, automation, and response (SOAR) and other orchestration initiatives need to support new process automations, and resources should have knowledge about and experience with the right processes to orchestrate.

Achieving the highest Zero Trust architecture is not sufficient. The enterprise must have the right services and policies in place to be able to manage and continuously improve Zero Trust adoption.

Seamless Adoption of Zero Trust Architecture for End-to-End Security

Zero Trust is ultimately about augmenting end-to-end security and managing risk. The right team should have the knowledge and expertise to identify and mitigate security risks effectively, which is crucial for the success of a Zero Trust strategy. This team must also be able to identify the critical business drivers and align the overall strategy to the enterprise business. With the right overall Zero Trust strategy and team, the enterprise transitions from a more static risk to a dynamically adapting risk-based policy, which gives the enterprise an edge over its competitors. Enterprises no longer need to focus entirely on security architecture and can focus on acquisitions, new products, or service enhancements. The Zero Trust adoption initiative will cover all needed requirements and will finally provide a tangible road map and strategy proposal, as covered previously in Chapter 6, "Understanding Zero Trust Maturity," with the help of experts and subject matter experts (SMEs) in their respective domains.

Strategy and Deployment: Two Sides of a Coin

Remember that in any enterprise there are going to be two verticals that permeate the Zero Trust concept. One vertical is the strategic leadership. Presentation to the strategic team must be in the language of risk, revenue and cost benefit. Coincidentally, most conversations with leadership relevant to Zero Trust typically begin with risk. The work and importance of Zero Trust can fully be appreciated only if we have a risk identified and a risk metric to work toward.

The second vertical is tactical leadership, which is mostly line managers and operations leads. Tactical leadership is usually concerned about how the day-to-day work and operations will function. They do not talk in terms of risk but rather in terms of performance and uptimes. They will have more detailed conversations about flows and how certain tasks get affected by the presence of this new policy. Some common concerns include "How will new users get access to the right applications?" and "How will new logging get affected?" The focus at this level will be technology and mechanisms rather than strategy.

In an enterprise, we are likely to see a lot of applications being accessed by subjects. It is imperative we insert a security mindset into application configuration and begin the tedious journey of application discovery. There are going to be multiple third-party integrations in the network such as chatbots, which will essentially need APIs to communicate to an application server or database. In today's world, the API is the currency of communication, and without securing these channels, we open a large backdoor for attackers. Someone with both networking and security skills can greatly impact the road map and development of a product and will be a key player in the development of the application as well.

Zero Trust focuses on changing processes rather than changing existing application infrastructure. There will be exceptions that will need immediate remediation if the gaps are critical; however, in general, Zero Trust does not mandate a full revamp of existing architecture but rather pivots toward designing applications and networks with the right security controls. A vulnerable web server is an immediate low-hanging fruit to fix, but as part of overall Zero Trust strategy, web server development will be improved to make sure the same vulnerable code is not deployed into production. Considering the criticality of the applications, the older applications may continue as legacy applications, and all that is needed is to identify the critical security mechanisms to protect them, as will be explored in Chapter 9, "Critical Security Mechanisms for Zero Trust Architectures." This is where security expertise is greatly appreciated and where practitioners are able to provide insight into what applications teams need to do to make sure applications and subjects have the right access control.

Security Ownership

Another aspect to bring out the importance of a strong team is to establish ownership of the strategy. The concern that still exists in most enterprises today is the fact that application teams do not consider maintaining security in the code as a performance indicator. Code reviews, security testing, and so on are owned by security teams, and application teams own only application validation and verification. A simple example would be coding an application to support only HTTP. Application developers code HTTP because it has fewer hassles with certificates, which ends up being a concern because most audits would highlight an HTTP-based service and direct it to be changed to HTTPS. DevOps and DevSecOps are catching up to make sure applications are coded securely and to the standard of the security architecture.

In the past, protections were built into the network environment and a moat was created. In Zero Trust architectures, workload is not just about controlling access but about the development of the applications themselves. Applications are still large threat surfaces. Every organization is a software company with internal applications, and all these applications need to have security baked in.

With this in mind, a strong Zero Trust team is needed to show that security teams definitely own the Zero Trust initiative, but they do not own the entire responsibility. Multiple business units, especially application and business owners, have a huge stake in the entire Zero Trust strategy. This is a huge shift from standard security initiatives, which are owned and deployed by security teams. When envisioning Zero Trust, even the end user is a key part of a Zero Trust strategy. An application owner must know that their application needs to support multifactor authentication. A compliance officer must know if specific security compliance is still achieved if the new Zero Trust model and its security controls are implemented. An end user must be able to gauge any possible hiccups when testing the strategy and provide feedback on how the strategy and implementation can be amended. Zero Trust is truly an organization-wide team effort.

Each business or function owner is the right entity to comment on or identify the impact of Zero Trust on their respective functions. Business owners will also be the right entities to add security into their applications. Security practitioners are the right entities to provide the strategy of how various business units can help the security organization achieve the Zero Trust vision.

Breaking the Barrier (Infrastructure, Operations, and Security)

Whereas infrastructure engineers are not covered under security projects, modern-day security projects cannot exist in silos from infrastructure initiatives. Every modern security project almost always requires conversations with the infrastructure teams. This is applicable when it comes to Zero Trust as well. You cannot implement Zero Trust without making major configuration changes in the network. Consider the vision of the infrastructure and operations teams. Their focus is to provide complete availability. If they are given an option to choose between downtime and security loss, almost always the answer is going to be tipped toward operational consistency. Their next focus would be to make access simpler and streamlined for users rather than have security initiatives break something and they have downtime. From the perspective of the enterprise security vision, Zero Trust fully satisfies both of these requirements. If you've done everything right, you should already know your assets and their flows, so infrastructure teams just need to make sure the security requirements and policies are implemented at the infrastructure level and monitor them for consistency. If infrastructure and security do not go hand in hand from the start and security comes in much later, the security strategy is bound to become an obstacle. Infrastructure teams provide their feedback to the team creating the strategy so that infrastructure and security are aligned early on.

Over time, there has been a major change with infrastructure teams. They are no longer just enforcers of policies but rather enablers of policy. Security teams give them the policy, but they not only execute it but also make sure there is some level of optimization by looking for automation opportunities. They no longer focus on controlling access to assets but rather facilitate access to the assets with the right direction from security. Infrastructure leads and engineers no longer primarily focus on stability but understand that with the right access control and visibility into the network flows, stability can be automatically achieved. This is why it is important that engineers understand at least the mission statement to which their Zero Trust initiative is aligned.

With Zero Trust breaking the barrier between the infrastructure, operations, and security, and with a common strategy between all three business units, it becomes easy to implement security at the beginning of a project and convince operations and the infrastructure teams to implement Zero Trust initiatives. Infrastructure teams understand certain aspects of why security teams request to control access to specific enterprise resources and try to support the organization by building automation and orchestration around it. Both the infrastructure and security teams help to reduce risk and improve compliance. It is increasingly clear that the security team is the best equipped to decide on how to design access control for a specific business requirement coming from the business owners, and the infrastructure teams are the best custodians to implement the control with guidance from the security teams.

Zero Trust is not about using a tool and achieving security. It's about developing and using the right process and making sure it is followed throughout the lifecycle of the Zero Trust deployment. The operational process definitions and strategy are owned by the COO. Making sure that the process and relevant technology is successfully implemented, is the role of the infrastructure (CIO) and security (CISO) teams. The CIO, COO, CISO and the CTO share an important partnership to achieve the broader Zero Trust vision. The Zero Trust strategy is typically owned by the CIO or the CISO.

DevSecOps and Its Relevance to Zero Trust

With the advent of DevSecOps, application owners are made aware of security flaws even before product release, and even if they have to release the product with security flaws, they know they need to add more stringent rules to protect it. For example, if an application is supposed to be integrated with single sign-on (SSO) and the SSO mechanism is not ready yet, the application can still be deployed with some form of MFA. If MFA is also not ready, the application could still be released as long as it's in a secure micro-perimeter with an enforcement point to protect it. Zero Trust does not state, "No it's not possible." It approaches a problem differently and tries to state, "Okay, will this method achieve what you need? Is there another way?" It breaks down silos between the various entities in the organization and helps bring together a collective responsibility by keeping the accountability of successful deployment with the security business unit. These are common traits that are adopted extensively in the DevSecOps strategy, where the core of DevSecOps is to incorporate streamlined secure application development without the traditional restriction of sequential design or siloed implementation.

DevSecOps can be visualized with the Department of Defense's illustration shown in Figure 7-1, which showcases how the dimension of security changes the way we approach application development and deployment.

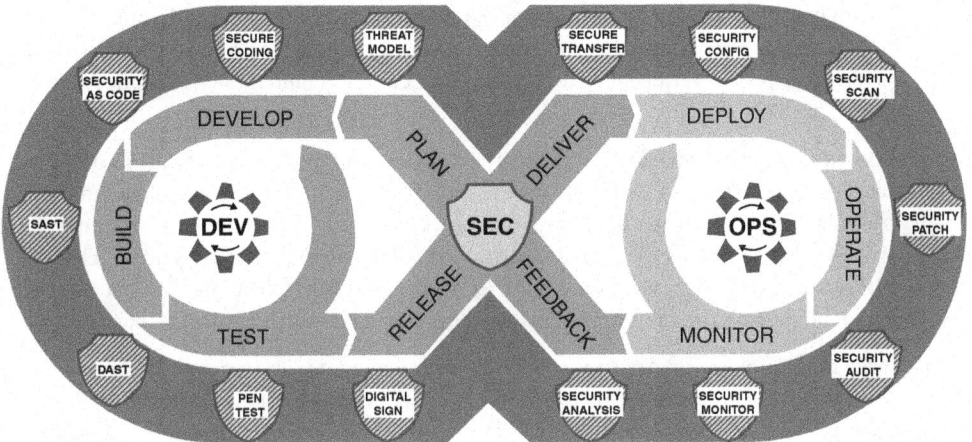

Figure 7-1   *The DevSecOps Cycle[1]*

Zero Trust adheres to the following strategies for the successful adoption of DevSecOps in the enterprise.

Secure Software Development (the Left Half)

The following are some key strategies and principles that most enterprises follow when incorporating security into their overall application development architecture:

- **Security SMEs:** Insert security staff into your enterprise's product design and development processes. This is a huge paradigm shift for many enterprises. Breaking that silo between development and security may look daunting, but once the security representation is complete and once security becomes part of the workflow, the benefits can be seen in the form of more secure and stable products. Security SMEs make sure that secure coding practices such as using digital signatures, external authentication, secure services, and so on are incorporated into application design from the inception.

- **Shared ownership:** Reset expectations around ownership for application security with the business owners. A key aspect of DevSecOps is to break the barrier between teams as well as to blur the lines concerning ownership of the application and its design. Application security is no longer just the security team's responsibility but instead more of a combined effort of business owners, application developers, and the security team. The aim is also to improve the efficiency of the application development process by breaking all barriers between development, security, and operations.

■ **Risk-based application development:** Use a risk-based approach to decide when security should administer technology. Security adds in a layer of risk mitigation when creating software or deploying technology. Without security representation early on in the design phase, multiple features get implemented without an understanding of the risk. For example, a developer might not understand the inherent risk of leaving a static password in the source code. A security angle must include threat modeling as a base requisite to make sure that common (and sometimes uncommon) threats are considered.

■ **Context-based classification at design inception:** Make sure classification of all data is completed and that functions are identified based on business, data, and asset criticality. An application developer will not know that the application being developed is used to process sensitive information; therefore, software might be released without appropriate kernel isolation or encrypted channels. The presence of security personnel during the operations and development phase avoids such scenarios.

■ **Agile methodology of development and patch management:** Improve the overall security posture of applications by developing in an agile methodology and fixing security vulnerabilities immediately rather than waiting for long timeframes. This is a common concern that we see in general product development. Unless the vulnerability is critical, most fixes to code do not go in before production. This, however, is suboptimal. Considering all the limitations of time and efforts for production teams, a security perspective could help optimize this by highlighting the concerns early on during development so that the fix can be incorporated immediately rather than being patched later. The best-case scenario would be to insert security personnel early on in the development to avoid this vulnerability from making it to the final release.

■ **Security as Code (SaC):** With more focus on automation, SaC allows for automating critical security compliance checks, application protections, and overall security requirements into the application development lifecycle. This reduces the time needed to independently evaluate the security posture of an application by automating most of the validation.

■ **Baked-in security:** Make sure all applications are developed with the principle of least privilege. Although this is easier said than done, it is extremely crucial to make sure that all applications, software or otherwise, deploy this basic security principle, which will take substantial testing and code review. Hence it is best done alongside a security subject matter expert. Source code access, including APIs, must be protected with least privilege and need-to-know access only. Static testing, dynamic testing, and penetration testing must be streamlined and automated for all applications.

■ **Identity control capability:** Identity and access management (IAM) mechanisms are a mandatory requirement to control access for subjects to their respective applications. Zero Trust is identity-specific, and hence any authentication or authorization (granular or otherwise) must be inherently incorporated into the software development. Developers must take into account all AAA protocols supported and, based on a larger study of the protocols used, deploy feature support to suit the generally adopted protocol. A product without an authentication mechanism is a house without a fence or a room without a lock.

- **Unified security controls across production, development, and staging environments:** Incorporate staging environments with Zero Trust. Once again, a common pitfall with most development strategies is to incorporate Zero Trust into production but not staging. This means when code is tested in staging, it will not be close to a real production workload because Zero Trust concepts might not have been considered during development. Subsequently, when security concerns are seen in production, the Zero Trust–enabling patched code needs to be retrofit into the product, which costs time and money.

Continuous Integration and Continuous Deployment with Security (the Right Half)

The following are some key strategies and principles that most enterprises follow when deploying and monitoring their existing applications:

- **Security analyst feedback:** Make sure security validations are a continuous check. Once again, similar to Zero Trust itself, the security validation, especially from the security practitioners and SOC engineers, must be continuous and must constantly provide feedback to the design and implementation via continuous integration/ continuous development (CI/CD) pipelines, preferably automated.

- **Validate the efficacy of security mechanisms:** This involves the continuous monitoring of data at rest and in motion to validate theft. Once products and software are released, developers must not just sit back and move on. Operations teams must run continuous monitoring to validate the efficacy of the security mechanisms in question with feedback from security SMEs. Have you enabled encryption? Yes. Is it strong encryption? Probably not. The development and monitoring happens in tandem.

- **Application anomaly detection:** Machine learning, artificial intelligence, and SOAR must work together to make sure applications are monitored for anomalous behavior and have automated responses configured for common incidents. Automation and orchestration must be the final goal of any strategy. The aim is to automate common tasks like anomaly detection, simple incident response, false positive analysis, and so on so that engineers can work on more complex scenarios. This is commonly referred to as *AISecOps*.

- **Effective application monitoring:** Make sure application logging is crisp and to the point but detailed enough to identify issues. Applications tend to get outdated or impacted by security advisories, and the continuous monitoring helps feed the CI/CD pipeline to automate patch management and vulnerability management. It is therefore important to make sure application anomalies and security events are logged accurately.

- **Continuous compliance checks:** Include constant Zero Trust–aligned compliance checks at each stage. Whereas there are no Zero Trust compliance checks in general, capabilities that exist today are specific compliance checks that align with Zero

Trust. For example, continuous validation of context is an important check that aligns with most compliance requirements. It is also important to make sure that the end users and devices are adhering to the overall governance policies laid down by the enterprise. This is another important check that the enterprise needs to build in as part of continuous compliance checks.

- **Secure configuration:** DevSecOps encourages automation for activities such as hardening. Secure configuration is not relevant just for Day 1 deployments; it is also relevant for constant improvement of security posture. CI/CD automates hardening activities, which is an important strategy toward effectively securing applications.

Key Stakeholders in a Zero Trust Team

It is fair to conclude that Zero Trust is not all about the technology being implemented but more about the people as well. It is not a simple security initiative but a broader IT strategy that has widespread impact on the people involved. Hence, technology proficiency is not enough when we are presenting Zero Trust strategies. It is imperative to know the audience and their stake in the overall strategy before we can present. This section will help identify some of the key stakeholders in the Zero Trust strategy's leadership. The subsequent section will showcase some technical skillsets needed to provide inputs into the strategy.

Overall, two types of roles need to be a part of Zero Trust:

- **Supporters:** Your supporters will be people who align with your overall vision and will usually be at a higher power scale. These include leadership stakeholders who believe in the vision and mission.

- **Enablers:** Enablers are specialists in their field who see how Zero Trust can benefit their overall operations as well as how they can improve the user experience based on these optimizations. You need to enable your enablers before they can enable your enterprise.

The sections that follow describe the roles that are usually relevant to creating robust Zero Trust leadership.

Executive Leadership: Supporters of Zero Trust

As is the case with all strategy, its success greatly depends on strong support from leadership. Without an enabler at the leadership level, any initiative will have challenges proving its value to the enterprise. The sections that follow describe some key leadership roles you must influence to enable your Zero Trust initiative and to make adoption smoother.

The Board of Executives

The board of executives is composed of the final decision-makers with respect to any decision made in the enterprise. This makes them key stakeholders in the Zero Trust team. You need to be able to convince the board that moving to a Zero Trust strategy will enhance employee and security experience. This will of course be viable only if you align the Zero Trust strategy to the business vision; however, if you follow the methodology mentioned in this book, it shouldn't be very difficult. What you need to remember is to be well prepared before proposing a strategy to a board of executives. They have very strong values, and it is important to be aware of each of these values before you begin your pitch.

The executives also drive the organizational culture. There are countless lessons in leadership where trust in employees and empowering them to make the right decision goes a long way to support the overall employee experience. When employees are trusted to make the right security decision, a sense of responsibility gets instilled into each individual to make sure they see why the enterprise is going down the Zero Trust path. The core of Zero Trust is not to remove trust altogether; instead, it's to give just enough access to those you trust and monitor that access continuously. Executives who see this as a blessing will align with you easily. Majority board support is mandatory.

Chief Risk Officer

When I talk at presentations regarding topics that I do not work with on a daily basis, a common strategy I use is to find a friendly face in the crowd and just speak to them. It greatly helps to boost my confidence and validates what I am saying. They usually nod and smile or just calm me down so that when the tough questions are asked, I am not rattled. The chief risk officer (CRO) is your cool influencer friend who will give your strategy a shoutout to all the other executives if you are able to convince them that your strategy reduces risk, which is their currency. In actuality, the CRO has a great say in a Zero Trust strategy. You will get to know the current risk in the enterprise, and the CRO can give you tangible examples and metrics to support your cause. Alternatively, the CRO can approve or validate metrics you have crafted. Without the support from the CRO, you might find it tough to push the Zero Trust agenda, or, worse, you might even get opposition.

A CRO needs to look at your proposal from multiple angles, and that is going to be critical when creating the final executive presentation. Zero Trust needs to provide an edge over competitors and must hence reduce the competitive risk. Implementing Zero Trust could also add a technical risk that needs to be mitigated. For example, it is possible that as Zero Trust implementation begins, certain technology might cause possible roadblocks to the deployment. The CRO also needs to confirm that with the strategy of moving to Zero Trust, the overall risk of compliance lapses is reduced. Hence, if you are able to convince a CRO to support your cause, that will be an added bonus when you're presenting to the board.

Chief Technology Officer

A chief technology officer (CTO) is a high-ranking official responsible for the technology vision of the enterprise. As the technology guru in the enterprise, the CTO is usually interested in using the best technology at optimal costs; hence, all product selection activities will be led by the CTO office. The CTO is also responsible for innovation in products that the enterprise owns and manufactures. A well-defined integration of products with the enterprise is usually top of mind for most CTOs; hence, aligning the Zero Trust strategy with the CTO is key to make sure that the strategy you propose and the products identified fit the overall technology vision.

Chief Operating Officer

A chief operating officer (COO) handles the overall vision for the enterprise—not to be confused with the business drivers. A COO covers topics like operational excellence in processes, employee experience, and overall operations of the enterprise. A COO has a sustainable vision for the enterprise. Security is definitely on their agenda, but not at the top of their mind. It is your job to make sure the Zero Trust vision aligns with the overall vision of the COO. Topics like financial management, people management, and even risk management to an extent must be aligned with the Zero Trust proposal.

Chief Information Officer

The chief information officer (CIO) and the CIO team are key players in the Zero Trust strategy. The CIO is responsible for the entire IT strategy and infrastructure. Data protection strategies and digital transformation are key when it comes to the CIO office. Zero Trust greatly impacts access to devices and how they implement access control. The CIO team is responsible for clearly articulating how they plan to achieve the Zero Trust vision. It goes without saying that the CIO team must fully understand your proposal and road map before they can consider other projects and finance to decide on their strategy. They will be responsible for strategizing and transforming their existing infrastructure along with the security team. It is also likely that conversations to approach Zero Trust will begin with the CIO or the CIO team. The CIO will dictate business drivers and the need to move to Zero Trust.

Chief Financial Officer

Every project or initiative needs money, and the chief financial officer (CFO) is your treasurer. It is especially tricky to get the CFO on your side because a CFO is going to look at how the proposal is going to save the enterprise from unnecessary spending. Support from the CFO office will rely on how optimized the proposed budget is. Some of the initiatives in the proposal might need an initial capital investment, which might not seem attractive at first, but it is important to showcase how the Zero Trust concept and overall vision is all about moving cost from capital expenditure to operational expenditure. Zero Trust might need some additional technological advancements, but in the long run it will be just managing the right control with the right skilled operators. This will save in overall capital investment each time a new technology is released and will keep the cost low at the operational level.

Chief Information Security Officer

The chief information security officer (CISO) is the leader of the project team. Consider the CISO as a highly skilled project manager from the context of Zero Trust adoption. The role of the CISO team is to support the CIO in realizing the Zero Trust transformation.

The CISO decides the security strategy and makes sure the overall strategy is implemented and monitored consistently to satisfy the overall business need of transformation. The CISO is the team captain and must be on board with the CIO to support the overall vision.

The Operations Team

Once the Zero Trust architecture has been successfully deployed, the CISO and their team must make sure the deployment is monitored to reach maturity; however, the story on the ground will not focus on strategy. The operations team will be firefighting customer and network issues every day. The strategy might involve migration of all users at a time, and the operations team may clearly disagree because it would be an operational nightmare when users call the IT help desk to complain about access loss. Operations teams provide valuable insights on how the implementation and monitoring will be taken forward. For example, operations teams will propose to use a simple monitor mode for the migration so that they can validate accesses from users and check how many failed access requests have been occurring. They use this to enhance and optimize the access control and then move the users per segment. The operations team must be onboard with the overall implementation plan; otherwise, there can be some serious pushbacks and delays in implementing the solution.

What you have seen are the important stakeholders, supporters, and promoters who need to be on board with the idea of Zero Trust. Selling the vision to senior leadership and multiple teams is a key step in making sure the motive and advantages are clear. You need to identify your important supporters and detractors and attempt to win the support of the stakeholders mentioned in this section. Of course, leadership is not the only group you need to convince. There are multiple teams that need to support your cause and, more importantly, give you more insight and feedback to add to your strategy and presentation.

Specialist Skilled Personnel: Enablers of Zero Trust

A Zero Trust–enabled team requires a wide range of skilled professionals to effectively implement and maintain the security model. The skills described in the sections that follow are highlighted separately because their role is twofold. They help strategize, design, implement, and monitor the Zero Trust architecture and at the same time provide valuable feedback to the overall architectural proposal to make sure that the Zero Trust vision aligns with the business drivers as well as operational needs.

Strategy and Design

The sections that follow describe the strategy and design roles that provide feedback to the overall Zero Trust strategy early in its lifecycle.

Business Owners

Business owners need to give feedback on their applications and its current road map for the business. If an application is going to be scrapped, it needs to be considered legacy and therefore the focus on its development may not be much. Similarly, if an application is time critical, application owners must work with software architects to design and implement security without impacting the performance. Business owners will assign an asset value and a risk to a specific application, and based on the overarching discussions of prioritization, the CISO will eventually decide how to categorize and label the applications and utilize those labels to design risk-based control policies.

Security Architect

A security architect (SA) designs and supports the implementation of the Zero Trust architecture in the organization's infrastructure. They analyze the organization's security requirements and design a secure architecture that can detect and prevent unauthorized access.

The architect's role is more technical in nature. The architect's role is to design the security mechanisms that are best suited for the Zero Trust strategy. This is different from the business analyst (BA) whose role is more oriented toward business drivers. For example, a BA may look at a flawed authentication process and provide feedback about the process itself, whereas an SA needs to look at the existing IAM mechanism and then provide improvement recommendations or direct the enterprise to an apt vendor. An architect will also identify key business flows relevant to the network and workload and help the enterprise create policies specific to those flows.

An architect is best equipped to validate the effectiveness of the deployed technology. Remember that a business analyst validates the effectiveness of the strategy and overall processes, whereas an architect is more technology focused.

Business Analyst

A business analyst helps to identify the organization's security requirements, analyze risks and threats, and assess the impact of security incidents on the organization's operations. The BA can also help to design and implement security controls that meet the organization's needs, along with the SA. If DevSecOps is a strategy being followed in the institution, the BA can help push for security configurations to be implemented into the applications as part of the DevSecOps initiative.

A BA also has a discovery role along with an analysis role. The BA must look at the current security posture and suggest solutions. Sometimes a BA can also help coordinate activities across multiple business units.

Finally, a BA gets involved at the end of a Zero Trust lifecycle to monitor the effectiveness of the Zero Trust metrics identified during the initial strategic discussions. Both architects and business analysts work together to translate the metrics, posture, and current state from the strategic perspective to tactical and operational, and vice versa.

Infrastructure Deployment and Operations

Infrastructure deployment and operations are important verticals in the implementation of the Zero Trust architecture. Once the architecture has been identified, implementation and monitoring commences. The personnel described in the sections that follow are enablers who support the implementation and monitoring strategy of the enterprise.

Implementation Engineers

A network or security engineer is responsible for implementing and maintaining the organization's network or security infrastructure, respectively. In some organizations, this role may be common across security and network infrastructure. In a Zero Trust model, implementation engineers need to configure the network to allow only authorized access, segment the network to reduce the attack surface, and enforce access control policies. They provide feedback and insights about deployment challenges and other factors that are otherwise not visible at a leadership level. It is imperative that the security architect translates any business drivers to the deployment engineer and at the same time translates the deployment challenges as risk metrics to leadership. It is a common segment where information may be lost in translation. A deployment engineer may not need to know the overall strategy, but they must know the primary purpose of what they are deploying based on basic network and security knowledge.

Identity and Access Management Administrators and Engineers

IAM engineers (AD admins) are responsible for managing the users' digital identities and the access they have to resources in the organization. In a Zero Trust model, IAM engineers need to ensure that only authorized users can access resources and enforce access control policies. Based on feedback from business owners and security architects, they will implement the required access control for the subjects and objects.

Security Operations

SOC analysts are responsible for monitoring the organization's assets for potential security incidents. In a Zero Trust model, SOC analysts need to have a deep understanding of the organization's architecture and access control policies to detect and respond to any security incidents.

SOC analysis is a key skillset in monitoring and maintaining a network irrespective of whether the organization has implemented Zero Trust. A SOC analyst must have a background in security and must have an inquisitive mindset and understand the architecture of what you are protecting in order to identify anomalies instead of passing them off as a one-time abnormality. It is important to make sure SOC has enough people to accommodate the number of alerts coming in. Invest in automation and orchestration and the right organizational processes to make sure engineers can process this information well.

A Zero Trust enabled SOC will be contextually aware of key assets and risky users; therefore, the alerts generated will be almost always relevant and the SOC will be much better equipped to recognize and respond to threats targeting sensitive data. SOC analysts will provide information around common threats the enterprise faces relevant to the critical assets being monitored, which feeds into the overall strategy of how the Zero Trust architecture is designed and implemented. With SOAR available to orchestrate processes,

SOC automation should be a priority skillset. This skillset will be able to bring in more automation for simpler tasks.

Incident Response (IR)

Another segment where Zero Trust gets feedback from is incident handling and response. Due to the strong granularity and segmentation requirement of Zero Trust, incidents generally can be caught much earlier, and if they materialize into a threat, they do not have larger impact on the enterprise. This is one of the larger advantages of effective segmentation.

Similar to a SOC, an incident response team needs to have the right people, processes, and technologies. Operations might look a little different from a SOC because in an IR plane, the focus is on ring-fencing an incident, which has much broader implications. Zero Trust has great support from IR teams because they see the largest benefits at various stages of incident response. According to NIST SP 800-61, Rev 2, incident response has the following stages, and Zero Trust strategies greatly enhance the experience for IR professionals along each of the stages:

1. **Preparation:** Before an incident occurs, an organization should have plans, policies, and procedures in place for responding to security incidents. This includes identifying critical assets, establishing response teams, and defining roles and responsibilities. This aligns with Zero Trust initiatives and activities like asset management, role-based user segmentation, and network segmentation.

2. **Detection:** The first stage of incident response is detecting that an incident has occurred. This can be done through various means, including monitoring systems, intrusion detection systems, or reports from users. Zero Trust demands central and dynamic visibility, which will enhance detection of incidents based on anomaly and heuristic technology.

3. **Analysis:** Once an incident has been detected, the next stage is to analyze the incident to determine its scope and severity. This involves collecting and reviewing data, identifying the cause of the incident, and assessing the potential impact on the organization. With the right visibility in the network, all information can be collected to identify the root cause of the incident, and this is facilitated by Zero Trust monitoring requirements.

4. **Containment:** Once the incident has been analyzed, the next step is to contain the incident in order to prevent it from spreading further. This might involve isolating affected systems or networks, shutting down compromised services, or blocking network traffic. Incident response personnel greatly benefit from Zero Trust at this stage because if the network is deployed with Zero Trust at its core, the blast radius of an incident is small and easy to contain. The second aspect is that the IR team can deterministically showcase which devices are affected due to enhanced visibility and effective segmentation. Finally, with the right automation, any incident can be rapidly contained automatically.

5. Eradication, recovery, and lessons learned: After containing the incident, the next stage is to eradicate the root cause of the incident. This involves identifying and removing any malware, patching vulnerabilities, or otherwise securing systems to prevent future incidents. Once the incident has been eradicated, the next stage is to recover any affected systems or data. This might involve restoring backups, rebuilding systems, or reconfiguring services. While embarking on the Zero Trust journey, SOAR initiatives will be key when it comes to eradication and recovery. With the right orchestration of specific processes, an incident can be thwarted, and recovery activities can be started without any intervention from analysts. As a constant feedback mechanism, the lessons learned will review the incident response process in order to identify areas for improvement. This will feed into the incident management lifecycle as well as provide more insights into possible SOAR enhancements.

In an incident management lifecycle, new and existing risks and threats are identified and analyzed. New policies and procedures are created, and existing policies are modified based on constant feedback. This is where IR leads provide their feedback to the overall Zero Trust strategy. The feedback from IR teams helps the policymakers understand the key risks and threats and how to support, automate, and orchestrate repeatable IR activities.

Zero Trust provides an additional layer of protection against cyber threats, as it assumes that attackers are already inside the network perimeter and therefore requires strong authentication and verification of every user and device that tries to access a network or application. This reduces the risk of successful attacks and minimizes the potential impact of any security incidents that do occur. The Zero Trust strategy aligns with what IR teams perform every day as part of their job and hence gets great support and feedback from them.

Governance Teams

Governance, risk, and compliance teams are instrumental in creating the overarching policy that governs all policy implementations. These abstract policies align with the compliance and risk of the enterprise and are subsequently translated to the actual policy based on the hardware implementation. Traditionally, the abstract policies were very high level and were considered a layer above the implementation step; however, Zero Trust aims to introduce a permeable layer between the abstract governance policies and tangible access policies so that the two layers are no longer distinctly separate and implementation engineers understand what governance teams are expecting, and they can provide more feedback on the implementation on specific products. The sections that follow describe some key personnel from governance teams.

Compliance Officer

A compliance officer ensures that the organization complies with the relevant laws, regulations, and industry standards. In a Zero Trust model, they need to ensure that the organization's security policies and practices are aligned with the relevant compliance requirements, such as General Data Protection Regulation (GDPR), Health Insurance

Portability and Accountability Act (HIPAA), and Payment Card Industry – Data Security Standard (PCI-DSS). They work with legal and regulatory teams to interpret regulations and ensure that the organization's security practices comply with these requirements. They also provides feedback on compliance requirements that the Zero Trust vision must cater to.

Another key role they fulfill is to work closely with the security team to implement policies and procedures that align with the organization's security goals and objectives. This involves identifying risks and vulnerabilities, defining access controls, and implementing security technologies that enable continuous monitoring and enforcement of security policies.

Audit and Governance, Risk, and Compliance Team

The audit and governance, risk, and compliance (GRC) team is another critical segment to consider when it comes to creating policies for the enterprise. All policy directives come from the GRC team, and it is an aspect of the Zero Trust architecture that is easily overlooked. There could be pushback and, at a bare minimum, questions on how you would implement the Zero Trust strategy and maintain policies. At this point, you should remember that the main motive of the Zero Trust initiative is to make sure that the GRC team understands the importance of the initiative, and your final goal is to change any detractors to promoters. In general, implementation of Zero Trust greatly helps GRC teams to implement security policies because Zero Trust inherently considers context-based policies.

Generally, you will notice that Zero Trust itself doesn't need to be called out under any compliance requirements. This is because, if Zero Trust is implemented right, compliance and governance are automatically taken care of. The main motive of the GRC will be to make sure that Zero Trust is aligned with the business and that all risks have been considered and regulations are being complied with. For example, compliance might require you to segment your network, and when you do so, you essentially protect yourself from the impact of certain attacks and inadvertently implement security controls as part of the Zero Trust Implementation roadmap. All the critical mechanisms we will discuss in Chapter 9, "Critical Security Mechanisms for Zero Trust Architectures" add to the overall compliance landscape of the enterprise. By segmenting the network as part of Zero Trust, you are also essentially restricting the audit scope, which makes the audit faster and more focused. The audit can help each segment differently and provide custom feedback per segment, which is better in the long run rather than performing a large-scale audit where the recommendations are not detailed enough.

One of the main drivers for an audit are IAM and data classification, and auditors are responsible for ensuring that organizations comply with regulatory requirements and industry standards. The best course of action is to make sure to have auditors involved early on in policy creation to make sure that all the requirements are fulfilled. Auditors will provide a maturity baseline and help the enterprise move in the right direction to set up and create a viable Zero Trust architecture. They will also provide the mandatory compliance and audit requirements, which will eventually help prioritize initiatives in the overall Zero Trust deployment. By implementing a Zero Trust model, organizations demonstrate that they have implemented appropriate security controls to protect their sensitive data and resources. This helps to reduce the risk of security breaches and potential financial and reputational damage that can result from a data breach. This will eventually align with what auditors are looking for.

Data Security Analyst

Data security analysts are responsible for securing the organization's data, ensuring that sensitive data is protected from unauthorized access. In a Zero Trust model, data security analysts need to ensure that data is encrypted, access is restricted, and data is monitored for any suspicious activity.

Zero Trust greatly depends on data classification and segmentation to make sure that all compliance requirements are in place. This is key because an IP-focused, perimeter-based approach is highly ineffective. With a perimeter-based approach, you cannot effectively add the right controls based on asset value or surrounding threat. Data and assets are unique based on their value, and protection must be tailor-made for them as well. Data analysts help facilitate this by identifying key functions and how the data is handled by those functions. Once the classification and segmentation is in place, the overall categorization can then be fed into the strategy, road map, and policy-making mechanism.

Project Manager

A project manager is responsible for managing the implementation of the Zero Trust model in the organization. They ensure that the project is delivered on time, within budget, and meets the organization's requirements. They also coordinate the activities of different team members to ensure a smooth implementation. The CISO team is the virtual project manager for the entire Zero Trust strategy and implementation, and they will delegate some of the management to tactical project managers for each initiative. All initiatives will eventually converge to the common Zero Trust vision. For example, the overall Zero Trust strategy is owned by the CISO, but the segmentation strategy and implementation might be taken up by one of the team members.

Legal Teams

The legal team helps identify the applicable laws and regulations that affect the organization's security practices and ensures that the security policies are aligned with them. They also assist in assessing the potential legal risks associated with the Zero Trust approach, such as data privacy and confidentiality, and develop mitigation strategies to address them. Furthermore, the legal team helps draft the necessary policies, procedures, and contracts to support the Zero Trust approach, such as user access agreements, data sharing agreements, and incident response plans. They will provide feedback about, for example, storing data in specific locations or the operational requirements of maintaining 99.99% uptime as per a service level agreement (SLA).

In addition, the legal team helps ensure that the organization's security practices are consistent with its contractual obligations, such as service level agreements and data protection agreements, and that the appropriate provisions are included in contracts with third-party vendors and partners. They also assist in responding to legal inquiries and investigations related to security incidents and ensure that the organization is compliant with legal requirements for breach notification and reporting.

Overall, a Zero Trust–enabled team requires a range of skilled professionals who can work together to implement and maintain a secure architecture that reduces the attack surface and mitigates the risk of security incidents.

External Feedback

External architects and consultants should be included to influence the Zero Trust policy based on their expertise. Their stake will be different when it comes to the other internal stakeholders. Independent consulting firms have a reputational stake where their feedback, along with the success of your initiative, gives them credit. Product vendor consultants are technological partners where feedback helps propagate the technology required to achieve the enterprise vision. Hence, the stake here would also be to maintain the credibility of the product and the strategy that the target enterprise is developing and consuming.

End Users

Finally, there can be no Zero Trust team without the end users. While it is unlikely that all end users, employees, or even customers might be security savvy, there must be representation of various demographics of end users to feed into the overall strategy. Security champions from each team or group must be able to articulate the needs, concerns, and possible feedback to the strategy representatives. End users are usually exposed to more sophisticated social engineering attacks, and when the security strategy is changing, end users need to have a say in what kind of user experience they wish to have, or what kind of automation they would like to see. Even topics like what form of authentication mechanism they are comfortable with are important, as some end users might not be comfortable with biometrics. Security-focused BAs are the most common security champions; however, if the teams are large enough to have their own security champion, the BA would usually be communicating with the security champion to make sure all requirements and feedback are captured.

Managing Your Stakeholders

With so many people involved and so many lenses to cater to, it will likely become difficult to manage the expectations of everyone. An easy method of focusing on the right group of people is to use the power grid illustrated in Figure 7-2.

Figure 7-2 *Power Grid of Supporters and Detractors*

Managing people greatly depends on how interested they are in your proposal and how much power they have. Members of leadership who are keenly interested in your proposal are your promoters or supporters. This is where you want most of your stakeholders. Of course, not everyone might support your idea, and you might find that leadership is not keen on the overall idea of Zero Trust or they are concerned with specific risks. They might be detractors, but you shouldn't completely give up on them. Leadership detractors need to be kept happy. These are folks you need to speak with, understand what they oppose, and try to align with them as well so that most of the detractors move to become leadership promoters. For example, if the CRO is not supporting the Zero Trust initiative due to the possible risk of delayed product releases based on vendor history, your role is not to ignore the concern but to work with the CRO to understand the actual risk and explain how Zero Trust could work around the risk or mitigate it. This will bring the CRO to your supporter quadrant.

People who do not have strategic decision power such as specialists might also be present in your stakeholder list, and your goal is to make sure that people who have high interest in your vision are aware of changes being made. They will want to know what is happening either due to personal interest or to propagate it within their business unit. Similar to leadership, there will be detractors in the enterprise via users and specialists. A well-known fact both in the enterprise and real life is that you cannot keep everyone happy, and from a decision-making perspective, you do not need them to fully accept and support the cause. That being said, you do not want to completely alienate them, so monitoring their feedback and interests is key to align with them and to move them into being promoters. If they stay detractors, keep track of their interests and constantly gather their feedback.

Friction is common in a healthy relationship, and the same applies to the Zero Trust initiative. Not all proposals are going to be a bed of roses. Any proposition will have opposition, and there might be many reasons for opposition, including the following:

- You did not or could not appeal to a business driver that everyone can accept.

- Your cost benefit analysis drove leadership to choose to accept the risk of not implementing the strategy.

- A traditional mindset where tangible hardware like a firewall is sufficient.

- Complacency that the current security controls are enough.

Irrespective of what is blocking the proposal, it is important to be patient, understand everyone's viewpoint, and take their key argument as feedback and modify the proposal to appeal to as many people as possible. The enablers and sponsors you made along the way will help you as well, and that is why pushing this proposal to leadership is not just a 30-minute presentation. It takes months of face-to-face calls, chats, analyses, and presentations.

The best course of action is to listen to and revisit the strategy to make sure everyone is aligned. This is where the power of the pack unfolds—when more people believe in your vision, you will automatically be able to bring the detractors to your side. They should be informed of your progress so they know that you are not just a one-trick pony.

Security Culture: The Last Piece of the Puzzle

The workforce of the enterprise is the final piece of the Zero Trust social puzzle. Strategy is not going to work if your enterprise workforce doesn't support you and understand what you want to do. Similarly, if users try to bypass security controls you put in place, it completely defeats the purpose of putting in those controls. There is a close relationship between enterprise leadership and the overall workforce, and like any relationship, communication is key. It is not enough that the strategic leaders and operational leaders accept and support Zero Trust; the workforce must support the strategy, too, because it is the strategy that will essentially translate into policy and then guidelines and procedures. If employees do not understand why policies are in place, the entire program will be useless. People are the weakest link in any security chain. There must be a strong sense of ownership in each employee to make sure security is maintained. Maintaining a program just to get a "yes" or "no" response from employees might be supporting the compliance initiative but not the enterprise in the long run. What we need is for people to understand the core motive for the paradigm shift and then propagate the best practice. People in a cybersecurity program must know why they are in that program and must try their best to add value, which then brings in a shared responsibility culture.

Security Champions

It is impractical to expect all of the workforce to understand the nuances of security. One strategy that has helped multiple enterprises to get the workforce on board is to identify the key stakeholders or security champions per team and then liaise with them. These security champions need not be experts in security but they must be passionate about it. These representatives will then help propagate the security directives from the top-down and make sure that the entire team understands what is expected out of them.

Tangible Campaigns

Create tangible campaigns for each of the communities identified. A community here means a diverse set of people (for example, HR teams, support teams, and so on). This also means targeted training and custom content based on the overall security training and practice. All courses must be personalized and customized.

Training must not be instructional, as no one likes to be told what to do. Make it interactive, funny, and engaging so that people listen and want to know more. Let them know issues closer to their heart than to the enterprise. Make sure your security campaign is not just a phishing campaign and that it covers all aspects of Zero Trust. This is a change for users, and you are giving them more power. With great power comes great responsibility, and they should know it.

Focus on Morale, Not on Blame

A team with higher morale has more chances of propagating the right idea to other teams. Try and craft your training to avoid blaming a user for making a mistake or being exploited by a malicious actor. Phishing campaigns are a common means of blaming the user and have usually been detrimental in creating a larger security consciousness. Avoid the blame game and enable them to make mistakes and learn with curiosity.

Drive and Lead Security in All Business Verticals

Business and technology leaders need to facilitate and inculcate security in their teams and products. DevSecOps and many such initiatives need to be adopted and engraved into day-to-day operations. Security must always be embedded early on in any social, procedural, or technological initiative and never as an afterthought.

Trust and Enable Your End Users

Humans are the weakest link, so end users must have security training and awareness. Consider end users as toddlers discovering new approaches to accessing data. Let them make mistakes and learn. It is common to provide training to end users, but what is not common is to give them power and trust them. Trust your users, train them well, and believe in them to make the right decision. Zero Trust is about giving the freedom to an end user to access only what they are supposed to, and if they don't, the enterprise puts in the apt controls around it. Zero Trust doesn't mean "no trust". It means "just enough trust" for just enough time. Don't offload security—own it.

Malicious Insiders: Trust No One

Often many enterprises consider campuses as a trusted segment and take security very lightly when connecting a campus segment to a data center. The logic here is that both are trusted, so what could go wrong? Similarly, employees trust each other, and once there is an informal friendship established, an implicit trust is created that clouds the possibility of anyone being a malicious insider. However, the bigger message that Zero Trust access models deliver is to remove any implicit trust we have not only in the network but also toward people. If you see your friend sulking around the cloud data center, remember to note it down, and if you see a pattern, report it. Do not assume it's alright because they are your friend, as that is the definition of implicit trust, which you must avoid. Employees need to know exactly what policies are around an insider threat so they can implement them too. Some common indicators are as follows:

- **Unusual behavior:** Employees who suddenly begin exhibiting unusual behavior, such as coming into the office at odd hours, spending time in areas they don't usually frequent, or acting secretive, could be potential indicators of malicious intent.

- **Disgruntlement or dissatisfaction:** Employees who are unhappy with their job, have a grudge against their employer, or are facing disciplinary action might be more likely to engage in malicious activities.

- **Accessing sensitive data:** Employees who access sensitive data that is not relevant to their job duties, or who access data outside of normal business hours, might be engaging in malicious activities.

- **Unauthorized access:** Employees who attempt to gain unauthorized access to systems, data, or networks, or who use unauthorized devices, can also be early indicators of malicious intent.

- **Social engineering:** Employees who use social engineering techniques, such as phishing or pretexting, to gain access to sensitive information can also be potential indicators of malicious intent.

It's important to note that these behaviors alone do not necessarily indicate malicious intent; however, they might warrant further investigation to determine if an employee is engaging in potentially harmful activities. Employers must take steps to mitigate insider threats by implementing robust security measures, monitoring employee activity, and providing training and education on security best practices. Zero Trust does not propagate implicit trust, and an insider by definition demands a basic level of trust. Hence, it is important for the Zero Trust policy to cater to insider threats. End users are most susceptible to social engineering and thus activities like tailgating and shoulder surfing must be avoided as much as possible. Awareness of the possible attack vectors such as phishing with targeted emails is an important step toward a more security-aligned workforce. Monitoring user behavior also warrants for a mature and automated SOC solution which will be the source of information for incidents involving malicious insiders which subsequently will be converted to automated playbooks.

Overall, work on not only the technology aspect of the proposal and strategy but also the people aspect, because Zero Trust as a journey is one the enterprise as a whole must complete together.

The Follow-Up

[Glenn sees mostly pensive faces across the table. He pauses for some time to let the information sink in and then continues.]

Glenn: Essentially, a brief summary of what we discussed is as follows:

- Zero Trust is not just about technology and metrics. There are intricate relationships that need to be capitalized upon to effectively propagate the vision.

- Zero Trust breaks all boundaries between security, operations, and application development. With DevSecOps strategies, applications are robust and the overall responsibility of a Zero Trust implementation is shared across all teams. Accountability stays with the leadership team, depending on who initiates the conversation.

- It is important to keep executive leaders as supporters and try to influence any detractors to become your supporters by understanding their requirements.

- Keep specialists on your go-to list. Utilize all segments of the human chain that will be affected by the Zero Trust strategy, including end users. Get feedback and improve the strategy before presenting it.

We have identified some of the key roles you will need to coordinate within your enterprise to create this virtual team.

[Glenn and the leaders have a conversation on understanding how certain maturity levels were ascertained, and they discussed in detail the dependencies on creation of the teams and potential challenges they would face getting other leaders on board. Subsequently the conversation continues.]

Mr. Smith: I think irrespective of our current state, this is very good work, Glenn. We have a more detailed and clear understanding of what we lack, and there is a lot of work for us to do with lot of people to speak to and convince. I will be pitching this idea and the metrics to specific people—especially Ali, who I know was on board with this initiative. As we gather the workforce to support this initiative, we will need Prolink Solutions' brief on our Zero Trust strategy and the important mechanisms everyone should be aware of—especially Jed's team. We need to be able to map out these gaps identified to mitigating solutions, which aligns with our strategic road map.

Glenn: That shouldn't be a problem. On the contrary, I am glad we had this discussion because the next step is to identify the right strategy that fits the business drivers and then identify critical security mechanisms. I will work with Jed's team and your CIO team to help you create your Zero Trust strategy.

Endnote

1. "DoD Enterprise DevSecOps Fundamentals," https://dodcio.defense.gov/Portals/0/Documents/Library/DoDEnterpriseDevSecOpsFundamentals.pdf

Develop the Target Zero Trust Architecture

Once the core team is identified, the next step is to understand the primary terminologies and concepts in Zero Trust architectures and how they overlay the existing networks. The following chapters are covered in this phase:

Chapter 8

Building a Zero Trust Architecture

After identifying the business drivers and Zero Trust strategy, crafting the right metrics, and assembling a strong Zero Trust team, the next step is to create a Zero Trust architecture for the enterprise to follow and to convert to policies, procedures, and products. To be able to propagate the Zero Trust vision, a blueprint is needed so that all subsequent initiatives follow a common design and implementation procedure. This blueprint must consider and include all the Zero Trust tenets. Zero Trust is an information security model with access control as a strong outcome, so strategies are not limited to a restrictive and fixed path. For example, some strategies may begin enhancement of object access with more stringent object-to-object control. An alternate strategy may focus on how the subject and object are separated, while other strategies may include augmenting subject context. Irrespective of the path being considered, a strategy and architecture blueprint is required as a reference for all policies and procedures to fall back to.

An important aspect of this chapter is to help the enterprise overlay the existing infrastructure with the target Zero Trust architecture. Zero Trust consists of architectural components such as policy decision points (PDPs), policy enforcement points (PEPs), and policy information points (PIPs); however, at an operational and deployment level, these capabilities are most likely existing mechanisms that are already performing some level of security control or information processing in your network. For example, the network most likely has an Authentication, Authorization and Accounting (AAA) server that performs identity and access control. Further, enabling multifactor authentication (MFA) or mobile device management (MDM) on this AAA server and automating and incorporating IAM practices into the AAA server aligns the security controls provided by this server along with Zero Trust strategy. Consider Zero Trust as a high-level security alignment of the existing architecture to the target architecture with some enhancements.

The best way to arrive at a viable strategy is to look at a typical enterprise architecture design and then implement or uplift it to the Zero Trust paradigm and explore how much change is needed or what additional security controls need to be incorporated to achieve that goal. Essentially, this chapter will help build some of the necessary background needed to craft a reference architecture or blueprint and an implementation road map for deployment.

A Typical Enterprise Architecture

Zero Trust has fairly detailed architecture documents via NIST 800-207, but as an executive leader or an architect, an effective way to take in a new paradigm is to place it as an overlay on the existing network. What is important to understand is that Zero Trust implementation does not mean a complete overhaul of the network, and it also doesn't mean that new hardware is always required. The Zero Trust access model augments existing infrastructure and is an overlay. Understanding the baseline infrastructure and overlaying it with the Zero Trust architecture is the strategy that this section covers. Consider a typical enterprise design diagram and map out some of the capabilities and architectural elements. Figure 8-1 illustrates what a moderately mature enterprise network looks like.

Most organizations have varied segments or routing domains in the network; however, the bare-minimum segments that exist across all maturities of networks, ranging from start-ups to mature enterprises, would be a server farm, a DMZ, and an Internet segment. This is, of course, a fairly basic level of segmentation, and more mature enterprises have multiple routing and security domains based on their functions. Macro-segmentation is an important aspect of Zero Trust, and enterprises that have considered this aspect will find that their Zero Trust journey is much more streamlined.

The architecture illustrated in Figure 8-1 is modular. There is a common routing entity called the *network core* that in the case of most enterprises will be a large throughput router or multifunctional switch. All segments route via this entity to each other, which makes this routing entity a common target for attacks. Most attackers will attempt to get access to this resource so that they can then route to any network without having to do additional port forwarding. The subsequent sections will list the other segments in the high-level architecture.

Untrusted Segments

The segments described in the sections that follow include the devices and services that are directly or indirectly connected to untrusted networks (namely, the Internet). In a pure Zero Trust environment, the existence of a semi-trusted or untrusted segment is strictly to segment traffic and services based on functional classification. They are not critical or mandatory to create access control. The purpose of this segment is to illustrate the contrast between traditional networking and security architectures and the Zero Trust architecture.

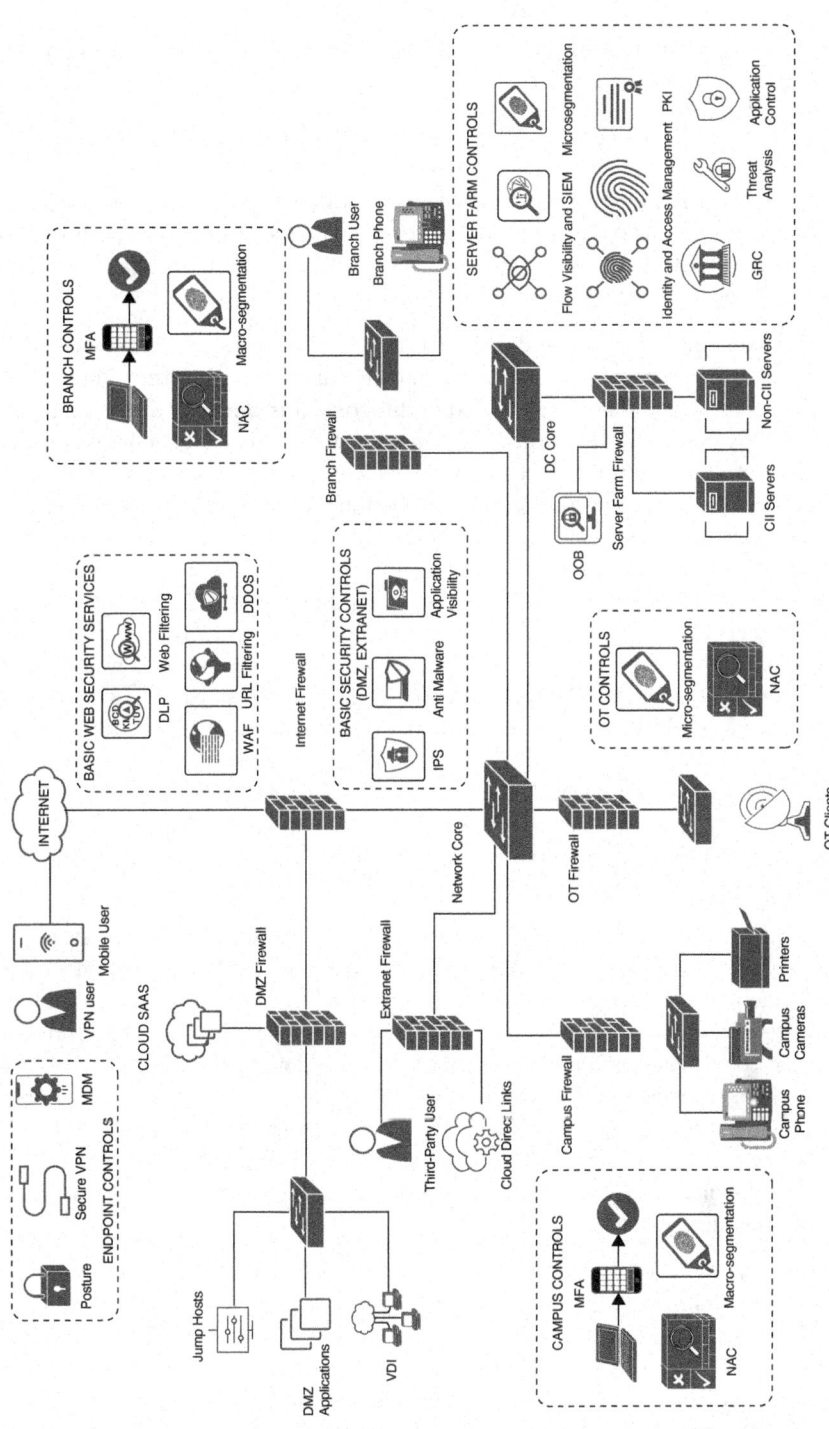

Figure 8-1   *A Typical Enterprise High-level Design*

Typical Untrusted and Semi-Trusted Functional Zones

The zone definitions are the logical divisions that are present in a typical enterprise network. In most enterprises, these are separate routing domains, but they might be virtual router forwarding (VRF) tables as well. The zones in a typical enterprise architecture are as follows:

- **Internet zone:** The Internet zone mostly has all Internet-facing network equipment (not applications) as well as network security capabilities that cater to protect from a wide range of attacks. This segment is the wider filter and is used to sift out close to 80% of all network attacks with the apt security controls.

- **DMZ zone:** This is a common demilitarized zone where Internet-facing web applications reside. Jump hosts, bastion hosts, and Virtual Desktop Infrastructure (VDI) are also some assets that can be found in this zone. It is generally a best practice followed by enterprises to harden their jump hosts well. The most common use case for the jump hosts in this zone is to provide secure access to the server farm applications. The concern here is the access control. Jump hosts are generally controlled by privileged access management (PAM), and once privileged users have access, they basically have full server farm access across a firewall because access control on the firewall is IP based. At a network level, jump hosts are given access to just the services needed to be accessed; however, it is not scalable to have one jump host per employee. When jump hosts are shared, the access opened on the firewall ends up being a subnet or a small IP range rather than specific hosts or users, which eventually becomes a problem as you cannot become more granular with the access you wish to provide for specific services. The onus of access control now shifts to the application developers who focus more on application validation than security control. VDI alleviates that concern to an extent, but granular context-based access is still not entirely achieved and is not commonly implemented. Chapter 4, "Always Start with 'Why,'" covered the details of the jump host flow in more detail.

  With a more mature organization or with organizations sending workloads to the cloud, direct connects and cloud integrations with the cloud service provider (CSP) may also be hosted in this zone. Some enterprises terminate the cloud-direct links to a separate cloud zone as well. The overall cloud entry point is a very enterprise-specific consideration. Some enterprises terminate cloud private links to the DMZ zone, though connecting the data center to a CSP over a direct connect link only extends the enterprise boundary and doesn't add security. The access control model in the cloud is very different from the on-premises model, and enterprises are usually keen to make sure that the access control both on the premises and on the cloud remain the same.

- **Extranet zone:** The extranet zone is a zone that most customers do not implement or tend to miss out on. Essentially, all third-party, vendor, and subsidiary connections or business-to-business (B2B) connections enter the network via the extranet. This is commonly not utilized and is combined with the DMZ segment. Some enterprises also terminate their cloud private links to this zone, which is usually more common than terminating the private links on the DMZ zone.

Typical Capabilities Deployed in Untrusted and Semi-Trusted Zones

In general, considering the architecture illustrated in Figure 8-1, most enterprises must have at least distributed denial-of-service (DDoS) protection, advanced threat protection, a web access firewall, URL filtering, and a web proxy. These are key security controls providing protection from untrusted networks for both inbound and outbound traffic.

An interesting perspective here is when the capabilities are explored in the sections that follow, you will notice that most of these capabilities are network focused. They access data in transit and rarely implement subject or object control. Hence, transferring some of these controls to the cloud is also an option, which is achieved with a Security Service Edge. A Security Service Edge solution does not store payload data, which makes it an attractive option for enterprises to save the costs of hardware and operational overhead.

DDoS Protection

Distributed denial-of-service attacks are some of the most common attacks an enterprise has to defend against. With the advent of malicious actors and hacktivists attacking various enterprises, DDoS is a key entry point into the network. One might think that DDoS is a denial-of-service attack that basically brings down the enterprise network and essentially blocks the attacker from accessing the network as well. This is mostly not the case, and most organizations have a fail-safe mechanism when it comes to certain aspects of the network. For example, enterprises might allow end users to pass through the network unauthenticated if the AAA server is down. All an actor needs to do is implement a DoS attack against the AAA server and they can access the network unauthenticated. DDoS protection prevents certain types of TCP and UDP packet–based DoS attacks, but modern attacks are mostly reactive and are not triggered by an access request. A DDoS mechanism is triggered when it perceives a certain level of packets being exceeded. This is manually influenced by an administrator.

Web Application Filtering with a Web Application Firewall (WAF)

Whereas a next-generation firewall (NGFW) can basically cover a large chunk of security controls, a WAF is a very specific functional firewall that caters to web traffic and L7 inspection. This is an important aspect of Internet security and is paramount in detecting and preventing web-based attacks such as SQL injection. One aspect to remember is that a WAF handles multiple types of users, including anonymous users on the web applications in the network. Presence of the WAF is a key security control protecting web-based applications.

URL Filtering and Security Intelligence

Most traffic facing the Internet is either inbound to services or outbound to the Internet. There are various reasons to police user traffic destined to the Internet, and one of the major reasons is acceptable usable policies. URL filtering can be used to block certain malicious URLs before a GET response is received. HTTPS traffic can be blocked based on the server name indication in the client hello packet. Most enterprises use this control to make sure users do not end up accessing fake websites or malicious URLs, which could expose the enterprise to web-based attacks.

Web Proxy

Although some enterprises do not use a dedicated device for web proxying, generally it is a good practice to keep all security functions separate. A web proxy is typically a good candidate to perform URL filtering, SSL decryption and deep-packet analysis on web traffic. Data loss prevention (DLP) is another security control used at the Internet layer, but it's not commonly used by a lot of enterprises unless there is a compliance requirement or unless the assets protected are extremely sensitive, such as credit card numbers or personal identifiable information (PII). At transitional segments (where traffic crosses security boundaries), additional capabilities are deployed, especially when traffic traverses the extranet and DMZ.

Intrusion Prevention

An intrusion prevention system (IPS) is a very effective security control when it comes to signature-based or anomaly-based detection. Multiple vendors provide strong IPS protection, and with updated signatures it is possible to quickly detect even zero-day threats over time. Of course, you cannot detect a zero-day on Day 0, but with the speed at which vendors research and release signatures, it is a good practice to place IPS capabilities at all transitional segments (Web to Application, Application to Database and so on). For example, if an extranet user wishes to access a DMZ service, they must undergo IPS inspection. Similarly, if the DMZ wishes to communicate to the server farm, it must undergo IPS inspection.

Although IPS is a strong inspection capability, it becomes a concern if a large amount of traffic starts passing through the IPS, consequently creating a bottleneck. Modern networks move most security capabilities to the cloud, and a user accessing the Internet has IPS capabilities in a Security Service Edge (SSE) and doesn't need to use an IPS locally or even physically. Capabilities like Cloud IPS are still being explored by enterprises, but until they are realized, most enterprises find that an on-premises IPS inspecting all traffic ends up overloading the device and impacting the user experience.

False positives are common with IPS as well, and enterprises struggle to identify legitimate traffic from the malicious or potentially unwanted traffic without the right visibility tools.

Malware Analysis and File Inspection

This control is specifically for controlling the blast radius and the spread of malware. Most malware will try to move laterally once it has a foothold in the environment. File inspection makes sure that any such file being transferred, even within a server farm, is inspected for malware. This is an added security control over the existing macro-segmentation and micro-segmentation strategy. Some enterprises do not enable Malware and File inspection in trusted environments, as it affects the processing power, but in general it is an important control to use and it's a best practice to enable it when protecting untrusted networks.

Application Visibility

With more complex attacks and sneaky attackers, it is now commonplace to disguise a malicious application as a legitimate one, especially as update files. Application visibility is a key capability that allows an enforcement entity to understand the true nature of the application and block it if it's potentially malicious. Application signatures are similar to IPS signatures and are statically defined based on behavioral research, which gets updated when there are changes in the behavior of the actual app. Like IPS rules, application visibility is also susceptible to false positives.

Trusted Segments

Whereas there is no implicitly trusted zone in Zero Trust, current enterprise architectures have zones that are generally considered trusted. From the context of a typical architecture, most segments are based on location and the service. Server farms are considered the most secure, followed by campus and branch segments. Most of the user traffic originates from the campus and branches and end at services hosted by server farms or private clouds. The concept being highlighted here is the *trust value* assigned to services just by being placed at a specific segment. An application in the server farm is still considered trusted instead of validating its posture continuously. Similarly, an authenticated employee is considered trusted when accessing services from within the campus simply because the traffic is from the campus. Headend OT segments have separate flows, and the server farm generally hosts all application services. Implicit location-based trust is prevalent in traditional networks and much more explicitly utilized when accessing server farm segments. The Zero Trust architecture, however, transforms location from a trust allocation mechanism to a trust attribute. This section explores the key implicit trust segments in a traditional network that Zero Trust transforms.

Typical Trusted Zones

The subsections that follow describe the commonly deployed trusted zones in an enterprise.

Campus

Often called HQ or headquarters, the campus is the zone where all users connect from. Users are most often authenticated via dot1x, and end machines are postured in some enterprises as well, though posturing is not a common practice. This is either because of concerns with an agent installation (which are also policy enforcement points) or due to the time taken to provide access, impacting overall user engagement. The Posturing agent must wait for the policy decision point (PDP) to respond before access can be provided, and if the requirements are complex, they are rarely used due to fear of delaying user access. With greater threats emerging, multiple enterprises are embracing MFA for authentication for all users, including the LAN/wireless campus users.

Branches

Branches are extensions of the main headquarters and usually process much less traffic than the actual head office. Most enterprises do not perform network access control for their branch offices. They do not segregate branches at a routing domain level and unfortunately do not provide direct Internet access to branches. Branches are considered extensions to the main data center, and a lot of resources are polled from the data center, forcing a lot of traffic to hairpin to the data center. This affects security control and impacts users. Enterprises, especially those with a large number of branches, accept the risk of a branch's controls and do not deploy the right security. They either extend all traffic to the data center or let traffic exist locally with minimal security controls.

Operational Technology Segment

Operational technology (OT) is often overlooked due to either the enterprise not catering to such devices or because the OT network, traditionally being a flat network, is isolated from the IT environment and is considered secure because it is physically isolated. With the advent of more efficient OT technologies, this is not the case, and there are designs for OT clients that need to connect to OT services on the server farm or the public cloud. OT clients might also be using shared services like DNS and DHCP from the server farm or DMZ.

Though not a common niche, many enterprises with a large OT presence have moved to adopt a common enterprise architecture where IT and OT networks coexist. Essentially, OT networks used to be specific to a Layer 2 domain, but that has changed drastically, and because the Internet is being shared between IT and OT segments, it is necessary to make sure the OT networks are protected in a similar method as IT networks. Most OT networks stay isolated from the IT network unless they need to use common shared services or if they need to access the Internet. Devices in this segment are not user-manned and therefore headend devices will need to be under a strong asset inventory management section and all headend devices must be profiled against a valid vendor product or a static list.

VPN

Although VPN is not an explicit zone per se, it is still a unique access mechanism with its own access control. The motive of the VPN is to make the subject appear to be on a trusted network. (Surprise! This shouldn't be the case as per Zero Trust.) Unfortunately, most enterprises consider VPN users as trusted users and provide full access to the entire segment without access control. A user from a VPN doesn't provide the right subject context without having a security posture agent configured on their endpoint.

A good trend observed in modern deployments is that most organizations integrate the VPN solution with an Active Directory or AAA server along with basic device posturing. This helps build the context and provide granular access to all VPN users.

Server Farm

The server farm is a common repository for all the services in the network. Most network infrastructure services are hosted in this segment. This is traditionally the crown jewel, where the enterprise's critical infrastructure as well as network-shared services like DNS, DHCP, and load balancers exist. Most of these entities have not entirely integrated with identity management, and there are enterprises that still do decentralized access management on application, where access control is performed by the local service rather than a central directory service.

Containers are a new form of workload where flows may generate as and when the container workload is up, since most containers are ephemeral. The IP that is used by the container might not be static, so if a decision point needs to really perform access control granularly, it is difficult to track IP, and hence control will have to be granted at a subnet level rather than at an IP (subject) level. Couple this with the general lack of flow visibility and you will have IT administrators groaning to create host specific FW rules.

A second aspect relevant to the Zero Trust environment is the presence of network access translation (NAT) or a proxy. With NAT in the picture, if static NAT is being performed, the original IP is lost and with it the context of the flow. Without the right visibility and flow monitoring tool to stitch the entire flow from various segments, this flow essentially gets interpreted as separate flows—one with the original IP and one with the NAT IP. If we are looking at a dynamic NAT scenario, it's even worse because Endpoints A and B will appear to traverse the network with the same IP address and randomized source ports; therefore, host-based access control or any form of granular control will not possible.

Typical Capabilities Deployed for Trusted Networks

The sections that follow describe capabilities that are expected to be present at most trusted segments. Unlike untrusted segments, controls are deployed based on the subject (users), network, or application (workload).

Typical Capabilities Deployed for Networks

Usual controls for networks involve visibility and contextual correlation of multiple log sources. Network segmentation remains the fundamental capability required for any network.

- **Flow and behavioral analysis:** Heuristics and behavioral analysis are used by enterprises to identify anomalies in the network. The primary purpose of these controls is to essentially provide visibility into not only anomalies but the legitimate traffic as well so that a baseline can be created to be fed into multiple other security solutions. Artificial intelligence (AI) greatly augments anomaly detection, which would otherwise have needed an initial manual definition of anomalies in the network. Enterprises are now seeing the value of having a solution that can identify and correlate flows to their identities. AI-based flow visibility is a rarity in today's network architectures, and in general flow generators and identity providers are not commonly integrated or do not have a middleware integration mechanism in place. This, however, will change with the adoption of Zero Trust, as Zero Trust is all about collaboration across multiple systems for more visibility. Though the main service is installed in the server farm, flow collectors collect data from all over the network to gather as much information as possible for the main flow processor to process.

■ **Network segmentation:** NGFWs are a good mechanism to perform macro-segmentation, but controls across a firewall will almost always be broad. They were helpful with a static perimeter; however, with a dynamically changing perimeter, the firewall falls short and can allow only broader access.

A decade ago, it was considered best practice to have multiple tiers of firewalls with different vendors. Today, it is an operational headache. Of course, various functions demand a different type of control. If you have a server accessing the Internet and you are performing application-based control and blocking YouTube at both your server farm firewall and the Internet firewall, both belonging to different vendors, it is likely that they do application detection differently. Hence, one firewall might block it and the other might not, leading to the age-old question of why one firewall is blocking and the other is not. What the enterprises lose is policy uniformity.

Typical Capabilities Deployed for Subjects (Users and Devices)

Security controls with respect to both subjects and objects become relevant because consideration now has shifted to service access and maintaining control to the services. Zero Trust is very identity-centric and follows the principle of least privilege The list that follows describes some of the common security capabilities implemented for subject control. It is important to note that the controls listed do not depict a hierarchy but rather merely serve as a list of controls that typical enterprises have in place.

■ **Identity and access management (IAM):** As the main source of user information, the IAM solution focuses on the subject aspect of the Zero Trust policy. IAM works in tandem with network access control (NAC). A policy server authenticates users with the help of an Active Directory and either uses a built-in NAC solution or a separate NAC solution to implement the access control policy. Because IAM is so widely used, most IAM solutions also provide NAC features. Unfortunately, a lot of enterprises do not have a truly mature IAM solution in place. The closest we see toward an identity solution is the integration of an IAM with Active Directory so that users can be authenticated. This of course is not nearly enough, and when it comes to complex AD and user hierarchies, the reluctance to move to a more segmented AD increases manyfold due to the general complexity involved. More mature enterprises utilize MFA and posturing, but the policies to provide access are broad and check only wider groups such as domain users rather than checking individual groups.

Another challenge enterprises face is the lack of integration of their IAM with the network infrastructure. It is very rare that we see Active Directory integrated with network devices like a firewall or a web proxy. Most access is IP-based. This takes away the visibility they gain about the identity associated with a flow, which is valuable when performing flow analysis because the identity of the source is interpreted differently at each siloed enforcement point.

Additional capabilities under IAM include user segmentation, multifactor authentication, and privileged access management. These are critical when it comes to understanding user-focused capabilities that greatly augment the security posture of trusted segments.

- **User segmentation:** Segmentation is key when classifying users and IoT devices. They are both different kinds of subjects as well as objects. Macro-segmentation is generally used to isolate vendors, guests, employees, as well as IoT devices. Within each macro-segment, smaller subsegments should be created based on how the Active Directory is segmented to provide more granular context-based control. All users should authenticate using dot1x or an equivalent mechanism, and it is important to maintain an asset inventory of personnel, corporate devices, and headend devices, not just the resources and services.

- **Multifactor authentication:** There is no need to detail multifactor authentication because any enterprise that has not deployed MFA is probably living under a rock. Multifactor authentication is the key to making sure users logging into the network are actually who they say they are. Depending on how and where users connect from, their authentication mechanism might differ. MFA with biometrics is also gaining traction when allowing users to the network, but the most common methodology still remains a password with a push notification or a one-time password (OTP). Products on the market today extend multifactor authentication to risk-based authentication as well to consider the overall risk of the user and the device from which they access resources.

- **Privilege access management (PAM):** PAM is another important aspect of identity management that handles all the privileged accounts and their access. User segmentation is a major enabler for PAM implementation, and almost all enterprises that have a working IAM solution have some form of privileged account management. The only challenge or concern is the maturity of that solution and its integration with IAM. It might be as manual as maintaining paper accounts or as automated as a developed CI/CD pipeline for onboarding administrative accounts, but PAM is mostly effective only if the context it uses and the access control it performs can be fed back into the IAM solution. User segmentation is also something that is not commonly implemented, as most Active Directory hierarchies are flat. PAM is also usually deployed for specific services only, based on the type of data they protect. For example, a bank may deploy PAM for a service that is processing PII but may choose to use standard AD authentication for other services, and this would depend on the architecture and the risk appetite.

- **Network access control and posture validation:** All users and devices connecting to the network are validated against an access control policy. Although this will become a key player in the Zero Trust Strategy later, most enterprises use broad policies to provide access. Posturing is an important factor to check for the presence of antimalware applications, disk encryption, and so on, and it's almost never implemented at the beginning of a network access control implementation. Enterprises may consider the time taken for posture checks and choose convenience over security, which may prove costly later. Network access policies are almost never granular enough, and rules are as broad as allowing domain users access to the entire network.

As mentioned earlier, posturing is not commonly used because of the time taken to allow access and general user experience, but it is gaining traction in modern networks due to the sophisticated nature of modern attacks and improvements in posture agent capabilities. Irrespective of the trust level, network access control and security posturing must be considered as key security controls for subjects.

Typical Capabilities Deployed for Application Services

Most protections in the application segment rely on service controls and application-level protections. The common capabilities utilized for applications and service objects are as follows:

- **Vulnerability and patch management:** The management of applications and their vulnerabilities is an important security control for applications. Without automated detection and mitigation of stale patches and vulnerabilities, there's a high chance that there might be dire consequences with large breaches occurring due to exploitation of an avoidable vulnerability.

- **Micro-segmentation:** It goes without saying that most enterprises haven't matured enough to embark on a micro-segmentation path. From conversations with multiple network leads and application architects, it is evident that the intent is clearly there but the application architecture is so old that moving to a micro-segmented path will need enterprises to convert their monolithic application infrastructure into micro-service architectures, and this will take a lot of time and effort, not only to design but to implement. Hence, we do not see a lot of micro-segmentation in most server farms. Server farm fabrics and applications are deployed as network-centric with a clear road map to application-centric paths, yet they do not see micro-segmented applications for a long time. Of course, with application dependency mapping or behavioral flow monitoring, it is possible to provide a mapping of the flows at least at a high level, which can be used to then segment the applications. This is usually the first step most enterprises take when embarking on a software-defined datacenter journey.

- **Process anomaly detection and runtime application security:** Process anomaly detection is a fairly new technique that is useful in detecting process anomalies in applications. For example, if data is being exfiltrated from the process via a specific port that the process is not supposed to be communicating over, then that anomaly is flagged and processed for information. Runtime application security is a similar anomaly detection technique, which is useful to detect runtime anomalies in applications.

A Zero Trust Architecture Overlay

This section explores overlaying the existing enterprise architecture with a Zero Trust model. There are no changes in the existing security controls. The strategy is merely augmenting and shaping the existing controls to better suit the Zero Trust model. The term "fine tuning" of the policies is more apt when describing the Zero Trust Network architecture, which makes sure access is more granular and controlled than it is in the existing scenario. A mature enterprise might find its Zero Trust journey a straightforward fine-tuning exercise. An enterprise starting off its corporate journey might find the transformation a herculean task, but the bottom line is that the security controls encompassed by the strategy need to be included, irrespective of the access framework being followed. The only difference is that with Zero Trust, it is more identity- and context-focused. If you think about it, this is something that is not new and has been in practice for quite some time but was never really adopted due to its relevance at the time. In modern security architectures, it makes more sense to incorporate Zero Trust strategy because you are not architecting designs with security and networks as separate silos but you are doing it as a combination of solutions. Zero Trust is all about making the security controls work for you rather than you as a security administrator working for them.

Before deep-diving into the extended Zero Trust architecture, it is time to use the NIST model for Zero Trust architectures. NIST 800-207 provides the diagram shown in Figure 8-2, which explains common Zero Trust architectures and some key terminology that will be utilized to understand the proposed Zero Trust enterprise architecture.

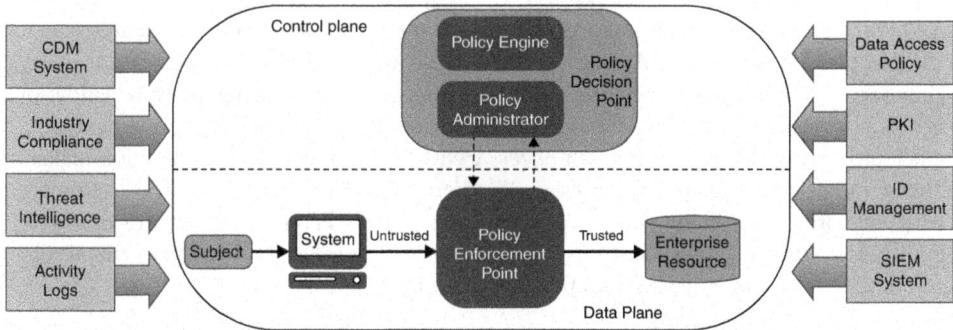

Figure 8-2   *NIST 800-207 Conceptual Zero Trust Architecture*[1]

By the core definition, one of the key outcomes of Zero Trust is to serve as an access model that restricts access of a subject over an object by ignoring any implicit trust in the middle and providing access based on context. Some terminology that will be commonplace when discussing Zero Trust architectures is presented in the sections that follow.

Subject

As has been briefly covered previously, the term "subject" covers the "who" part of the equation. When a subject is being referred to, the context is usually about any user, application, or even headend non-user device that can communicate over the enterprise network. This means your mobile phone, your laptop, the CCTV camera, and an IoT fridge are all subjects. Subjects are entities that consume an object. The key here is consumption, and therefore access to "consume" an object is a privilege and must be validated against policy. A subject is a consumer.

Object

The object is the "what" aspect of the equation. It covers what can be accessed and to what extent. For example, applications, headend devices such as printers, and even other users can be objects. For example, when users communicate over voice, it is usually an inter-user communication similar to how peer-to-peer communication works. Any resources, or "providers," or services will be considered resources and must also be part of the asset inventory of the enterprise. An object could literally be a text file on a server. It's any entity to which access can be controlled. The object "provides" controlled access to the subject based on certain attributes.

Policy Decision Point (PDP for Policy Engine and Policy Administration)

A PDP is a central or synchronized policy engine that decides if a subject has access to an object and, if it does, to what extent. It is usually a central NAC appliance or an Active Directory instance that can perform authentication, authorization, and posture validation. In modern networks, it is unusual to see a single decision-maker in the environment, so it is important to clarify that it is not necessary to have a single physical decision-maker. The common practice is to have multiple decision points that synchronize their verdicts with each other, thus maintaining overall end-to-end access control verdict. For example, if an Active Directory considers a user to be malicious, the AAA server syncs this verdict and blocks all access requests from that specific user.

Circling back to Zero Trust concepts, the assumption is always to expect that a breach has occurred already, so a subject asking for access to an object is considered a malicious request at the very start unless proven otherwise. Think of this as "guilty until proven innocent." That is where the PDP becomes critical. It is the central logical authority that gets a full viewpoint of the entire network's asset and flow requirements. The PDP will have full visibility of the network flows and will get feedback from application owners to create service policies so that they can be referenced when access to an object is requested.

A final aspect of the PDP is to be able to cater for independent configurations from multiple administrators to make changes to the same policy without affecting the integrity of the policies. Along the same lines, there must a receiving entity that receives change requests for the policy based on external factors like incidents, SOC playbooks, SOAR initiatives and so on. The Policy Administration aspect of the PDP is basically the owner of any changes to the policy, irrespective of whether it is manual via administrators or automated via playbooks. This separation of role on the PDP is necessary to make sure that there is clear segregation of information flow as well as accountability of configuration changes.

Policy Information Point (PIP)

A PIP is an entity in the network that gives information to the PDP so that the PDP can make more granular contextual decisions. Currently, if we had to take a medium enterprise segment, a PIP would be an Active Directory, which is the most common PIP for segments when considering users. Note that it is not necessary to separate PDP and PIP roles, and an Active Directory acts as both a PDP and a PIP. If an Active Directory is well segmented, it will give much-needed information to segment your access policies. Similarly, a unified SOC solution can be a PDP as well as a PIP where the information collected from the enterprise will drive actionable outcomes based on heuristics and automated playbooks. Another growing information point is a flow collector, which collects network flows to give more visibility and understanding of traffic in the network. Information security policy, as part of the governance initiative, is another information point that feeds into decision-makers like AAA servers and Active Directory instances to drive the access control for specific flows. Finally, threat intelligence is an important source of threats to the enterprise and feeds both to the PDP as well as the unified SOC solution.

Policy Enforcement Point (PEP)

A PEP is the laborer in the network that enforces the policy between the subject and object. In the NIST model, the only entity that actually controls traffic is the PEP. All the elements like governance, directory lookup, and so on are all PIPs, and these information sources influence the access control but do not directly control access between subjects and objects. Each PEP has the capability to control access but does not have the reasoning or the "brains" to do so, and hence the PEP must have frequent bidirectional communication with the PDP. When an access request comes in, the PEP must communicate to the PDP and ask for a verdict. It would be counterproductive if the actual communication to the PDP happens over a data plane; therefore, as per best practices according to the NIST document, communication from the PEP to the PDP must be secure and should be separate from the data plane. Specifically, they must communicate over the control plane. Loss of the data plane must not impact the capability of a PEP to query the PDP for an access verdict. The following are the three major types of PEPs:

- **Application-level enforcement points (A-PEPs):** A-PEPs are deployed close to each application or system and are focused on protecting the application or host system itself. They run specifically for the host they are on and provide the required contextual details to the PDP to be able to make a decision specific to access control relevant to the application. These enforcement points function at an OS level on the machine or are external to the machine like a PAM or an NGFW, but the common factor is that they serve a specific function for a specific application or system. In monolithic environments, this type of in-app enforcement is rare and generally relies on an external entity like an NGFW to perform these controls. A-PEPs provide for much more granular control based on the micro-segmentation strategy and are usually local to or around application systems with brief communication to the PDP.

■ **Network-level PEPs (N-PEPs):** These are the most common enforcement points and are likely already in the network. Firewalls, IPS devices, proxies, and so on are all part of network PEPs because they control access to a network and not to a specific application. These are the easiest to fit and are commonly the low-hanging fruit to be implemented first. They constitute coarse-grained control and are used to provide access control at OSI Layer 3 and OSI Layer 4, though modern security products provide much more context-based network enforcement.

■ **Endpoint-level PEPs (E-PEPs):** E-PEPs are usually agents installed on endpoints to provide contextual information on the security posture of the devices being used. Mostly these are either Zero Trust agents or posture agents that gather information from the endpoint itself and can block or allow access based on policies directed by the PDP. As with all PEPs, these agents must communicate with the PDP to understand if the security posture of the endpoint, along with other attributes and conditions, is enough to provide access to the desired object.

With these definitions in mind, a simple Zero Trust architecture is illustrated in Figure 8-3, which covers basic components in the Architecture with an added flavor of flow. Figure 8-4 showcases the entire Zero Trust Architecture that has been created to provide end-to-end coverage of all aspects of the Zero Trust strategy.

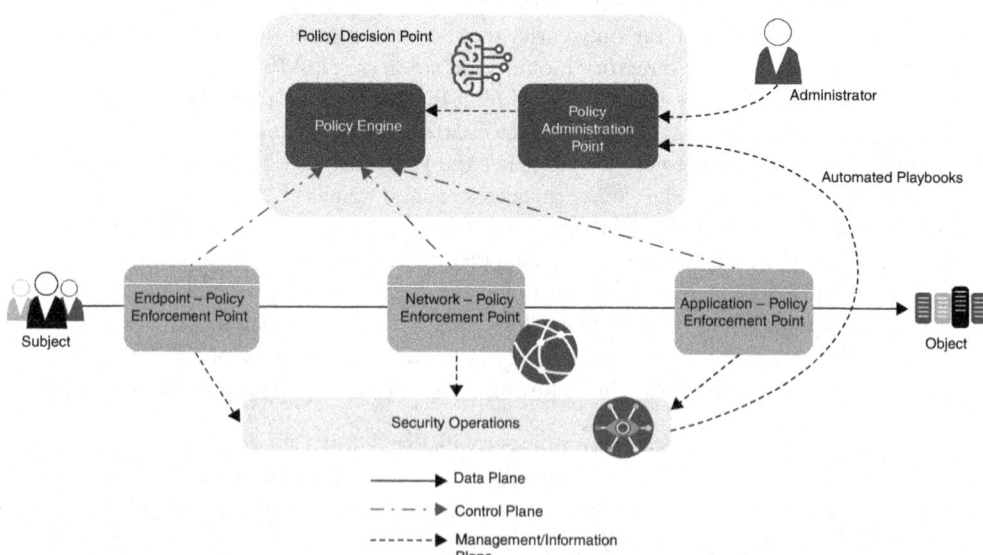

Figure 8-3 *A Simple Zero Trust Architecture depicting a user to application flow*

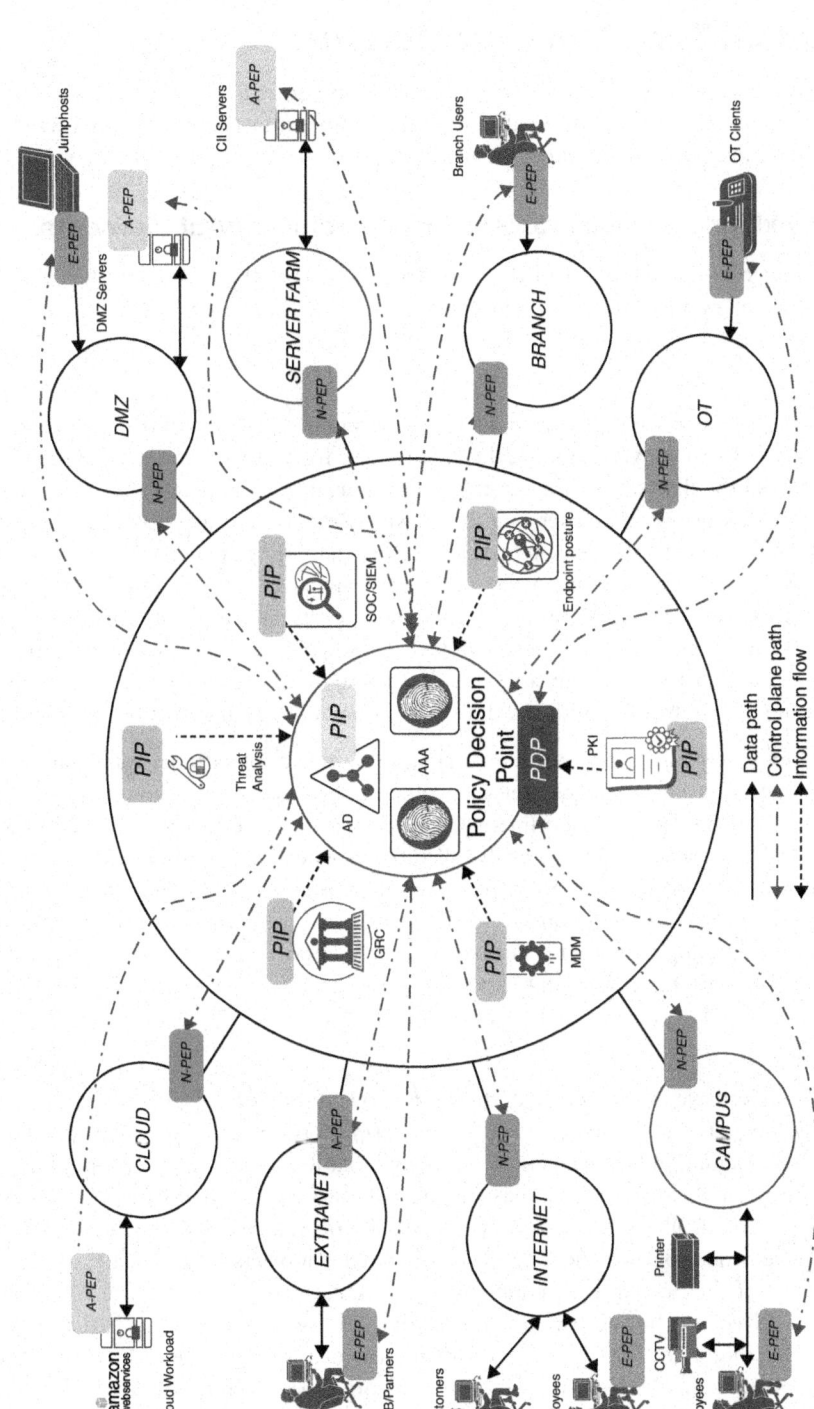

Figure 8-4 *A Conceptual End-to-End Enterprise Zero Trust Architecture*

Conceptual Zero Trust Architecture

There is a lot of ground to cover in the illustration in Figure 8-4, and each aspect will be explored in detail to help you visualize how they map out to the typical enterprise architecture discussed in the previous section.

Mapping Existing Entities to Zero Trust Architectural Elements

As a first step of making sense of the Zero Trust blueprint, the approach taken will be to map the Zero Trust architecture components to the existing architecture.

Subjects

One of the first steps to start crafting an architecture is to list out all the subjects or "consumers." Who are you building the architecture for and how do they get the right access? From the illustration in Figure 8-4, a campus user, a VPN user, and a branch user might very well be the same user, and that is what Zero Trust is all about. The same user must be provided the same access irrespective of where they are accessing the object or service from. Similarly, the same users when accessing a web service might have different access as compared to when they are accessing corporate applications. A third-party or B2B partner user is another unique subject. They will be part of the same Active Directory but a different organizational unit (OU) with different controls. Alternatively, there could be a federation with the third-party domain to facilitate extended trust.

Applications are subjects as well when they access other applications. For example, a web service accessing a database service would imply that the web service is a subject. Similarly, headend devices like printers, CCTVs, and operation technology (OT) endpoint clients like PLCs, sensors, and so on are all subjects when they request access to their respective servers, which may be in the server farm or in the cloud. By extension, any entity that needs access to an object or is a consumer of a service is a subject; therefore, when you're defining zero trust access policies, it is important to remember the basic definition of subject access at all times.

Objects

When subjects have been identified, the next step is to identify what is being accessed and how. Subjects can be objects too, and the term "applications" has wider implications. If micro-segmentation has been taken seriously, isolation would be incorporated even between web application and database application servers. Irrespective of the segmentation implemented, all applications or workloads are objects because they are providers of a service, while some of these applications may be subjects as well, depending on the context of the flow. If users communicate with each other (which in an enterprise constitutes voice traffic at a bare minimum), then all users are objects as well as subjects. Generally, it might get confusing to understand the concept with the lines between subject and object being blurred so easily, but one core aspect to remember is that a subject and object are specific to a flow. A user accessing a web service showcases the web service being an object, whereas a web service accessing another database service showcases the web service being a subject.

Network devices are also objects when accessed by subjects for management access. Remember that objects are any entities that "provide" a service and are accessed by subjects. In modern networks, workloads are likely unsegmented or broadly macro-segmented, and hence objects would likely be IP-based subnets rather than applications themselves. The motive of Zero Trust is to overlay the existing workload infrastructure and at the same time pave the way for a new application segmentation path so that objects can be as granular as possible (like a service group rather than an IP subnet).

Policy Enforcement Points (PEPs)

Every security control that was discussed in the existing enterprise architecture is already a PEP. The subtle difference here is that in a typical enterprise architecture, the decision point and the enforcement point are local to each device. For example, a firewall has rules that are manually created on it to be the decision point to block traffic. At the same time, an IPS has its own rules, and a web proxy has its own rules as well.

One of the best aspects of Zero Trust is the ready integrability with existing security controls, which reinforces the notion that Zero Trust doesn't need new technology. Note that the architecture diagram in Figure 8-4 does not visibly showcase network boundaries but treats them as additional add-ons to the actual subjects and objects. When an assessment is performed for the current Zero Trust readiness, the findings will almost always be that Zero Trust architecture is an overlay for the existing technology and can be easily achieved mostly with architectural uplifting performed on PEPs or a fine-tuning exercise for certain controls. Eventually, PEPs enforce what the PDP decides, and that is where the main brain is. Traditionally, PDPs reside in each enforcement point, and one main tenet in Zero Trust is to maintain policy uniformity with a centralized decision-maker or have the key decision-makers synchronize the verdicts. A practical example is the migration of edge firewalls to the Security Service Edge (SSE).

In the conceptual architecture illustration in Figure 8-4, each macrosegment is allocated a PEP and each PEP could very well be a combination of multiple security mechanisms. Each PEP constitutes a **protection policy**. It is important to understand the flexibility that uniform policy provides us. You maintain the policy at a higher level and selectively choose what enforcement each type of flow needs.

Network Enforcement Protection Policies

This section describes some of the common protection policies in place in modern networks.

- **Untrusted protection policy:** The untrusted or Internet protection policies are useful when placing protection for untrusted networks. Overall, the following are minimum protections needed when considering objects and subjects from untrusted entities:

  - DDoS protection

  - Network-based firewalling

  - IPS inspection

- Web application firewall

- Malware analysis with sandboxing

- Application visibility

Third-party entities, B2B, extranet flows, or even cloud workloads constitute untrusted flows and therefore need a specific protection policy. Depending on the type of flow, additional controls can be added. For example, for an outbound Internet flow, a web proxy should be designed as well.

- **Semi-trusted policies:** It is important to highlight that semi-trusted does not mean more implicit trust. The only difference here would be the use cases. For example, web service access, jump host access, or even in some cases VDI access and so on would constitute a semi-trusted profile. These entities will have communication from untrusted as well as trusted sources. Overall, the following are minimum protections needed when considering semi-trusted networks:

  - Network-based firewalling

  - IPS inspection

  - Malware analysis with sandboxing

  - Application visibility

Depending on the type of flow, more controls could be implemented. For example, in a DMZ network, a web service access might need an additional WAF to protect from web-based attacks. For applications or jump hosts, additional application protection may be implemented. VPN users should be considered as semi-trusted as well, and endpoint checks must be coupled with network checks.

- **Trusted policies:** Unlike the previous policies, most enterprises will have a policy where there is far less protection. There would be free flow, flat networking, or at a bare minimum VLANs and a couple of VRFs. This, however, is not enough to protect a trusted segment from lateral movement. Overall, the following are the minimum needed when protecting from trusted networks:

  - Network-based firewalling

  - IPS inspection

  - Malware analysis with sandboxing

  - Behavioral analysis

  - Heuristic IPS

  - Micro-segmentation

  - Application-specific protections (if any)

Trusted policies generally get overlooked in most enterprises due to implicit trust propagation.

Endpoint Enforcement Protection Policies

Similar to network enforcement, endpoint enforcement has various policies based on the type of segmentation implemented for the subjects and the use case in question:

■ **Untrusted subjects endpoint protection policies:** This usually constitutes end users the enterprise doesn't control. Usually customers, end users, and sometimes even other vendors are covered under this protection. The bare minimum posture that should be checked must be the presence of antimalware.

■ **Semi-trusted subjects endpoint protection policies:** This usually constitutes end users who are partners or vendors who need limited access to enterprise applications. For example, when a vendor needs access to certain devices in the network to configure them as part of a project, they might be given temporary access as a local user via a VPN or by adding them to a separate OU on the AD. The enterprise commonly enforces posture checks like drive encryption, minimum OS, and so on. Some checks include the presence of the following:

   ■ A minimum approved OS version

   ■ OS critical patches

   ■ Full Disk encryption

   ■ Antimalware

   ■ Presence of Zero Trust agents

■ **Trusted subjects endpoint protection policies:** These are usually for employees and other trusted subjects or high privilege accounts. These users must be part of the domain and must use certificate-based authentication. Basic checks for security posture include the presence of all of the following:

   ■ A minimum approved OS version

   ■ Full Disk encryption

   ■ Antimalware

   ■ Being domain joined

   ■ Automated patch updates

   ■ OS critical patches

   ■ Presence of Zero Trust agents

With this in mind, consider the use cases outlined in Table 8-1.

Table 8-1   *Example Use Cases, Subjects, and Objects Driving the Type of Protection Policy Needed*

Subject	Device	Object	Endpoint Protection Policy
Employee	Personal device	Company website	Untrusted policy
Employee	Corporate device	Company website	Trusted policy
Partner	Personal device	Company Website	Untrusted policy
Employee	Personal device	Employee self-service portal	Semi-trusted policy
Employee	Corporate device	Employee self-service portal	Trusted policy
Partner	Corporate device	Employee self-service portal	Deny
Employee from another branch	Corporate device	Employee self-service portal	Trusted policy
Partner	Partner device	Select network devices	Semi-trusted policy

Note that deciding the endpoint protection needed depends on how "trusted" the subject is and how critical the "object" is. This is a good point to introduce risk-based access control because a typical endpoint posture adds to the overall context of the subject. A well-postured endpoint reduces the overall risk of the subject. If the object is critical and the subject is not trusted (vendor, partner), then the endpoint protection deployed is more advanced to reduce the risk. If network protection policies were designed without considering endpoint protection policies, then an untrusted subject would trigger an increase in the protection desired, which will translate to more stringent network protection policies due to the high risk of the subject. In the case of an integrated network and endpoint context, the network rules can stay the same as long as the risk of the untrusted user can be reduced with stronger endpoint protection policies. Zero Trust is flexible because the dependency is on the asset, flow, and risk.

Application Enforcement Protection Policies

Application enforcement protection is unique. There are no untrusted applications. The overall idea of applications is that they are the core functionality of the enterprise. Workloads have a minimum segmentation of web, application, and database. With this context, web workloads are mostly considered semi-trusted workloads. Applications may be trusted or semi-trusted based on the criticality of the application. Web apps or customer-facing apps may have a lower trust than critical enterprise-specific apps. The protection policies are as follows:

- **Semi-trusted application protection policies:** In semi-trusted policies, workloads like web applications, web services, and so on do not process critical or sensitive workloads. They extract the data and pass it on to the application layer, hence protection is limited to the following:

- Application visibility

- Workload monitoring and vulnerability scanning

- Application firewalls

- Endpoint detection and response

- Micro-segmentation

- **Trusted application protection policies:** In all likelihood, close to 85% of the work-load in any enterprise would be trusted. It would be expected that trusted policies are more stringent specifically when databases are involved, but they are rarely so, and it is usually the opposite with trusted policies being very broad and not capable of blocking lateral movement. Hence, based on various functions of the applications, specific security controls may be additionally implemented. At a bare minimum, the following controls must be included:

  - Micro-segmentation

  - Application-dependency mapping

  - Process anomaly detection and forensics

  - Continuous vulnerability scanning

  - Process change and configuration management

  - Automated patch management

  - Mature DevSecOps (CI/CD pipeline) strategy and implementation

  - Runtime application security

If the workload involved databases, then database monitoring may be added as an additional control. This goes to show that each profile can be further improved based on the type of workload being protected and subject accessing it.

PEPs "enforce" and are the entities that implement the policy access defined by the PDP. At this point, the enterprise has identified the subjects, objects, and enforcement points, which means the enterprise has gained visibility into **who** accesses **what** as well as **how** the access is controlled.

Policy Decision Point (PDP)

A PDP is an entity that controls the access a subject has over an object. It is essentially the brains of the Zero Trust ecosystem. The PDP considers all the information available to it and builds a context relevant to the subject, and based on the crafted context, it dynamically decides a verdict regarding the access of a target object. The PDP collects information from multiple information points. Note that the PDP cannot reprocess the context if it was not asked for access reevaluation again, so it is imperative that there is a mechanism with which a PEP has continuous access-verdict querying built into its system so that it doesn't just accept a previous context. What this means is that your AAA server (a common PDP and PIP) wouldn't decide to check whether access is legitimate unless it can force the enforcement point to retrigger authentication. That is why there are timeouts for authentication and authorization sessions. The PEP must query the PDP each time the session times out to validate and confirm whether the context is still valid. The best approach is to use session based access that terminates the session once a flow is complete.

A PDP is a logical construct of a group of decision-makers making key decisions for access control. It would be counterproductive if this entity was as single device that processes the entire set of access requests. All that an attacker needs to do is to perform a DoS on this entity. Hence, in all practical sense, the logical PDP is usually a combination of several synchronized decision points. Here, "synchronized" is they key word because this is what keeps the policy uniform throughout the end-to-end use case. For example, an Active Directory is the most common source of truth when it comes to user information. When an Active Directory makes a decision to block a user, that attribute must be synced to all other decision-makers in the logical unit so that when the same request appears elsewhere, it is blocked accordingly. Another example is the SSE, where the CASB, DLP, NGFWaaS, and Zero Trust Network Access is combined into one enforcement point with its own decision point that is centralized and accessible from all edge segments.

If we had to compare this to an enterprise, consider the PDP as the board of directors. A decision is not made by just one member but is agreed upon by all members. If one member is not around or missing, the decision the others make is still relayed to make sure that the decision is unanimously implemented. A PDP works in a similar manner. It could be a cluster or a common policy server, but there will always be a common decision for a specific flow. Usually in existing architectures, AAA servers and Active Directory instances can provide the PDP services. With their increasing upliftment and capability enhancements, SOC solutions are also a good candidate to be a centralized PDP. PDPs are used to create the context of flows and provide verdicts to block or allow specific flows. Each application will consume this common context and build its own context based on local information, which it will then use to allow or block a user. In a mature Zero Trust architecture, local attributes and contextual information from workload is sent back to the PDP in a continuous feedback loop.

Policy Information Point (PIP)

The final aspect of the Zero Trust architecture is the policy information points, or PIPs, which are basically any source of information that can help add more context to a subject or an object. In an enterprise architecture, there are some common sources of information that add great value to the context of a flow, as described in the sections that follow.

Governance, Risk, and Compliance Systems

Governance refers to the processes, policies, and procedures that define how an organization operates and manages risk. Compliance refers to the adherence to industry regulations and standards that govern security practices. Governance and compliance are important components of a Zero Trust policy, as they help to ensure that the organization's security practices align with industry regulations and best practices. To implement a Zero Trust policy effectively, organizations must establish governance structures and compliance frameworks that align with their security objectives. Governance and compliance play a critical role in a Zero Trust policy by making sure their inputs feed into the overall strategy and policy implementation. These include the following:

- **Defining policies:** Governance establishes policies that define what needs to be considered for users, devices, and network traffic management and monitoring. These policies help to ensure that security practices align with the organization's security objectives. Strategies on various approaches for Zero Trust are greatly influenced by the overall user, device, or application policies.

- **Enforcing policies:** Compliance frameworks help to enforce the policies defined by governance. Policy enforcement usually dictates how a policy is enforced and with what security mechanisms. Compliance ensures that the organization adheres to industry standards and regulations. For example, PCI-DSS might mandate the presence of a WAF. This will feed into the Zero Trust strategy, which will include the WAF based on the specific type of flow and protection policy desired, such as untrusted or semi-trusted policies.

- **Auditing and monitoring:** Governance and compliance help to monitor and audit security practices to ensure they are effective and aligned with the organization's objectives and feed into the business drivers and overall vision of Zero Trust. It is also important to measure the extent of adoption of the governance policies across the enterprise to make sure that enforcement of policies is being followed across all segments and not just formally documented for compliance purposes.

- **Risk management:** Governance and compliance frameworks help to identify and manage risks associated with security practices, including the risk of data breaches and cyberattacks. Risk management activities build a risk profile for each application and subject, which then feeds into the overall Zero Trust policy, which involves crafting the subject's access to the object based on risk.

Overall, governance and compliance help to ensure that security practices are effective, aligned with industry standards, and properly managed and monitored.

Identity and Access Management Systems

Identity and access management (IAM) plays a critical role in Zero Trust policy, as it helps ensure that only authorized users and devices are granted access to services and data. IAM solutions provide the necessary capabilities to verify the user's identity and ensure they have the appropriate level of access to services. IAM is a valuable source of key attributes about subjects as well as objects.

IAM solutions help establish a Zero Trust policy by providing features such as multifactor authentication, identity federation, and access control policies. Multifactor authentication requires users to provide additional verification methods, such as a fingerprint or a one-time password, to access an object. Identity federation allows users to use their existing identity from another trusted source, such as a social media account, to access an object or service. IAM enriches all decision points with valuable user identity context. A typical enterprise will have a basic IAM in place and will be a constant source of user context information for multiple enforcement points.

Mobile Device Managers (MDMs)

Modern networks are no longer restrictive to devices. With more BYOD stories emerging and enterprises opening their networks to mobile phones and other personal devices, MDM is a key information point for decision-makers in the network. A common deployment setup is to have the AAA server (PDP) sync with the MDM server (PIP) so that it has full visibility into the unmanaged devices. Of course, this will also need a strong BYOD policy in place as part of the overall Zero Trust Governance capability.

User and Device Identity from Certificates

With password-based access becoming less secure for enterprises to adopt or retain, certificates for devices provide much needed information about overall context of the device or the user. For example, the location of the user might be Singapore but the access request might be from Japan. That might mean the user has moved or there is an endpoint proxy application being used. Worse, it could be that the certificate has been compromised and the access is indeed from Japan. Certificates and PKI in general provide much needed information to a PDP to make decisions for access. Authentication via methods like EAP-TLS are considered more secure than EAP-PEAP and MS-CHAPv2. In this case, even if the user is considered authenticated, if the device cannot produce an authentication for itself (typically a certificate), the user is not provided access. This access control is controlled by governance and compliance policies as well as risk appetite, and this is how all information points tie in with each other.

Threat Analysis

A key (almost overlooked) aspect of security in an organization is the threat analysis. Threat analysis helps Zero Trust security by identifying potential security threats and vulnerabilities in an organization's network and applications. By analyzing potential threats, Zero Trust security can create a more comprehensive security strategy that provides more robust protections.

When a specific attack vector or malicious subject has been identified, the end-to-end target flow is captured and fed into the Zero Trust policy, which adapts itself to block out such flows. As part of application and flow discovery, a threat modeling activity is commonly performed to identify threats and critical flows so that security controls can be identified accordingly. Threat analysis focuses on risk of exposure and the criticality of an asset. Therefore, when the list of threats are analyzed, the one with greater risk to the assets poses a larger threat, and this threat provides the starting point to generate a Zero Trust policy. Threat analysis also drives continuous monitoring due to the need to qualify the threats in real time. This enhances the Zero Trust capability and complements the existing flow discovery efforts.

Security Operations, Security Information, and Event Management and Flow Collectors

Generally, a security operations center (SOC) capability provides a single pane of glass from all sources of information. The SOC will have a security information and event management (SIEM) component that processes all security logs from various segments. Syslog and NetFlow are protocols used to send traffic from various sources. User logs, NetFlow logs, application logs, server logs, and so on are types of logs sent to a SIEM in a SOC. A common pitfall in SOC architecture is to have various architectural components retrieving and processing data but no component correlating the data. For example, user logs are sent to the SOC directly, and network flows are sent to the NOC. Security logs are sent to the SIEM, and application logs are sent to a specific log server, such as Elasticsearch, Logstash, Kibana (ELK). There is no correlation between these sources to get an end-to-end picture.

SOC solutions, SIEM systems, syslog, and flow collectors are all valuable sources of information for a decision point. SOC analysts process events and incidents and separate false positives from the actual incidents. This feeds into the PDP so that further access requests or flows from common false positive sources can be processed efficiently. Zero Trust aims to enhance the SOC by extracting data from all the PIP and processing it at a central PDP. A SOC is the centralized information aggregator that deduplicates repeated logs and builds a context that the PDP can utilize.

A PDP uses information about the user, the device, the network, and the object being accessed to determine a verdict for access control. This can be provided by SIEM, syslog, as well as flow collectors via NetFlow. The SIEM also provides real-time data monitoring information as well as behavioral information to the PDP. User and endpoint behavioral analysis (UEBA) is a key information source for the PDP to make context- and risk-based decisions. Usually, an enterprise already has a SIEM or at a bare minimum a syslog server, so the effective work to be done is centered around uplifting the SOC to a more dynamic proactive monitoring rather than the traditional reactive monitoring.

When viewed holistically, one will observe that most of these architectural components are usually already in place or are part of the common enterprise road map. Zero Trust does not always force new technology. Emerging technology like SSE, however, help to make the Zero Trust journey much more streamlined. Additionally, the Zero Trust journey will also drive new policy that is based on contextual information collected along with the relevant regulations, standards and governance policies. Historically, corporate endpoints have always been part of an enterprise architecture, but a Zero Trust architecture will make the enterprise consider extracting more information from an endpoint device to add more context to a flow. This will allow the decision-makers to select the right security controls for a flow without having to generalize the controls across the enterprise.

What Does a Zero Trust Policy Look Like?

Now that it has been established that Zero Trust is not about adding new technology, this section will focus on what a Zero Trust policy is really about and what it looks like. At a bare minimum, a Zero Trust policy changes the way we perceive access control. An employee should have the same access whether they access the network from the VPN, branch, or headquarters. Similarly, access to specific services that process critical information must be restricted based on role of the subject and not based on IP of the packet. Zero Trust Network Access can be broadly identified as Secure Access Service Edge (SASE) and Secure Private Application (SPA) Access.

Access control is no longer about source IP or destination IPs and ports. It is all about context. What is the first thought that comes to your mind when the topic of rules is brought up? If you thought firewall rules, then you're not alone. Firewall rules are the most common type of simplistic access control, which lists the basic five tuple:

- Source IP

- Destination IP

- Source port

- Destination port

- Protocol

They exercise the bare-minimum access control and cover almost 90% of the traffic. Firewall rules use you, the security administrator, as a decision-maker. You create the rules, and the firewall does the enforcement. If you observe the trend, you have been in Zero Trust mode from the beginning. You just did not focus on context and never really synced your rules across various security controls. That is what Zero Trust aims to enhance. Zero Trust wants to take away the decision-making load from the human administrator and wants to help provide the same power to specific devices in the network and then move away from five tuple or even nine tuple and create subject-to-object intent based on context.

From the original definition, Zero Trust is all about controlling access to an object. A subject requests access to an object; everything else is context that we build around that specific flow, based on attributes of subject, object, and the access method. This gives us enough information to decide on what a policy looks like.

Policy Components

A Zero Trust policy should consist of the components outlined in Table 8-2.

Table 8-2 *Policy Components in a Zero Trust Architecture*

Component	Definition
A subject	The source or consumer who is requesting for a service. This is usually checked against a subject database like an Active Directory or an asset inventory.
An object	The service or provider of the service that is being accessed, also validated against an Active Directory or an asset inventory.
Action	The activity that the subject wants to perform on the object. This is generally one of the five actions: **Create:** For example, create a configuration file, create a rule, and so on. A POST is another example. **Read:** A GET is a common example in the API world. Read basically means requesting for access to read a specific configuration and so on. **Update:** Make changes to existing configuration, policy, rules, and so on. Killing an application, for example, is updating its current state. **Delete:** Remove configuration, policy, and so on. One of the more dangerous actions would be to delete a specific configuration, users, or even security controls to allow access. **Exchange:** Whereas exchange is more of a form of read, it is a common type of information flow, especially when it comes to APIs. Application actions are key as well when it comes to considering actions for the verdict.
Characteristics	This constitutes the information that has been collected regarding the subject and the access requirements for the object. Characteristics or attributes are the core of creating a context because context is what allows the PDP to select specific verdicts.
Conditions	Conditions are specific arrangements of attributes in a unique format to achieve a specific outcome. For example, department is a characteristic. The department being HR is a condition.
Verdict	After evaluating a set of conditions, the policy engine sets a verdict. It is usually a simple allow or deny, but modern systems allow a monitor option as well.

Consider the following example to understand these components:

> *A web service admin accessing a database service to add a new web admin.*

When we dissect this activity and identify the components of the policy and how various aspects of the architecture fit in, the overall concept of a Zero Trust policy becomes evident, as Table 8-3 illustrates.

Table 8-3   *An Example Policy for Zero Trust*

Component	Explanation
Subject	Web service admin (user: granola).
Object	Database service (y.y.y.2) (MAC: aa:bb:cc:dd:ee:ff).
Characteristics	**Posture:** Compliant.
	Endpoint: Company owned.
	User: Authenticated with MFA.
	No anomalous traffic toward the DB server cluster.
	No sudden increase in SSH traffic or other common exfiltration mechanisms.
	Location: Over VPN.
Action	Change and edit the database table to add a web admin account (update).
Condition	A database service can be accessed only by a database administrator and only while on campus. It cannot be accessed over VPN or from a branch, even by a database administrator.
Verdict	Denied.

This is a very simple example, but it shows how versatile policy creation can be and how granular it can be made based on available information. It also showcases how important it is to have accurate policy information points. Without a full view of the subject and object requirements and a centralized PDP, enforcing an accurate policy is very tough. The central PDP is the easiest way to be able to deterministically create policies because all information can be fed to the PDP to process and create an accurate context. If the PDP is logically distributed, broken into multiple PDPs, or just simply local to every enforcement entity, then it is possible that some PDPs might not have the full view of the flow and hence will impede you from creating an accurate end-to-end policy without compromising identity. An SSE is the best example to explain the challenge.

Referring to the illustration in Figure 8-1, if a VPN user has to access a service in the server farm, the user has to be controlled by the Internet firewall, the server farm firewall, possibly a DMZ firewall, and many other security controls, which usually are isolated from each other. If the identity context is extrapolated at all of these enforcement points and the access control provided is the same, you have essentially achieved uniform policy. Couple this with continuous logging from various sources with some correlation happening at the SOC, and Zero Trust policies become easier to visualize. The entire exercise is cumbersome and in all practicality very difficult to implement, however. Instead of multiple firewalls, the SSE allows you to combine various security services into one platform that is also the decision-maker. Hence, there is one logical decision-maker and one enforcement point that caters to all the requirements of a Zero Trust flow.

The high-level trust policy must be deployable on all the enforcement points in the end-to-end flow. If an edge firewall doesn't have policies for a flow where a VPN user is accessing a database service, it would essentially allow or deny the flow based on IP access lists, and this flow would be denied only at the database service. If a policy was then created to block VPN access to the database service, it could easily be circumnavigated by accessing the web service and subsequently creating a session from the web service to the database service. This is a key factor where, irrespective of where you access the service from, a web admin must not be able to access the database. Access control must be uniform across all enforcement points.

It is important to highlight that Zero Trust policies help augment and enhance coarse-grained policies that were originally at the network layer. They do help applications control access, but, in general, finer decision-making is preferred to be retained locally on the application. As long as the Zero Trust policy can be propagated and implemented at the application layer, fine-grained policies will likely be implemented at the application level with constant communication with the PDP.

Information Flow in the Zero Trust Architecture

In a Zero Trust architecture, there are no silos. Everyone communicates with everyone else. Note the various communication flows in the diagram shown previously in Figure 8-4. An important set of communications toward the PDP are the **control communications** between the PEP and PDP. All PEPs must be able to communicate with the PDP to be able to get real-time contextual updates for access requests. This either can be triggered by time-based synchronizations or can be triggered by the endpoint reauthenticating.

The solid lines in Figure 8-4 connect all the PEPs. That is the data plane, and that is the actual communication between the subject and object. Note that these data flows are separate from the control path.

There is another important flow—the communication between the PIPs and the PDP. This might seem to be an inconsequential flow, but it is equally important when it comes to the Zero Trust architecture. Without the right information or context about the user, service and the data being served, you might as well be creating just a plain and simple firewall rule. Zero Trust policies are much more versatile and, if used correctly, can be utilized to create extremely granular access.

Basic Flows in a Zero Trust System

Considering the varied methodologies of how Zero Trust systems are implemented in the market, it is important to start somewhere when it comes to identifying the types of flows in the network and then correlating them to a strategy for access control. This is especially essential when it comes to enterprises that do not know where to start flow processing and application visibility. Most enterprises struggle to visualize the flows in their own network and are rarely able to get a full picture. As part of strategy building, the logical starting point is to discover subjects, objects, and flows.

Zero Trust greatly revolves around the idea of a subject and an object. The subject is likely either a user on a machine or a service. The object is almost always a service, except when we're considering P2P communication. Therefore, if we reflect retrospectively once again, Zero Trust is simply access control for a subject over an object. Considering the subject-to-object communication at a broad level, we have four types of flows, and each flow has a different type of protection. This will help provide enterprises with a good starting point to identify their end-to-end flows. The types of flows are as follows:

- User-to-user

- User-to-service

- User- or service-to-internet

- Service-to-service

All of these flows have PEPs, PDPs, and PIPs as part of the overall architecture, but the difference will be where the PEPs are placed and how the PDPs will make decisions for specific traffic. When diving deep into each flow, consider the proposed Zero Trust architecture discussed in the previous section and explore each flow and its deployment model with the architecture in mind.

The flows, however, are very generic and are purposed to serve as a starting reference for enterprises. As more flow discovery takes place, granular flows like voice communication, shared services communication, monitoring communication, and so on are identified to support the larger business flow discovery exercise. Cisco offers a Secure Architecture For Everyone (SAFE) architecture template that provides a guideline for all aspects of enterprise design, ranging from campus to data center, as well as Zero Trust architecture. The SAFE architecture provides key business flows that can be utilized as a reference point for additional flows relevant to the Zero Trust architecture. A key value-add of the architecture framework itself is the consideration of threats and required security controls for a flow rather than defining security controls specific to a location. Once the theoretical flows are clear, a practical discovery maps out the intents and existing flows. Consultants make sure the flows are aligned before a Zero Trust strategy can be formulated.

User-to-User Flows (Nonstandard Object Flows)

Note that the term subject-to-subject is not used but rather the flow is classified as user-to-user. The reason is because in a flow there will always be a subject and an object. There will be a provider and a consumer. Flows in this category are generally user-to-user flows and will likely not have any PEP to protect the flows from some of the lateral movement threats. These flows are usually between subjects and objects sitting behind a network PEP like a firewall. A classic example is user-to-user communication within a campus or Branch. Users are usually placed in the same VLAN and by default can communicate with one another. It is of late that wired and wireless users are designed to be placed on separate VLANs to avoid unfettered communication. These are actually fully allowed flows that ideally must be blocked or only certain flows must be allowed across.

In Software-Defined Access (SDA) networks, technology like TrustSec is used to control inter user traffic at a network level.

User-to-user flows are critical flows that may be allowed by default due to implicit trust. Two users in the same corporate VLAN might seem harmless, but as we have explored multiple times, implicit trust can lead to avoidable breaches with larger scope to perform lateral movement. Figure 8-5 illustrates what a typical flow would look like.

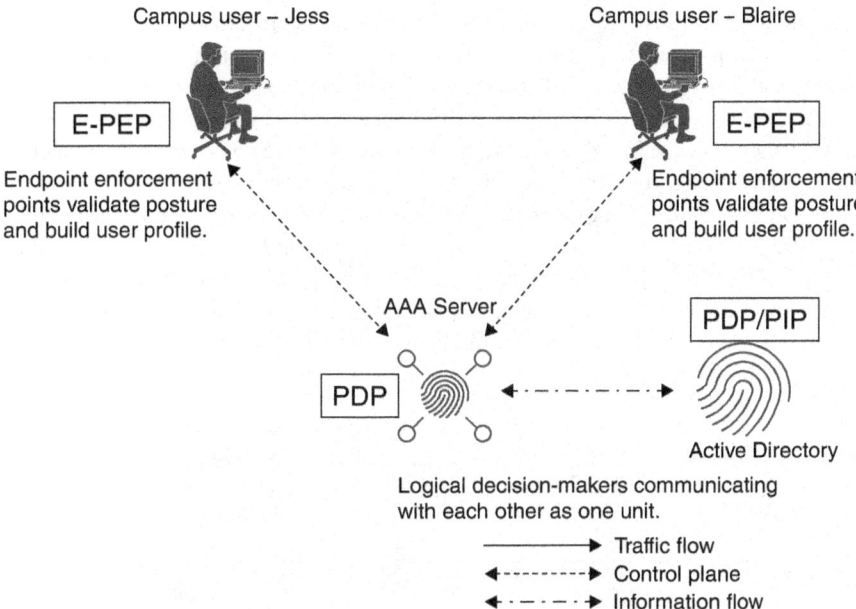

Figure 8-5 *Subject-Based Zero Trust Access*

In this type of flow, the access control is specific to the endpoint, which means essentially the PEP is at the endpoint. This usually translates as a posture agent for the end-user machines or as a Zero Trust agent installed on both end-user devices. There are advantages to this kind of connectivity, the main one being the capability to be able to granularly control flows to a much deeper grassroots level. You can create groups and allocate resources into those groups and control traffic, granularly specifying which ports need to be opened across which groups. The disadvantage of this methodology would be the scale of endpoint connectors that would need to be connected or deployed.

The implicit trust boundary in this case still is the actual service object, and enforcement points can be extended to cover controls such as end host firewalls, host NGIPS, and so on. The key factor here is that the PEP exists at the subject and object endpoints rather than at a network level, so most endpoint PEPs come under this type of flow. Lateral movement within a "trusted" segment is a common threat in this type of flow.

User-to-Service Flows (Common Macro Enclaves)

User-to-service flows are the most common type of flows in the network, as these are the common infrastructure services visible in the network. In all practicality, around 85% of all traffic in an enterprise are user-to-service flows. An end-to-end application access may include several user-to-service flows. A user from untrusted segments like the Internet may access the web service, and subsequently the user's identity may be utilized to access an application in the server farm. Flows would also include users accessing public SaaS. The crux of these flows is that the subject is an identifiable user who can be authenticated and authorized by an identity provider, and the object is a service that may be physical, virtual, or containerized. This is slightly different from user-to-user communication because, in this case, we have PEPs protecting a service macro-segment rather than specific service objects. The PEPs are most the commonly used network PEPs such as network firewalls. The end-to-end flow will use a hybrid combination of the user-to-user and user-to-service flows, but essentially PEPs for a user remain tied to the user and the endpoint. The service PEPs, however, are usually server farm firewalls and/or a combination of NGFW, IPS, and malware inspection. In the grander scheme of things, a service will essentially be protected by a combination of its application PEP (service-to-service model) and an NGFW at the server farm boundaries (user-to-service model). Figure 8-6 provides a visualization of the flow.

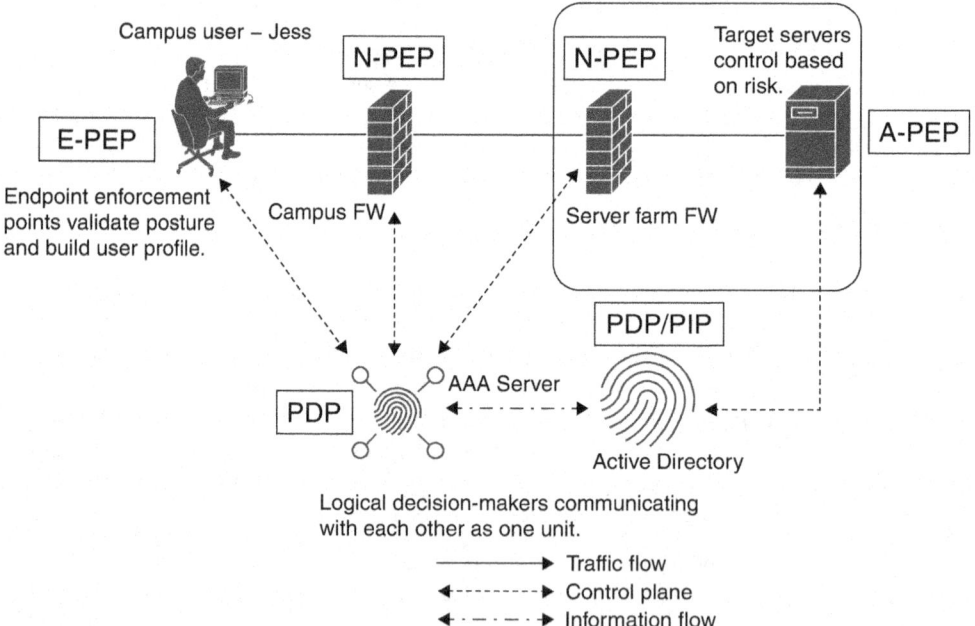

Figure 8-6 *User-to-Service Flow Model for Zero Trust Flows*

The network PEPs in this model protect a larger segment than a smaller host scope. This is advantageous because if the workload has dynamic IP addressing (for example, containers), the policy does not need to change. The disadvantage of course is the broader scope of policies rather than specific subject-to-object flows. Hence, it is always better to have a hybrid deployment with a combination of any of the other models.

User-to-service flows are also useful in grouping legacy devices together to achieve the right security posture. Legacy devices might not fully support modern microsegmentation, so grouping them into a larger enclave helps support the overall Zero Trust strategy. Creating a legacy segment, however, creates implicit trust within the segment and therefore is not desirable in the long run.

The Internet is generally not preferred for user-to-service communication because this means sending all internal traffic to the cloud and then sending it back on the premises. However, as part of the modern Secure Access Service Edge Architecture, the SSE component integrates with the SD-WAN such that when user-to-service traffic traverses the SD-WAN, it is inspected by the SSE. The SSE, being a modern security construct, encompasses a wide variety of security services, including DNS protection, Firewall as a Service, secure Web gateway, cloud access security broker, and identity mapping. The endpoint connects to the SSE via a secure channel (either via its own agent or via a network gateway, which could be an SD-WAN router or an NGFW). This is the more common use case that will be encountered in the current network and security landscape, where user-to-service traffic is sent to the SSE for inspection via the SD-WAN or any edge device capable of sending encrypted traffic to the SSE. Over time all the SD-WAN traffic will also get inspected by the SSE thus completing the SASE architecture.

The other, less common alternative is to allow for secure private access of applications where both applications and users are onboarded to the SSE. The endpoints have agent-based deployments where services in the server farm are exposed to the Internet without a public IP using application connectors and strong encryption capabilities. That specific flow involves an end user (hence an endpoint PEP) communicating with a service in the server farm (with an application PEP), which makes the overall secure private access flow a user-to-service flow, with the Internet as a transport medium, and does *not* get covered under an outbound-internet-specific flow that is covered in subsequent sections.

As an example of this concept, in Figure 8-6, the N-PEP in this case is the on-premises campus firewall. This firewall will transition to the SSE in the future, which is also an N-PEP. The traffic is inspected and controlled at the SSE. If the traffic is allowed and is destined to the on-premises network, then the traffic is redirected to the on-premises network also via a secure channel. Usually, the transport for endpoint Zero Trust access clients is DTLS and over the Internet, thus removing the dependency on technologies like IPsec and SSL VPN on endpoints. If gateways are used to access the SSE, however, IPsec may be utilized to connect securely to the cloud. Cisco's Security Service Edge capability (as part of Cisco Secure Access) along with many other vendors like Zscaler (Zero Trust Exchange) provides this form of access control and visibility.

User- or Service-to-Internet (Common Outbound Internet Flows)

The third use case is when a user or service accesses the Internet. Here, the relevance for protection is toward the subject and not the Internet. When services and users access Internet, the PEPs are usually on-premises Internet firewalls, but this can be extended to the cloud, and essentially a cloud firewall can control access to the Internet as well. In modern-day networks, both options are possible, but with the advent of Secure Access Service Edge (SASE), on-premises access control is no longer desired. With the scalability of the cloud, Security as a Service and Firewall as a Service is the direction the security community is moving toward.

Most Zero Trust systems provide cloud capabilities to perform secure Internet gateway controls, and as an outbound-internet use case, secure internet access is something most customers expect. Traffic moving toward the Internet traverses a PEP entry point into the cloud, and once inspection is done on the cloud, it is sent out to the final destination on the Internet. Protection is for the consumer and not the provider. The advantage of this option is the common acceptable usable policy (AUP), which can be enforced across Internet access from all locations. For example, a VPN user, a branch user, and a campus user all get enforced with a uniform policy. Currently, enterprises have a situation where branches are deployed with SD-WAN technology and direct Internet access, and branch users get subjected to cloud-based policy, but when the same user accesses the Internet from the campus, the traffic passes through an on-premises firewall with a different implementation of the policy. This breaks the basic tenet of Zero Trust, which is to maintain a common security policy. There is no need to look at having multiple firewalls and multiple tiers/vendors of firewalls for Internet connectivity when all Internet control can be performed on the secure Internet gateway on the cloud. The key attribute of this flow is that the service being retrieved is on the Internet, so it's not in the jurisdiction of the enterprise. Figure 8-7 provides a visualization of the flow.

The immediate advantage of this type of deployment is the operational simplicity and reduction in total cost of ownership, especially with hardware like firewalls and proxies. These flows are the first flows that most enterprises will utilize to transform to the Zero Trust model.

Service-to-Service Flows (Micro-Segmented Flows)

Finally, the last type of flow modeling is service-to-service communication—the key here being the fact that services have reduced identity, which means they cannot do MFA and greatly rely on certificates to authenticate. Similarly, application server endpoints may not support the installation of a PEP on the server itself, so PEPs get implemented just "in front of" the objects as either gateways (like NGFW or NSX) or application connectors (to the SSE). These usually translate to inter-service access control with the help of micro-segmentation. An example is VMware NSX.

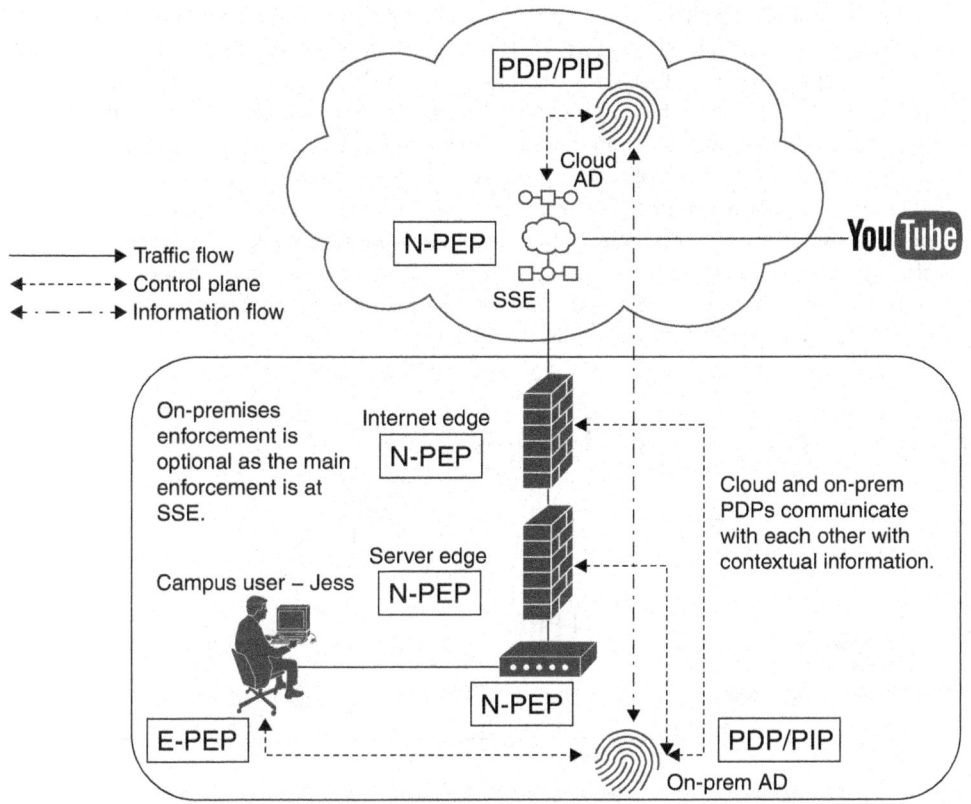

Figure 8-7 *User- or Service-to-Internet Flow Model for Zero Trust*

In the network considerations explored in the "User-to-Service Flow" section, a network PEP sits overarching the entire server farm architecture or at least in one specific routing domain (like a VRF). In contrast, the service-to-service flow specifically deals with workload control, where access control is not at a larger segment level but at a smaller function level. If a set of services need to communicate with each other on common ports like DHCP and DNS, then they get placed into a consumer and provider group and access is controlled between them granularly. This means, similar to IPS rules, there will be a learning period where these flows will not be enforced, and once administrators monitor and validate the flows, they will enable the access control to be enforcing rather than monitoring. With constant monitoring of flows, these rules may be changed and enhanced, which reflects the core tenet of Zero Trust. Specifically, dynamic policy adaptation based on the changing nature of traffic flows via flow visibility is the key highlight of this type of flow.

The service-to-service flow model has a PEP for each function rather than a specific service object. For example, a database cluster will be a specific function that has multiple service objects in the same VLAN or segment, with network PEPs protecting the cluster. Applications could be in a separate segment, and communication to the database segment would need specific controls that are enforced either at the service (object) level or at the macro-segment level (network). The key here is to avoid implementing too broad of policies at macro-level flows or too granular of host-based policies at the workload level. Micro-segmentation-based models focus on control based on function and will fit in well in a service-oriented architecture environment, where micro-services are the norm. Following the same hybrid approach, application PEPs enforce granular decisions on applications, and network PEPs control more coarse-grained access across various workloads. Figure 8-8 showcases applications communicating with each other across the SSE. This is not a common flow and is unlikely to be adopted into modern network and security architecture; however, the aim of this book is to look forward into the future where SECaaS will control all access centrally.

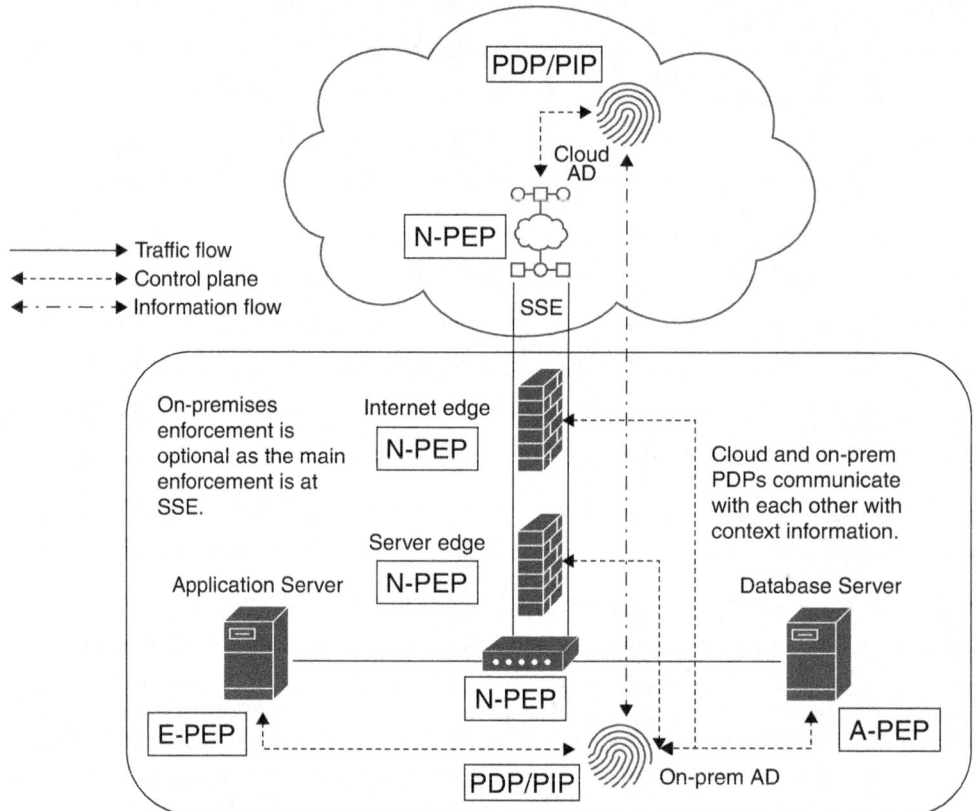

Figure 8-8 *Service-to-Service Flows Across SSE*

In all practicality, enterprises will not be sending any internal communication or specifically east-west communication to the cloud. Hence, the local control will be on the premises with a firewall or other security mechanisms like private VLANs or Cisco TrustSec. Figure 8-9 illustrates one such example where east-west control is via TrustSec.

When access segments cannot provide network layer protection, mechanisms like TrustSec that are derived from strong segmented group-based policies can be leveraged to restrict traffic across segments.

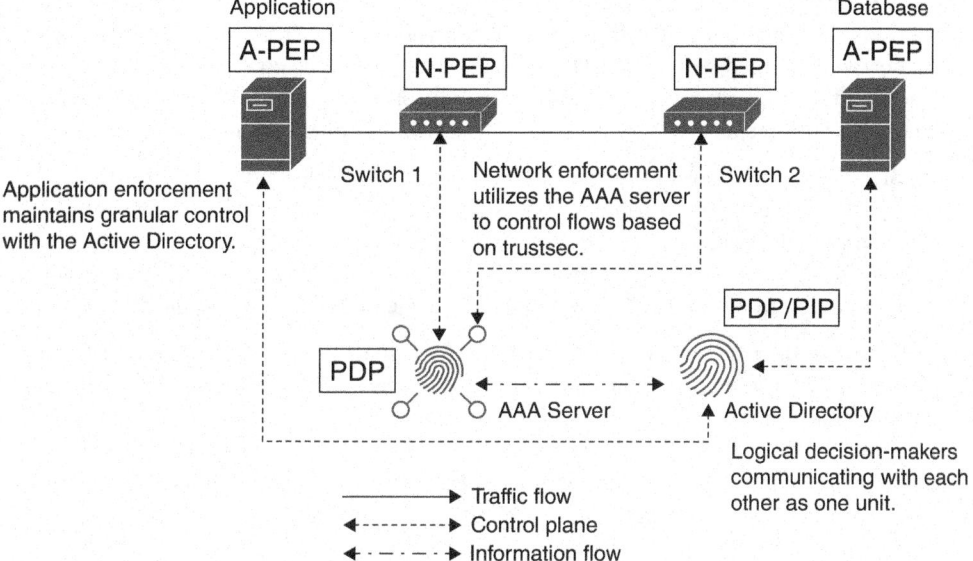

Figure 8-9 *Service-to-Service Flows on the Premises Protected by TrustSec*

In the service-to-service model, only specific similar functions of services can communicate to each other. Different functions and systems communicate via the network PEP. This is essentially facilitated via micro-segmentation and its implementation. The cons of microflows remain the same as user-to-user flows, wherein the smaller you make each functional honeycomb, the less scalable it becomes. The advantage, though, is the hybrid approach that utilizes architecture for both user-to-service flow models as well as service-to-service flows. For example, in Figure 8-9, you can observe that applications communicating to databases may be controlled via NSX or any other Layer 2 protection mechanisms like Cisco TrustSec (service to service via L2 control). As long as the applications and database can be isolated at an asset categorization level, control can be implemented. If an application needs to communicate to web services, it gets controlled by multiple N-PEPs, including the overarching NGFW, which is a common practice to have in a server farm as the server farm firewall (service-to-service via gateways), thus keeping the approach hybrid. The enforcement points are deployed specifically for the purpose of the function. Managing requirements and creating policies for these functions becomes easier to operate when planned well in advance with robust asset inventory, data classification and micro-segmentation.

Software-Defined Perimeter

As part of understanding the high-level Zero Trust architecture and its components, it is important to understand how the Zero Trust architecture is most practically implemented into a Security Service Edge solution. The principles implemented may also be utilized in a traditional environment with the right combination of security controls and policies. The software-defined perimeter (SDP) is an extremely valuable concept that will help understand the underlying concepts of the various products and solutions that exist in the market with respect to Zero Trust network architecture and secure access.

The SDP framework, developed by the Cloud Security Alliance (CSA), is a security concept that focuses on securing devices and resources in modern IT environments, particularly in the context of cloud computing and remote work scenarios. SDP is designed to enhance security by implementing a dynamic and identity-based approach to network access, moving away from traditional perimeter-based security models.

SDP does not draw a parallel to NIST SP 800-207. Instead, it derives inspiration from it. Consider SDP as an implementation plan of the architecture that NIST 800-207 provides. The advantages and field implementations of SDP are beyond the topic of discussion in this book; however, we will cover some aspects of the concept, including its tenets, the similarity to Zero Trust tenets, the overall deployment models that SDP provides, and how it aligns to the NIST framework.

Tenets of SDP

Similar to Zero Trust as an access model, SDP has the following tenets:

- **Assume nothing and always confirm the identity:** This maps to the assumption of implicit trust that is across most of the enterprise networks today. This also aligns with removing or reducing any implicit trust in the network.

- **Trust no one or nothing:** Always request for constant authentication. Do not trust once authenticated. This maps with the Zero Trust tenet of avoiding trust once authenticated.

- **Validate everything:** Constantly monitor for change in security posture and automate remediation actions.

These align with Zero Trust tenets where the primary principle is to assume a breach and *not* assume trust or identity knowledge. Verify and validate the identity and access at each request and not just at the perimeter.

SDP Principles and Methodologies

SDP removes implicit trust by implementing a deny-all firewall rule. Though it sounds more like a simple and commonly used network-level concept, the core of the concept is that a user cannot have access to an application that is not explicitly defined in policy. The difference here is the extension of an IP segment to an entire application context.

Rather than allowing IP segments, the solution allows users to applications. SDP validates dynamically the endpoint and makes sure that the endpoint can access only the application it is allowed to, thus maintaining a uniform policy irrespective of where the user is accessing the resource from.

SDP is independent of the communication medium. That means irrespective of whether workload is in the cloud or on the premises, the connection is what drives the security. There is an IP-based connection from the user to the SDP perimeter and from the perimeter to the application in scope. Because the communication channel is inconsequential, it is absolutely critical to separate the control plane, data plane, and management plane. The usual practice would be to isolate the control plane, where all the authentication takes place, and keep it outside the SDP perimeter. Once the authentication is done on the control plane, then and only then is the data plane revealed to the requesting host.

SDP focusses on uniform policy with integrated security controls. The main access policy will cover all controls in a single plane of glass rather than configuring them across various network devices as is the case in traditional networking.

Mutual TLS (mTLS) is an important aspect of SDP. mTLS enhances the security of communication between devices and services. When mTLS is implemented, both the client and the server (or, in the context of SDP, the user/device and the protected resource) have their own digital certificates. These certificates are used for authentication and to establish an encrypted communication channel. Hence, when an end-user device wants to connect to an application that supports mTLS, it initiates a connection to the secure device perimeter. This is the control plane that exists outside of the data plane. Once the authentication is complete, it gets access to the perimeter and sees the application the user has access to. If observed carefully, SSE is an implementation outcome of SDP and many parallels can be drawn with respect to implementation design of SSE and core tenets of SDP.

Another important mechanism utilized in SDP is single-packet authorization (SPA), which is a security concept and technique that aims to enhance the security of networked systems by only allowing authorized access based on the reception of a specific, carefully crafted packet. This approach is in contrast to traditional access methods that rely on open ports, which can be exploited by attackers. SPA (different from Secure Private Access) is designed to be a complementary security measure and can be used alongside other security mechanisms like firewalls, intrusion detection systems, and VPNs. The core idea of SPA is to require an external entity, often a client or user, to send a specially formatted packet to a target system before access is granted. This packet serves as a form of authentication and authorization.

Architectural Components

The architectural components described in the sections that follow are integral to understanding the overall flow of information in SDP. Primarily, the following architectural prerequisites must be deployed before being able to utilize the DP architecture:

- **Identity-aware applications:** Onboarding applications to the SDP is a key step in controlling access to applications. Implementations could start from IP and slowly move toward more context-specific identification of applications.

- **Client-aware devices:** All devices are capable of authenticating themselves either with a hardware chip (TPM) or with a user certificate. Usually an end-user agent will be able to initiate secure control plane communication to the SDP.

- **Network-aware security controls:** These include on-premises firewalls, gateways, and SSE.

Once the prerequisites are met, the following components are onboarded to the SDP ecosystem.

Initiating Host (IH)

The initiating host is the user device that is initiating a communication to the server via the SDP. An initiating host can be anywhere inside or outside the perimeter.

SDP Client

This is analogous to an endpoint policy enforcement point in the NIST architecture. This is a software client that initiates a communication to the SDP and is installed on the IH. All control plane communications are initiated by the SDP client toward the SDP controller.

Accepting Hosts (AH)

These are the applications or workloads that receive the connect requests from the IH. The SDP protects the accepting hosts either with per-application connections or via a gateway.

SDP Controller

The controller is the brain of the architecture and handles all policies and authentication requests. It isolates services from the initiating hosts and is analogous to the policy decision point (PDP) in the NIST architecture.

SDP Gateway

The gateway is a method of creating a secure perimeter around applications that cannot initiate or cannot communicate securely to the SDP. The gateway makes sure that all authentication and authorization is performed before the applications and workloads can be accessed.

With this clear understanding of the architecture, let's look at Figure 8-10, which illustrates the overall flow of communication in an SDP segment.

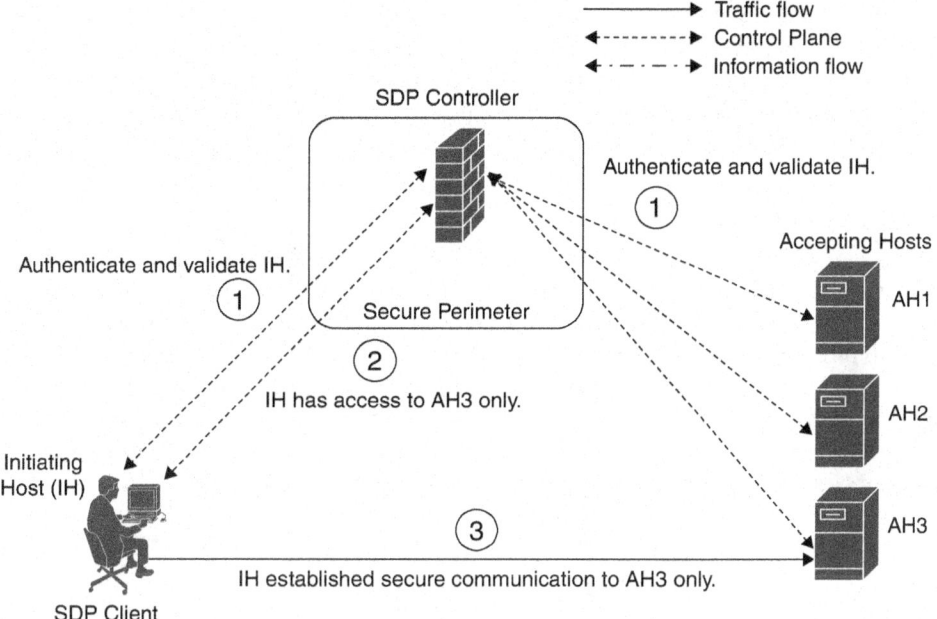

Figure 8-10 *SDP Architecture and Flow.*

The flow can be summarized with the following steps:

1. The SDP client on the IH initiates an mTLS session to the SDP controller. Similarly, the AH also is onboarded to the SDP controller. The controller validates the certificate and sends its own certificate for authentication. mTLS is established, and control communication is complete.

2. Once the mTLS is established, a login is attempted and the authorized accepting hosts (AHs) are exposed to the IH.

3. The IH initiates an mTLS session to the AH directly.

Once the communication channel is established, the information can flow. Careful observation reveals that these principles are already incorporated into a lot of security products like SD-WAN and SSE.

Deployment Models

There are six deployment models in the SDP architecture. These are discussed in brief so you can understand the key implementation methodologies of the concept of ZTNA. Most products like Zscaler Zero Trust Exchange and Cisco Security Service Edge utilize the concept in a similar manner. This section will attempt to categorize these SDP implementation models into the four basic architectural models discussed earlier in this chapter in the section "Basic Flows in a Zero Trust System." This section will overlay the six SDP models over the four NIST models to showcase the alignment and overall similarity of concept.

User-to-Service Flows

The sections that follow describe the models that involve an authenticated user accessing an enterprise service. The service could be a physical server, a virtual machine, or a container. The models are the most common deployment models and are aligned to current perimeter-based controls.

Client-to-Gateway Model

This communication model is usually utilized when there are many services that need to be protected at a time, and these services/workloads do not fully support per-application flows. These flows are typically covered with legacy applications, and the gateway is used to provide the necessary protection controls for the application or applications. Within the gateway's boundary there is implicit trust. From an NIST perspective, this would be covered under the "user-to-service" model, where the service is not capable of performing secure communications to an SDP. Figure 8-11 illustrates the concept, and you can clearly draw a parallel to a VPN implementation.

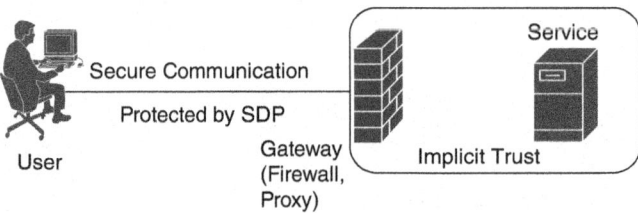

Figure 8-11   *Client-to-Gateway Model*

Client-to-Service Model

This is the ideal model of deployment where the gateway to connect to the SDP exists on the application itself, which makes the communication per application. As an extension of the client-to-gateway model, this specific model suits legacy applications as well; however, the gateway is not for a group of applications but a more specific application. This is an important model to consider as this is the general final state most applications should be at. Gateways are restricted to the specific application, which means that micro-perimeters for applications are clearly defined. Zero Trust Network Access (alternatively referred to as secure private access) is an outcome of the client-to-service model and is commonly achieved with an SSE implementation. Figure 8-12 illustrates the client-to-service model and draws a parallel to an SSE deployment.

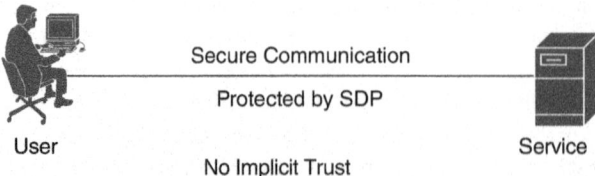

Figure 8-12   *Client-to-Gateway Model*

Service-to-Service Flows

The service-to-service flows are where both the subject and object are enterprise service assets that might not be able to identify with a user associated to an identity provider. Usual authentication methods would be certificates, secure private access, and mTLS. Models can either be across gateways (like SD-WAN) or across applications (SSE).

Service-to-Service Model

This model maps to the service-to-service model discussed in the "Basic Flows in a Zero Trust" section earlier in the chapter. These are perfect when services need to communicate to each other (peer to peer) like Internet of Things devices or virtual machines. Though not many enterprises might embark on a "service to service across Internet" paradigm, it is still a possibility that has many advantages—the main one being that it greatly reduces overhead and increases the security posture with all the security controls configurable at one universal policy administration and decision point. The long-term advantage is the reduction in total cost of ownership. The cost of the Internet link versus the cost of having multiple firewalls, firewall administrators, and processes in place is a common comparison that most enterprises are having. This is similar to the client-to-service model, except the originator is another service. This might be similar in concept, but service-to-service communications are more complex to architect than north-south Internet flows. Much time must be spent discovering applications and their dependencies to be able to design an apt per-application policy. Similarly, Internet bandwidth is a consideration, along with logging costs when traffic is sent to the cloud for inspection. Figure 8-13 illustrates a service-to-service model where communication is via an SDP, typically the SSE.

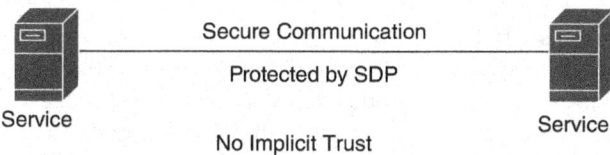

Figure 8-13 *Service-to-Service Model*

Gateway to Gateway Model

In this type of deployment model, the IH and AH need to sit behind a gateway that exists outside of the application itself. This is a more legacy network insertion methodology, where the actual end clients and servers might not support the overall capability to install SDP clients, but they still need to be protected by the gateway, which likely is a firewall, router, or proxy. Hence, the implicit trust exists within the gateway but not between gateways. This could manifest as a user-to-service or a service-to-service flow; however, the key here is the extension of trust, which is the least desirable model but the most practical one during the early stages of implementation. Enterprises might deploy this model and slowly move toward a more application-centric security control as they improve their micro-segmentation strategy. In Cisco SD-WAN, there is gateway-to-controller DTLS control plane communication followed by gateway-to-gateway IPsec communication, which can be modified to make the communication per application as well. Figure 8-14 illustrates a gateway-to-gateway model that draws a parallel to SD-WAN. This is one of the primary reasons why SD-WAN is the first step toward a unified SASE architecture.

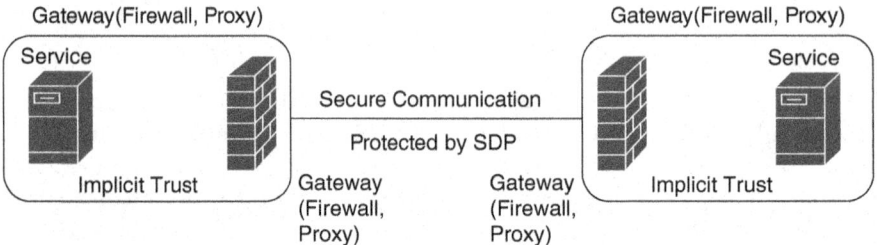

Figure 8-14   *Gateway-to-Gateway Model*

User-to-User Flows

The sections that follow cover the peer-to-peer communications that exist between users. Users in the context of this flow are identified and authenticated with an IDP based on certificates, biometrics, or other MFA-powered mechanism. Usually, users communicate directly over the SDP, which is not common in modern networks. The second option is when users communicate via a gateway. These models map to the user-to-user flow.

Client-to-Service-to-Client Model

This deployment is analogous to the user-to-user voice flow, where users have to communicate peer to peer via a voice gateway. Here, there is no mandatory requirement that the clients need to have direct communication with each other. Both clients connect to the SDP, and the SDP basically provides a logical communication path for both clients. There could be an intermediate server/enforcement point in the path that communicates to the main controller and would terminate the mTLS. A communication model that needs a middleman controller is an example of client-to-server-to-client flow. An example is when a VoIP phone needs to communicate to another VoIP phone via a controller (Cisco Unified Commutation Manager). Figure 8-15 briefly illustrates the model.

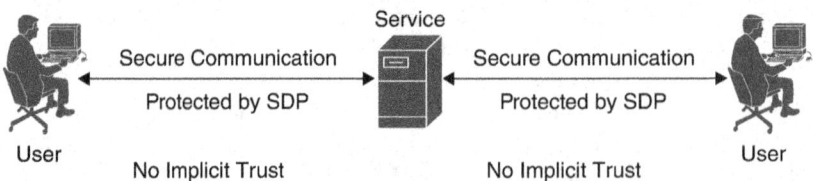

Figure 8-15   *Client-to-Service-to-Client Model*

Client-to-Gateway-to-Client Model

This specific model is similar to the client-to-service-to-client deployment model but with the unique requirement that the clients need to have direct communication to each other. The intermediate gateway does not take part in the mTLS but communicates to the controller to verify if the flow is allowed. This gateway is usually a firewall or proxy. Figure 8-16 illustrates the concept briefly. It can draw a parallel to how DMVPN works; however, the communication is not router-to-router but instead user-to-user, with the intervention of the controller to validate principle of least privilege and need to know without taking part or terminating the authentication process

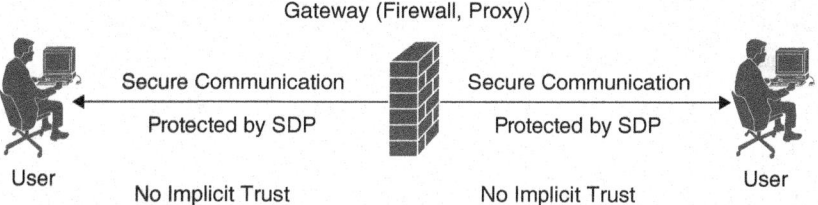

Figure 8-16 *Client-to-Gateway-to-Client Model*

User or Service to the Internet Flows

SDP as a concept exists to provide a secure perimeter for a subject to access an object. If the object is on the Internet, the secure perimeter ends at the SDP. Hence, the flows are similar to the client-to-service model; however, the service in this case would be the SDP itself. Traffic will enter the SDP and get subsequently processed and then exit the SDP to its target on the Internet. Figure 8-17 illustrates this concept and is similar to the secure internet access flow (user- or service-to-internet) when the endpoint Zero Trust agent initiates an encrypted tunnel to the secure perimeter. Rather than being one user-to-internet flow it is broken down to a user-service flow(client to SDP) and a user-to-internet flow(client to internet after SDP authentication, authorization and inspection)

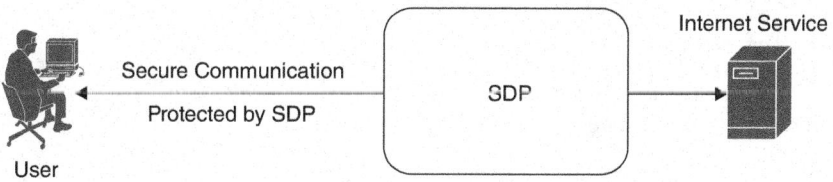

Figure 8-17 *Secure Client-to-Internet Model via SDP*

The second option is when agents cannot be installed on endpoints and a gateway is required to initiate an encrypted tunnel to the SDP. The SDP will terminate the tunnel, and subsequent encryption from the SDP to the target on the Internet will depend on whether the target supports application-level encryption. Figure 8-18 illustrates this concept.

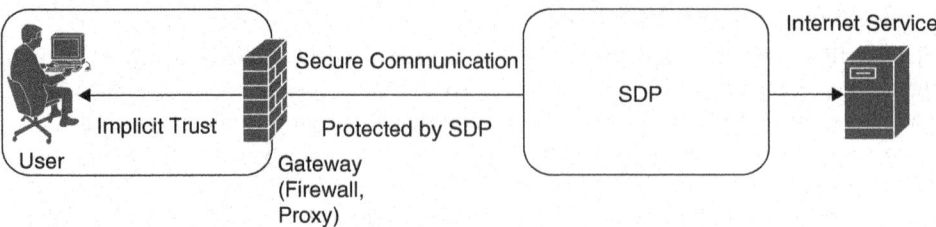

Figure 8-18   *Secure Client-to-Internet Model via Gateway*

Most products today deploy ZTNA in a very similar fashion, so understanding SDP is an important part of understanding the implementation aspect of Zero Trust and how the architectures and principles are applied practically.

The Deep Dive

[After a month of discussion with Jed's team as well as the CIO team, Glenn helped craft the strategy for the overall Zero Trust architecture, with feedback incorporated from the tactical road map he had created. He then presents and explains the architecture components to the entire virtual Zero Trust team.]

Glenn: Zero Trust does not require a full overhaul of your network. If designed correctly, it can be implemented with very simple and straightforward changes in the existing network. In summary:

- A Zero Trust network architecture has a subject, an object, and a context complemented with a secure communication channel.

- There is a decision point that decides what access to provide and enforcement points that enforce this policy.

- Finally, there are multiple information points that provide the decision-maker much needed information to evaluate context to make the right decision.

- Zero Trust is never a project that needs to start from scratch. It is almost always an overlay on the existing architecture.

- There are multiple types of business flows in a network, but Zero Trust Architecture broadly categorizes them into four flows:

  - User-to-user

  - User-to-service

  - Service-to-service

  - User- or service-to-Internet

Mr. Smith: Alright, I want to understand where we are at right now. We have got our key business drivers, the measurement metrics based on performance, and risk. From what I hear we are not in a good position, and I would like to get this ball rolling ASAP. We understand how the overall architecture will look like, have mapped out our tactical road map, and you have presented the strategy to my Zero Trust team. The team here with me is now Zero Trust educated and ready to get their hands dirty with the technology. So, am I right in assuming that the next step is the implementation? I do not want to wait any more knowing there are so many gaps in the security architecture that are not patched.

Glenn: Not really, you need to identify the vendors you wish to partner with to achieve your Zero Trust vision. For that you need to understand the critical security mechanisms relevant to your specific use cases. Once you identify the key security mechanisms that align with your business drivers and tactical initiatives, the product selection stage will commence and you would start releasing your request for proposals to your selected vendors. Subsequently, you will need to prepare the relevant budget and present all the findings to the board. Once the strategy, metrics, and budget are approved, then the various projects and initiatives can kick off.

Mr. Chen: *[visibly unhappy]* So the next step is to identify the critical security mechanisms and the product vendors, correct? Let's get started since this is going to be something I need to get deeply involved in.

Glenn: Yes. The focus is to identify the critical security mechanisms and select some vendors so that you can send out a request for proposal with the features and capabilities you are interested in. Let's dive in.

Endnote

1. NIST Special Publication 800-27: Zero Trust Architecture, https://doi.org/10.6028/NIST.SP.800-207

Chapter 9

Critical Security Mechanisms for Zero Trust Architectures

A common misunderstanding most enterprises have is that they need to start identifying Zero Trust–capable products before they begin their Zero Trust journey. As part of the overall Zero Trust Journey, a product-level selection or critical mechanism update happens much later in the fourth phase. It is important to spend time to understand requirements, build metrics, bring together a strong security team, and finally create the strategy. With the tactical road map and a strategy in place, the next logical step is to identify the mechanisms required for the desired initiatives, and this is where product selection is performed. Previous chapters focused on the overall architecture and logical placements of various security controls. This chapter explores some of the security mechanisms that are critical when embarking on a Zero Trust journey. These mechanisms will be covered in the same context of how we define Zero Trust, which is from the perspective of the subject, device, transport, and, finally, data and workload.

Zero Trust Mechanisms for Subjects (Users and Devices)

Here are some important aspects of a subject we should consider when selecting security mechanisms:

- They are mostly uncontrolled users with varied profiles. User A might be connecting from a corporate laptop, and two hours later the same identity might be connecting from a personal iPad.

- A subject can usually be mapped to a tangible entity that can think on its own. This means if you deny access to a subject, they could potentially look for other ways in.

- A subject's traffic patterns might be different. A subject might start surfing Netflix more often after the release of a new series. Patterns are not traffic-specific but more external to the enterprise.

Zero Trust is all about identifying the right context so that a decision point can decide whether a subject can access an object. In comparison to objects or services, the context of a subject is more dynamic with more characteristics to help determine this access. Keep these characteristics in mind as you look at the following common mechanisms in Zero Trust specifically for adding more context to a subject:

- Identity and access management (IAM)

- User segmentation

- Multifactor authentication (MFA)

- Profiling of endpoints

- Device posture checks

- Mobile device management (MDM)

- Bring your own device (BYOD)

- Privileges access management

- DNS and URL filtering

- Email security

- Remote browser isolation (RBI)

The sections that follow cover these mechanisms in more detail.

Identity and Access Management (IAM)

Most enterprises already have their IAM in place before they consider embarking on the Zero Trust journey. As part of basic enterprise operations, there must be a framework in place to provide authorized access to only those who need it. Simply put, the access model must provide the right access to the right person with the right context at the right time. The more mature the IAM solution is, the less cumbersome the Zero Trust journey will be because IAM is the heart of Zero Trust. However, it is important to stress that having an IAM does not mean your system is Zero Trust ready or that in some way most of your gaps are implemented. It implies that with an effective IAM in place, the enterprise is on the right track toward a secure Zero Trust implementation and will greatly benefit from more enhancements to the overall IAM architecture.

Of course, there are challenges in the real world when it comes to IAM, and a poorly implemented IAM system is still a critical gap that needs to be addressed. Enterprises rarely have simple Active Directory (AD) structures, and many customers stay apprehensive that Zero Trust strategies will make IAM deployments more complex. Another scenario includes customers who do not implement a centralized decision point but rather let applications handle access control. These customers usually view Zero Trust as a hindrance. This, however, is not true. Zero Trust is an access framework—an overlay across your existing security infrastructure, not a hindrance. Hence, Zero Trust not only appreciates the IAM infrastructure currently installed, but it also helps augment the security strategy and vision from an IAM perspective.

Zero Trust supports and enhances the access model for various aspects of IAM. Circling back to the survey "CISO Perspectives and Progress in Deploying Zero Trust" conducted by the Cloud Security Alliance,[1] close to 40% of enterprises have implemented MFA and around 40% have implemented cloud-based Active Directory instances, which is much more (almost double) than the mere 25% that have deployed on-prem ADs. The distinction is clear: enterprises are moving identity and access management to the cloud and acquiring better scale.

Common Roadblocks to Enhancing Identity and Access Management

Some of the common issues or roadblocks that customers face when enhancing IAM are described in the sections that follow. Although these are not new, listing them together emphasizes how important IAM really is to achieve the broader Zero Trust vision.

General Complexity of AD

As is the case with entropy, enterprises expand. The workforce is increased when the enterprise is growing and decreased when there are cuts. Such is the dynamic nature of a workforce in general. Things get complicated when there is a merger or an acquisition, where different identity technologies now need to work together to make sure both entities are at the same security level. With technology like Active Directory Federation Services (ADFS), OpenID (OID) and OAuth, there is now dependance on external identity providers to provide identity and authorization capabilities, along with creating diverse forests of identity trees across multiple ADs. Overall, ADs are rarely static, but operations teams would prefer to keep them static and hence tend to stay away from information security models like Zero Trust.

As an extension of operations being impacted from Zero Trust, it can be noted that AD administrators need to manage all of the AD complexities. Yes, they are experienced engineers to support the deployment, but with increased complexity comes a larger scope of errors and more hesitance to change. Segmenting users into smaller granular groups can be a large overhead for administrators, as it will involve the creation of new Organizational Units (OUs) or security groups.

Hesitance to Give Up Control

Most enterprises do not want to give up decision-making capabilities to a central server/entity. Application stack developers prefer to let the applications decide what kind of access control the application can deploy. This leads to local control and inevitable utilization of local accounts. In the standard enterprise architecture, most server farms will have a common policy enforcement point (PEP) like a next-generation firewall (NGFW), and the application will then decide what access to provide, assuming that if the connection was allowed by the firewall it is likely trusted. The application itself is another trust boundary where access control checks are performed. This, however, means that the application needs to integrate with all of the ADs in the network, and if the ADs are not synced, this would be a nightmare. Developers eventually implement local accounts for administrator and read-only access, which defeats the purpose of removing implicit trust. Enterprises do not perceive value in a hybrid network and application access control approach simply because they fear that by delegating some control to a central server, they are exposing the application. This also leads to distributed decision points, meaning each application and enforcement point is a singular decision-maker for itself. The decision-making is not uniform when considered from an end-to-end perspective.

For example, one of the earliest access models for subjects to a workload would look like Figure 9-1, where the AAA mechanism is embedded in the application. However, as time passed by, this was not scalable to map out to all applications, and with old monolithic applications, access to the application credentials essentially gave access to the entire web and database as well.

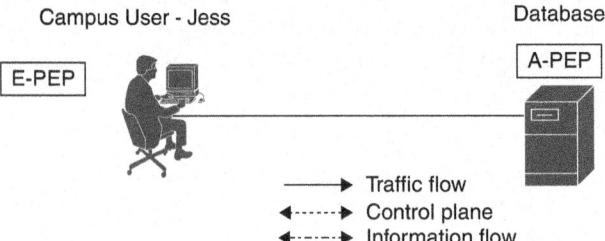

Figure 9-1 *Original Application-Based Control (Access Control at Application Level)*

The focus then shifted to control access to the application externally. Modern networks understood the importance of scale and moved the control to an external source but still kept the capability to do granular decision-making on the application. Access to the device is validated against an AD or RADIUS server, both of which can support MFA. This is by far the most common deployment, and additionally all servers will have a network PEP protecting them, which will get policy information from a policy decision point (PDP) in the server farm. Figure 9-2 illustrates this concept. Figure 9-3 illustrates how policies are disjointed when multiple decision-makers are involved in the network in a siloed access control environment.

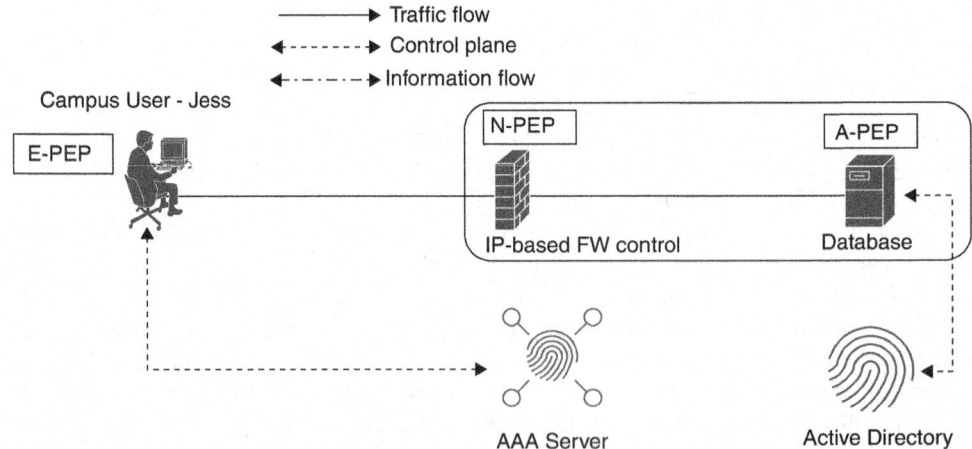

Figure 9-2 *Access Control with an External AD*

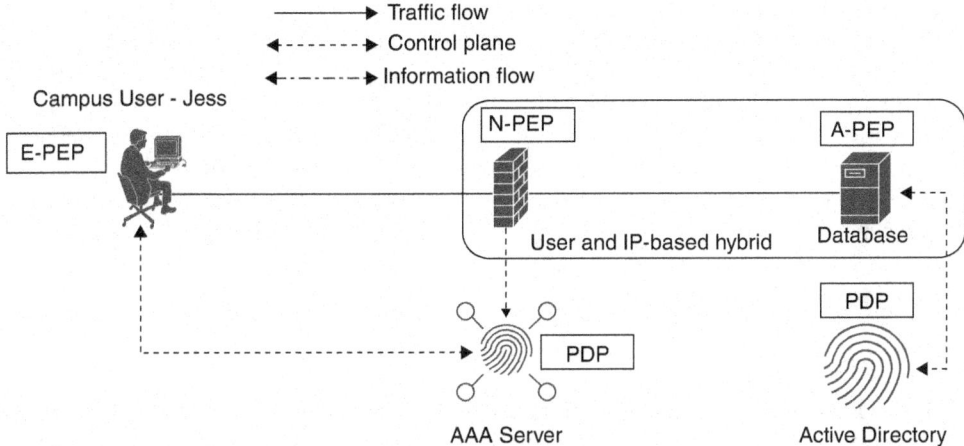

Figure 9-3   *Access Control with Disjointed PDPs*

Although all the highlighted scenarios are waning methodologies in modern enterprise security architectures, they show the purpose of what an administrator wants to achieve. The administrator wants to retain the authorization capability on the application and use an external identity source for authentication only. Zero Trust helps delegate some more features to the PDP, including complex context-based authorization, so that the policy is truly centralized and uniform.

Shifting to Certificates

Microsoft, Cisco, and many other companies are now advocating for password-less authentication via fingerprints and so on. Certificates are a direction that has great value in the overall Zero Trust road map but just never gets implemented due to technical issues or an organization's lack of strategy. Enterprises may embark on a VPN deployment project where the deployment is done using username and password; however, because the AD might not be mature enough to be able to do granular control, an informed decision might be made to move to certificate-based authentication to achieve nonrepudiation as well. Of course, that activity is easier said than done because multiple prerequisites like an enterprise Certificate Authority (CA) server and a means to deploy certificates at a large scale like Group Policy Object (GPO) policies are just not in place to support the broader initiative. Due to some of these limitations, certificates get overlooked and password-based authentication takes us back to the stone age; however, this doesn't mean that passwords will get replaced soon. The motive is to get the enterprise to start looking at better options (fingerprint, certificates, and so on) and transition in time toward a more secure method of identification.

Slow Movement to Attribute-Based Control

Zero Trust is all about attribute-based control, but in all practical sense, it is not easy to get the entire context of a flow, especially from a subject's perspective. Operation and

SOC analysts are constantly bombarded with logs, which could possibly give more information regarding a subject but more often than not there is no automation or orchestration in place to process this quickly. The attributes readily available at the time of rule writing are usually only the IP address as part of the usual five tuple (source IP, destination IP, source port, destination port, and protocol). This slows down the overall movement of the enterprise to Zero Trust, and operations engineers are not ready to make changes to the AD or the IAM in general to process more context and provide access, based on risk or context. Modern Zero Trust products utilize attributes to calculate a risk score that can be used by administrators to create risk-based control; however, the implementation of the feature in production is rare.

Presence of Legacy Devices

Most organizations will have a set of legacy devices that are too critical to migrate to new infrastructure and hence will not have support for modern authentication mechanisms. For example, a chip manufacturing company might have Windows systems that are so old that they do not support dot1x. This, of course, is a challenge when implementing IAM because now an identity specific to that user accessing the network and a legacy device cannot be deterministically identified. Gateway-based models are useful to deploy in such scenarios, the details of which are covered in Chapter 8, "Building a Zero Trust Architecture."

Similarly, when security is built around this application, the policy is very crude and does not reflect the actual criticality of the legacy application. Enterprises will not change the application because it is critical and will place a firewall before it and perform IP-based control to achieve pseudo-security. There are obvious flaws with this logic—the main one being that the data criticality and application availability are not entirely taken into consideration when deploying policies on the firewall.

Zero Trust Enhances IAM

If you have encountered all of the aforementioned roadblocks, whether in your enterprise or in an enterprise that is taking your consulting services, then do not fear, Zero Trust is here. Zero Trust will not completely rewrite your security control, but it will overlay your security control with the right policy and the right security mechanisms.

Consider the example of a common architecture for a simple application illustrated in Figure 9-4.

In a Zero Trust model, the existing infrastructure need not change; an application most likely will still need AD access to provide more granular application access, but a user's access to the application will be controlled by a AAA server and enforced with MFA. That means the application can check if User A is allowed to perform a specific action on itself. For example, "can the user change files on the server?" "No." "Can the user READ a web page on the server?" "Yes." At an application level, the question the PDP will answer is, "Is User A allowed to perform a specific action?" This can be fulfilled under specific conditions like "accessing the application from a postured device."

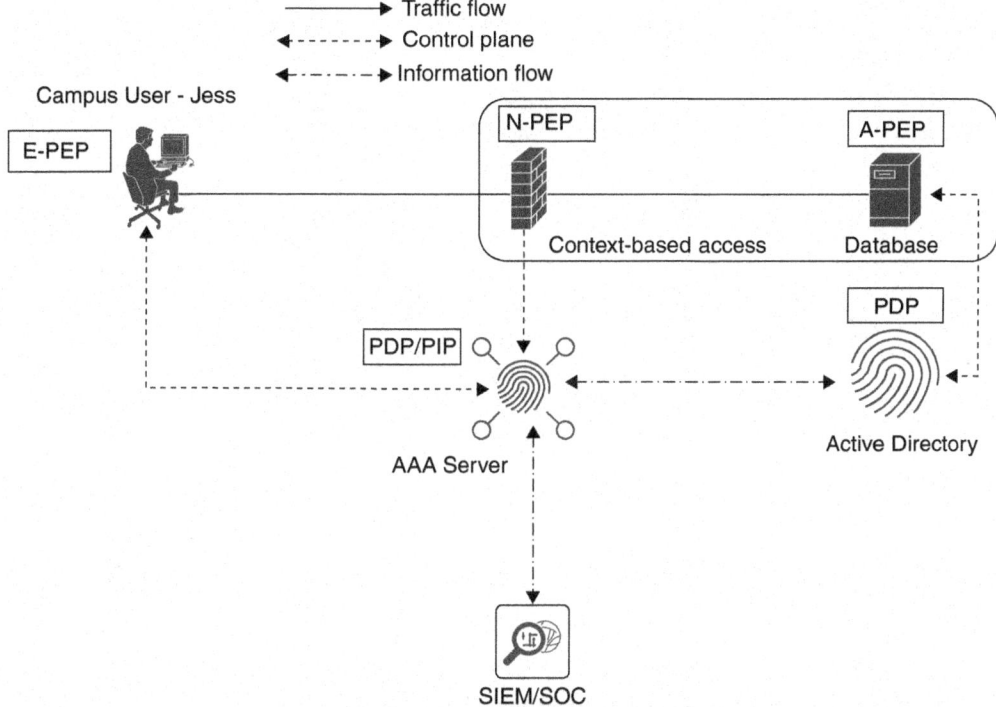

Figure 9-4   *A Full Zero Trust Application*

With the right segmentation, access to the network, specifically where this application is deployed, will also be controlled at the decision point. This is essentially helpful in a legacy environment where the application itself might not be capable of providing AAA services. A network PEP and an access control AAA server will help restrict access to the legacy network, thus mitigating the risk of a breach. Essentially the N-PEP like an NGFW can answer the question, "Is the subject allowed access to this specific network?" One must appreciate how this hybrid model helps not only to simplify policy creation but also avoids large changes to existing infrastructure, especially where legacy applications might not have integrations in place. Figure 9-5 illustrates the hybrid model concept. This architecture is the most common starting point for enterprises beginning their Zero Trust journey for access control enhancements.

Note here that the application itself has no common integration with an AD or a AAA server but the user can still be controlled at the N-PEP or E-PEP.

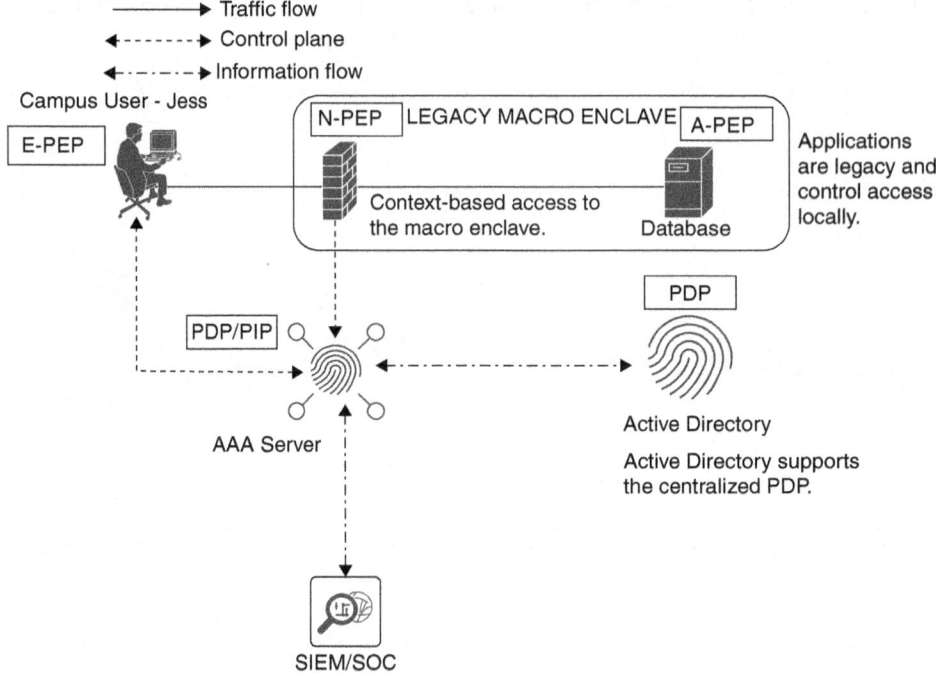

Figure 9-5   *Zero Trust in Legacy Environments*

Zero Trust Allows for Effective User Segmentation

It is important to clarify that Zero Trust as an access model doesn't mandate user segmentation but definitely benefits from it. On the other hand, Zero Trust helps identify specific flows and then assign subjects that need access to the flow. This helps the enterprise strategize its segmentation. For example, if a lot of voice traffic exists in the network, phones and users will be common subjects. Within general users, there might be privileged and unprivileged users as sub-segments. If the flows and devices are known, a segmentation strategy can be formulized. Figure 9-6 illustrates a small portion of a larger segmentation strategy.

User segmentation greatly benefits from the right segmentation strategy that is supported by the overarching Zero Trust strategy.

In summary, Zero Trust improves the IAM. It connects siloed enforcement and decision-making points to one common universal policy enforcement mechanism, which has its own information point, decision-maker, and enforcer. By combining or at least integrating these silos, Zero Trust provides a single plane of visibility and makes operational tasks much simpler. Simplification is key in any Zero Trust model, and that is what the framework aims to achieve, which is to simplify the operations and access control to make sure the right policies are implemented based on the right identity. A large percentage of enterprises embark on the IAM enhancement aspect of Zero Trust to show a quick win and get more funding for other critical mechanisms like posturing and NAC.

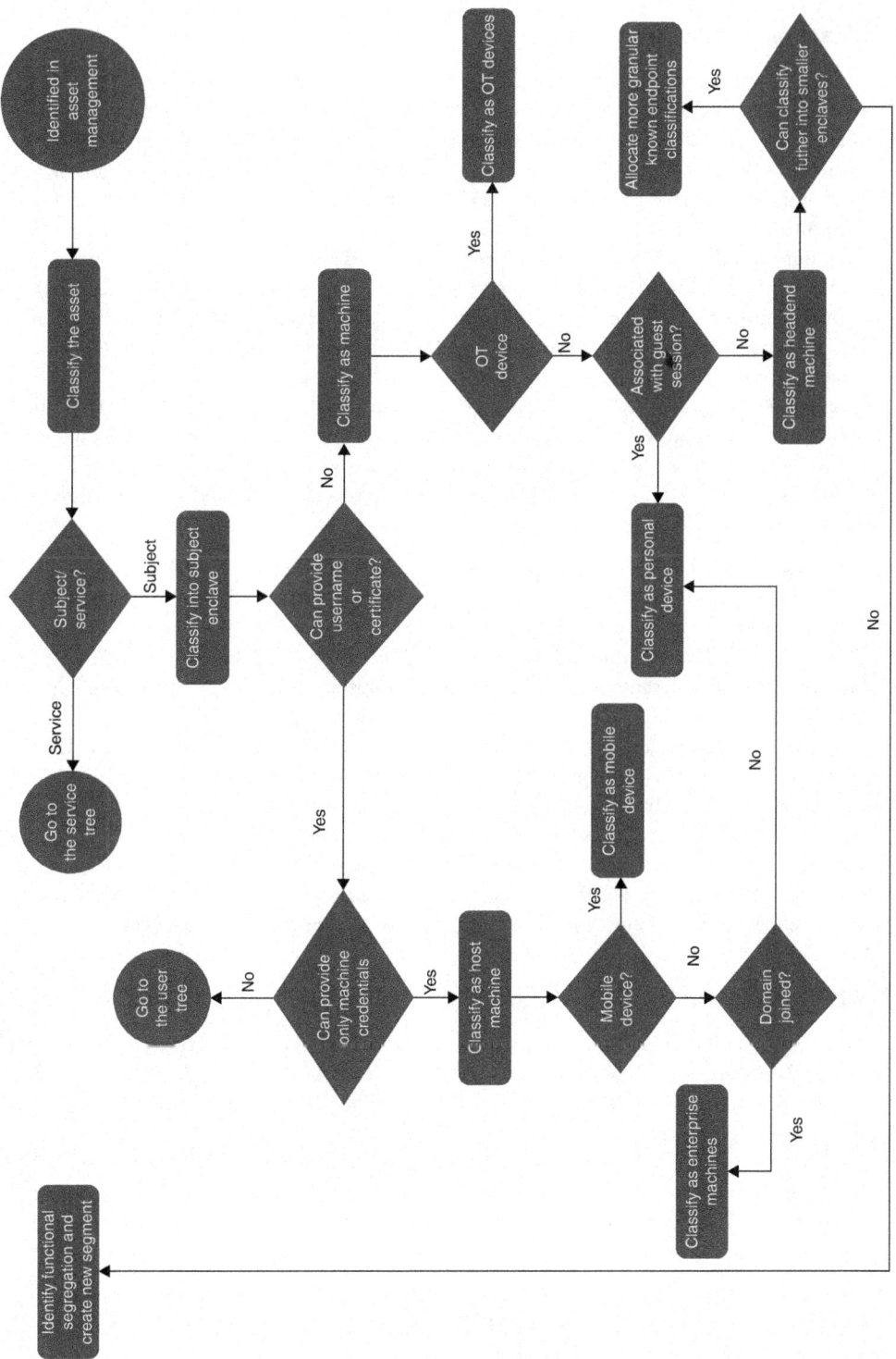

Figure 9-6   *Segmentation Strategy for Users (an Example)*

Multifactor Authentication

Multifactor authentication (MFA) is a security measure that requires users to provide multiple forms of identification to verify their identities. It adds an extra layer of protection beyond traditional username and password combinations. Typically, MFA involves something the user knows (password), something the user has (such as a physical token or mobile device), or something the user is (such as biometric data like fingerprints or facial recognition). Of course, many enterprises are moving toward password-less authentication, but until then MFA is how enterprises protect password-based users. The purpose of MFA is to minimize the risk of unauthorized access to sensitive systems or data via techniques like password-guessing, rainbow tables and so on. By requiring multiple factors of authentication, even if one factor (for example, password) is compromised, the attacker would still need access to the other factor(s) to gain entry. MFA significantly enhances security and reduces the likelihood of successful credential-based brute-force attacks. The key phrase here is that it *reduces the likelihood* but is not a silver bullet. MFA moves the vulnerability of guessing a weak password toward the user. Attackers must now socially engineer people to force them to reveal the second factor with modern tactics like push phishing.

MFA plays a crucial role in implementing the Zero Trust access model. Because Zero Trust assumes that no user or device can be automatically trusted, MFA becomes an essential component to verify the identity and trustworthiness of users attempting to access resources. By combining MFA with other security controls, such as device health checks, network segmentation, and continuous monitoring, Zero Trust helps organizations detect and mitigate security threats in real time. Cisco Duo and Silverfort are prime examples of forerunners in providing secure identity services. The relevance of MFA to Zero Trust lies in its capability to provide a strong authentication mechanism that aligns with the Zero Trust philosophy of continuous verification. It ensures that even if an attacker manages to bypass other security measures or compromises one factor of authentication, they still face additional hurdles to gain unauthorized access. By implementing MFA as part of a comprehensive Zero Trust architecture, organizations significantly enhance their security posture and protection against advanced threats.

Profiling of Endpoint Subjects

Profiling of endpoints is a key aspect of Zero Trust. It helps organizations to identify and authenticate every device that is accessing their network, regardless of its location or ownership. Most modern networks implement profiling by default to make sure that the decision-maker is aware of the types of devices in the network. You could statically profile a set of endpoints as corporate or IoT devices manually, and it completely depends on the enterprise strategy. Usually this is not done for user endpoints, and static mapping based on MAC addresses is implemented for devices that cannot do dynamic authentication via dot1x. Devices like printers, old IP phones, and so on fall under this category.

Dynamic profiling is when the decision point sends a set of probes or monitors certain types of traffic such as DHCP, SNMP, RADIUS, and so on to glean more information from the packets. This is especially useful in capturing critical information from endpoints, as this information can be used as additional context when creating policy. By discovering and identifying all assets on the network, organizations gain a better understanding of the overall security posture of their infrastructure and identify potential vulnerabilities or threats. This contextual information helps PDPs in organizations to make more informed access decisions and enforce more granular security policies, such as limiting access to sensitive data and to specific devices or users.

One important point is that profiling can never be fine-grained, which goes to say that profiling is an information source to the decision-maker, which considers other attributes as well.

Device Posture

Device posture encompasses validating the security posture of a device before providing access to a user. As part of CISA's Zero Trust maturity model, devices are key aspects of the network that were not traditionally included as part of standard AAA dot1x deployments.

In a dot1x deployment, a user is authenticated via certificates or username/password. It is quite recent that machine authentication was considered an important part of the authentication process. Enterprises still do not perform additional machine authentication along with user authentication. User authentication alone is no longer considered secure. Machine authentication is also possible via the same hostname identification or with certificates. Yet after all the breaches, the question still remains. Is machine or user authentication alone or a combination of both sufficient to decide the security posture of a subject?

The answer is quite simply no. That is where device posture and all the endpoint policy enforcement points (E-PEPs) come into play. They get connected to a decision point, which gives a set of conditions that need to be fulfilled to consider the device as compliant. In the end, the E-PEP returns a posture compliance status that the PDP uses as a criterion for access approval. If the device is not compliant, the subject is not given access. This is especially useful when the enterprise does not allow unmanaged devices for their employees. General device posture conditions include the following:

- Antivirus/malware applications installed, running, and up to date
- Latest OS versions installed
- Critical patches installed
- Integration with an external MDM (such as Microsoft Intune) complete and active
- Full Disk encryption enabled
- Domain joined
- Host firewall enabled
- Endpoint detection and response (EDR) installed, running, and up to date

Mobile Device Management

Mobile device management (MDM) is a security solution that is designed to manage and secure mobile devices (such as smartphones, tablets, and laptops) that are used within an organization. MDM enables IT administrators to control and enforce security policies, manage application deployment and updates, and monitor device usage.

MDM is relevant to Zero Trust because it provides a way to enforce security policies and monitor device usage, which are essential components of the Zero Trust model. MDM can be used to enforce device encryption, require strong passwords or biometric authentication, and restrict access to certain applications and data based on user roles and permissions. By using MDM in conjunction with other security solutions, such as identity and access management and network security, organizations can implement a comprehensive Zero Trust security model that provides a high level of protection against cyber threats.

The final goal is to empower the end user or an employee to use the device of their choice as long as they adhere to enterprise security policy and still maintain the best user experience. MDM bridges that gap.

Bring Your Own Device

Zero Trust is all about flexibility. Therefore, it would be hypocritical to say that Zero Trust allows anyone to access a resource from anywhere as long as there is a policy allowing them, and then provide access to only corporate devices. Enterprises are keen to restrict wireless access only to guest users so that any wireless access is isolated from the enterprise segments. This, however, is not the ideal scenario to be in, and employees should be able to connect to an enterprise SSID as well. Enter bring your own device (BYOD), where you can bring any device and connect to specific services in the enterprises as long as there are minimum posture checks. This might sound counterproductive, but enterprises that are embarking this journey will allow, for example, any laptop with a minimum of Windows 10 installed with the latest Bitdefender access to the network. The use case is also applicable to mobile access where a BYOD services provider or an MDM might help provision your device to meet the minimum requirement to access the network.

If the use case is about allowing access with the least intrusion to endpoints, a guest user is an option; however, the concern with guest segments is that you will not be able to get the level of granularity you can get with BYOD or MDM, and you need to be in the corporate network to get the guest SSID. Alternatively, the enterprise might issue a device certificate for the endpoint (like a mobile device) and then mandate endpoint protection software that will then allow a guest user to access the corporate network on a restricted VLAN. They would still, however, be able to communicate to other devices in the same VLAN, which will be controlled based on the workplace segmentation strategy.

With Zero Trust, the entire process is more streamlined and granular. The application will deploy MFA or single sign-on (SSO). Either the endpoint will have a Zero Trust agent to allow it to access private IPs in the enterprise or it might just allow access to public services of the enterprise. Device posture will be key, and preferably authentication will be done using certificates because they are more secure than a username and password. In the future, biometrics will also play a key role in supporting overall Zero Trust and BYOD strategy. The user doesn't have to connect to a corporate network but can connect from anywhere, making network access truly location-agnostic.

Privileged Access Management

Privileged access management (PAM) is a security practice that focuses on controlling and monitoring access to sensitive data, systems, and applications by privileged users, such as system administrators, database administrators, and other IT staff with elevated permissions. PAM is designed to reduce the risk of insider threats, external attacks, and accidental or intentional misuse of privileged access.

PAM typically involves the following key components:

- **Privileged account discovery and inventory:** Identifying all privileged accounts within an organization, including system-level and service accounts, and maintaining an inventory of these accounts.

- **Privileged account management:** Controlling and monitoring access to privileged accounts, including password management, session monitoring, and auditing.

- **Privileged session management:** Controlling and monitoring privileged user sessions, including session isolation, monitoring, and recording.

- **Privileged session monitoring:** PAM solutions often include session monitoring capabilities that record and monitor activities performed during privileged sessions. This control ensures accountability and allows for the detection of any suspicious or unauthorized actions.

- **Privileged access governance:** Establishing policies and procedures for granting, monitoring, and revoking privileged access as well as ensuring compliance with regulatory requirements.

- **Access request and approval:** A PAM solution will have a formal process for requesting and approving privileged access. This process typically involves submitting access requests, verifying the need for elevated privileges, and obtaining proper authorization from appropriate stakeholders.

- **Privileged threat analytics:** Analyzing privileged user behavior to detect anomalies and potential security threats.

Most PAM solutions also enforce policy using a variety of procedural controls. Some of these include:

- **Just-in-time (JIT) privilege elevation:** JIT privilege elevation is a procedural control that enables users to obtain privileged access on an as-needed basis for a limited time. It reduces the risk of prolonged privileged access by granting temporary and specific privileges only when necessary, and PAM solutions provide this capability.

- **Privilege escalation and delegation:** PAM solutions provide controls to manage the escalation and delegation of privileges. These controls determine who can elevate or delegate privileges, under what circumstances, and for what duration.

- **Password management policies:** Procedural controls around password management define the policies for password complexity, rotation, expiration, and storage. These policies ensure that passwords used for privileged accounts are strong, regularly updated, and securely stored. PAM helps implement these controls.

- **Separation of duties:** The principle of separation of duties ensures that no single individual has complete control over a critical process or system. PAM solutions enforce this control by defining and enforcing role-based access controls (RBAC) that distribute privileged tasks among multiple individuals or teams.

- **Audit and compliance reporting:** PAM solutions generate comprehensive audit logs and compliance reports. These reports help organizations demonstrate compliance with regulations, track privileged access activity, and identify any anomalies or policy violations in line with the continuous validation aspect of Zero Trust.

- **Incident response procedures:** PAM solutions include predefined incident response procedures for handling security incidents related to privileged access. These procedures outline steps to contain, investigate, and remediate any incidents involving privileged accounts or activities and support threat hunting and threat modeling efforts.

- **Continuous monitoring and review:** Regular monitoring and review of privileged access controls are essential to ensure that they remain effective and aligned with evolving security requirements. Procedural controls include periodic audits, assessments, and reviews of the PAM solution's configuration and usage.

PAM solutions typically provide centralized control over privileged access, with features such as password vaults, session recording, and access controls to help enforce security policies and reduce the risk of unauthorized access. By implementing a PAM solution, organizations can improve the security of their IT systems and applications, reduce the risk of insider threats and external attacks, and maintain compliance with regulatory requirements.

With basic PAM in consideration, it is important to understand that PAM itself serves as both a PDP and a PEP. It has capabilities to decide and enforce access control to privileged accounts. When you add Zero Trust to the mix, it can serve as a policy information point (PIP) as well and provide important context and even device posture conditions to the centralized PDP so that when specific privileged users request access, the PDP knows that special conditions need to be achieved.

Similar to how applications keep some control of subject access to themselves, PAM is capable of decision-making as well. PAM uses information from a PDP, or a PIP connected to the PDP, to enforce PAM-specific use cases for privileged users. For example, if a PEP in the cloud is validating a specific user and this user is allowed to access a database server, the access verdict request goes to the PDP. The PDP will ask PAM if this user is allowed access to the database, and PAM will then give its verdict that it's not. This then gets propagated via the PDP to the PEP, and access is denied.

PAM use cases are similar to user-based AD access. The only difference is that the PAM system has manysub-systems that are integrated into a larger system, contrary to the AD, which has a single system managing all sub-functions like key management and access management. PAM may use AD to get more identity-based context, but overall it works in a similar way.

Generally, when enterprises are looking for products, they will be more inclined to get a product that is specialized in more than two of the aforementioned mechanisms. For example, Cisco Identity Services Engine can provide a one-stop shop for a majority of mechanisms discussed, including AAA services, guest services, profiling, posturing, and even rapid threat containment based on dynamic threats. ADs like Microsoft and Azure AD seamlessly integrate with Cisco ISE to provide end-to-end identity solutions.

DNS and URL Filtering

DNS filtering is an important mechanism that can be implemented by most enterprises, even if they are in a beginner maturity phase. It is a simple addition to most security mechanisms but has a much wider impact. DNS filtering involves blocking access to a website by preventing the user's device from resolving the website's domain name to its corresponding IP address. This is done by configuring a DNS server to block requests for specific domains or categories of domains. This approach is often used by organizations to block access to websites that are known to be malicious or inappropriate. It is more proactive than URL filtering and is generally capable of blocking malicious traffic, even if it is encrypted. URL filtering involves blocking access to specific web pages or URLs within a website. This is typically done by analyzing the URL of each page requested by a user and comparing it against a list of blocked URLs or categories of URLs. This approach is more granular than DNS filtering, as it allows organizations to block access to specific pages or sections of a website while still allowing access to other pages.

In general, DNS filtering is easier to implement and can be more effective at blocking access to entire domains or categories of domains. However, URL filtering can provide more granular control over access to specific pages or content within a website. Both approaches have their strengths and weaknesses, and organizations may choose to use one or both, depending on their specific needs and goals. As part of the overall security mechanism portfolio, DNS and URL filtering are key edge-facing controls that need to be in place. Cisco's Umbrella and Security Service Edge (SSE capability of Cisco Secure Access) are market leaders in DNS security, and all Cisco products come with threat intelligence and URL categories powered by the research group TALOS.

Email Security

Email security is an important aspect of workforce security, as it protects against socially engineered phishing. Phishing is an extremely common attack vector, and these attacks can be particularly effective in a non–Zero Trust environment because users typically have implicit trust in the sender's spoofed identity. Implementing email security measures, such as spam filters and sender authentication protocols, can help prevent such phishing attacks and help move the enterprise toward Zero Trust.

Email security is also instrumental in implementing data loss prevention mechanisms to avoid sensitive information being sent over email. Cisco's Email Security solution provides a strong email security solution for most customers with features that include a spam filter, malware checks, as well as outbreak control.

Remote Browser Isolation

Remote browser isolation (RBI) is a security technique that isolates web browsing activities from the endpoint and moves them to a remote server. RBI is relevant to Zero Trust because it helps to address the security risks associated with web browsing activities. By isolating web browsing activities, RBI can help to ensure that the user's device and the corporate network remain protected from web-based attacks, and it can help to strengthen the overall security posture of a Zero Trust environment. With RBI, web browsing activities are executed on a remote virtual browser, and only the rendered display is transmitted to the user's endpoint device. This approach helps to prevent web-based attacks from reaching the user's device and the corporate network. With ransomware and phishing being common security attacks targeting the workforce, RBI is an important feature to explore, though it is generally implemented at a more advanced stage of Zero Trust maturity.

Cisco Umbrella and Security Service Edge (SSE capability of Cisco Secure Access) provide strong RBI capabilities.

Zero Trust Mechanisms for Networks (Workplace)

After exploring the overall Zero Trust tenets and strategy, it is clear why a location shouldn't be a mandatory deciding factor on trust. A server farm or a campus network does not mean implicit trust. What it does mean is that locations are not completely obsolete. Locations are still relevant in deciding specific types of access. A campus user might use dot1x, whereas a guest user might use a guest portal. The access varies based on various segmented networks. Segmentation is the key word here that we will be discussing in this section. Here are some common capabilities for networks:

- Cloud access security broker (CASB)
- Macro-segmentation and related security controls
- Direct Internet access for branches
- SD-WAN for branches
- Encrypted traffic visibility
- Flow visibility and flow stitching
- Network access control

Cloud Access Security Broker

Cloud access security broker (CASB) is a security solution that provides visibility, control, and protection for cloud-based applications and data, especially SaaS. CASBs act as intermediaries between cloud service providers and cloud consumers, enabling organizations to apply security policies and monitor user activity across multiple cloud platforms.

Similar to email and web proxies, a CASB can provide mechanisms like MFA, data loss prevention, and can even be utilized for posture and compliance checks. CASB is usually utilized to provide access to a SaaS application based on multiple factors. Being cloud native, CASB is a critical component of SSE. Whereas a CASB does identity verification, authentication, and authorization, it is also a gatekeeper between you and the application, which makes it a network-level control. The application will have its own authentication mechanism (to a cloud LDAP, for example), which is different from the CASB. It is analogous to buying a ticket to a carnival and then needing a separate ticket to enter the House of Mirrors.

CASB is an important tool to detect misconfiguration on the cloud configuration or even for monitoring for other anomalies. Cisco SSE is a strong partner solution for CASB.

Macro-Segmentation and Related Security Controls (NGFW and NGIPS)

A primary prerequisite and technical outcome of Zero Trust is segmentation. Macro-segmentation involves segregating the network into broad enclaves. If you consider the standard enterprise architecture, there are already common enclaves as part of years of segmentation. Some of these include the Internet, DMZ, extranet, server farm, and so on. Of course, segmenting to a macro enclave basically means implicit trust within the enclave, and this is something Zero Trust doesn't endorse. That is where security technology like virtual router forwarding (VRF) and NGFWs come into the picture. Both NGFW and VRFs are common security mechanisms used to achieve logical segmentation.

Consider an example of how an NGFW and VRFs help serve the macro-segmentation purpose illustrated in Figure 9-7.

A common division of the segments in a campus network will be to use a wired and wireless VRF so that wired and wireless clients do not communicate to each other without security control. Apart from these two VRFs, which will most likely constitute the entire employee base, there are two more critical VRFs, which are IoT and guest. IoT devices will constitute endpoints that cannot perform dot1x or are not directly mappable to a real person. For example, printers, video conferencing equipment, IP phones, and so on can be considered as part of this VRF. Some of these devices might support dot1x but can provide only machine identities. These devices are likely going to be identified by their machine ID, which is a certificate (if installed) or their MAC address. IoT devices generally follow a hybrid model, where IoT will have further subdivisions like printers, IP phones, and so on, which might get assigned a separate functional group. These can communicate to each other with a specific security control (IP ACL, TrustSec, and so on), and all these functional segments are protected by a coarse-grained NGFW that controls north-south traffic.

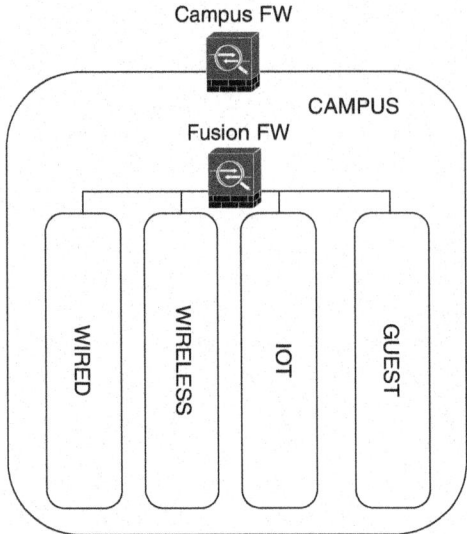

Figure 9-7   *A Typical Campus Network Segmentation*

Finally, the guest user is a common non-employee who is provided access either with a guest self-registration portal or with a sponsor. The general use case of a guest VRF is to isolate all the guest users from internal users and provide only Internet access. Devices will not be managed, and posture checks will be minimal. Usually, any device that cannot be managed or cannot provide corporate credentials (vendors, partners, and so on) ends up being under the guest VRF. For example, a research institute might mandate all their guest scholars to use the guest VRF. Depending on how frequently scholars show up, they can be long-term and short-term guests. Some key concerns with overall guest VRFs communicating with other VRFs in the network are as follows:

■ Overall lack of posture for endpoints

■ Lack of traffic encryption unless the server being accessed encrypts the traffic in HTTPS

■ Lack of user access and session limit, which would allow a guest to essentially use the Internet for more than the stipulated time, and if this user is compromised, access will not be revoked

As per Zero Trust principles, an employee must be able to access enterprise resources from a guest network in the campus or even from an airport Wi-Fi. The security controls must be in place for any such identified use case.

Any communication between these VRFs must have NGFW control along with IPS and other important security controls like malware inspection and application visibility. This is to ensure that inter-VRF (or segment) east-west communication is well protected from lateral movement. This is where IPS and IDS, along with several other security controls, will be discussed.

Apart from macro-segmentation, an NGFW or an IPS can provide the following security controls as well:

- **Advanced threat detection and prevention:** NGFW can detect and prevent advanced threats such as malware, ransomware, and zero-day attacks. This helps to prevent attackers from gaining unauthorized access to the network.

- **Identity-based access control:** NGFW can provide identity-based access control by enforcing policies based on user identity and role. This helps to ensure that only authorized users can access sensitive data.

- **Application visibility and control:** NGFW can provide visibility and control over applications and services, enabling organizations to control the flow of data and prevent unauthorized access to sensitive applications.

- **Continuous monitoring and analytics:** NGFW can provide continuous monitoring and analytics to detect anomalies and potential security threats in real time. This helps organizations to quickly respond to security incidents and prevent data exfiltration.

In a standard enterprise, security controls are generally placed just behind a segment edge device. For example, in the Internet segment, the NGFW or an IPS would be placed after a common routing entity. This is why protection policies are important when it comes to identifying security controls. As discussed previously in Chapter 8, "Building a Zero Trust Architecture," in modern architectures, location drives the creation of protection policies. An Internet protection policy will have a distributed denial of service (DDoS) capability inline before any routing entity is encountered, but a server farm protection profile will have more controls related to data exfiltration, advanced malware protection, and application security.

Observe what is common between all security controls that have been discussed so far. IPS, NGFW, web application firewall (WAF), DDOS capabilities, and so on are all enforcement points, but the key here is that they can make independent decisions based on policies that an administrator configures. For example, an NGFW has firewall rules, an IPS has IPS signatures, and so on. When relevance to the Zero Trust architecture is considered, the mechanism that binds these enforcement points together is the PDP. When there is a common policy creator and decision-maker that can influence all of these enforcement points, that's when the architecture has transitioned toward a software-defined policy model and adopted true Zero Trust–aware enforcement points.

The end goal is to introduce some level of orchestration and automation into policy creation with network devices so that the entire access control is risk-based and dynamic. In traditional networks, a change in application flow requirement or a change in security posture for a specific endpoint would need a manual change on all the network devices. This is cumbersome for security operations and usually is avoided due to its operational overhead. Zero Trust adds in the element of policy orchestration where a products like AlgoSec and Tufin provide policy orchestration capabilities to support the broader automation vision. These products integrate with the PDP, and when there are changes in security posture for an end user or an entire network segment, the PDP updates the orchestrator and subsequently all changes get pushed automatically to enforcement points like firewalls and IPS. The advantages include overall ease of end-to-end policy operations as well as end-to-end adoption of dynamic risk-based policies. The administrator achieves an end-to-end insight into flows, intents, and policies.

Let's consider the Zero Trust–aware WAN and campus diagram illustrated in Figure 9-8.

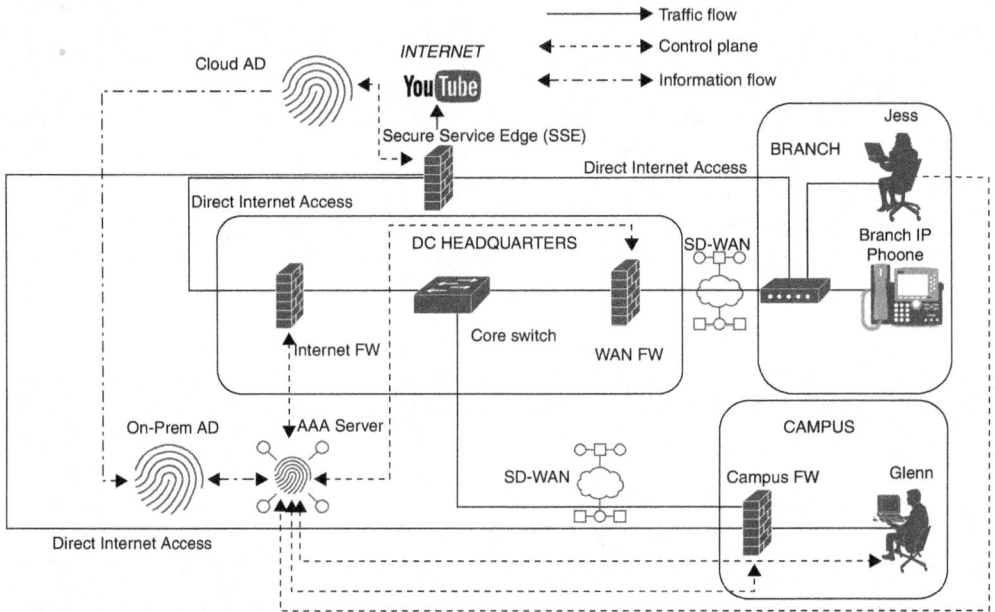

Figure 9-8   *A Common WAN and Campus Network*

The common factor here is the PDP (in this case, a AAA server coupled with Active Directory). Together they maintain the policy of the entire enterprise, which is distributed to various enforcement points. The most common enforcement point at a border that also is performing the macro-segmentation is the NGFW. This is the key aspect we want to drive home. Any PEP in the network becomes "Zero Trust aware" as long as it can consume identity, context, attributes, conditional requirements, as well as policies from the PDP. Over time, if there are multiple branches, having an on-premises firewall for each branch might increase total cost of ownership, which is where cloud solutions like SSE seamlessly integrate with the SD-WAN to provide a scalable and sustainable security solution.

Most Internet gateways provide firewalling and IPS capabilities. This is an important aspect to extend or maintain uniform policy both on the cloud as well as on the premises. Threats exist both on the premises and in the cloud, and NGFW and next-generation intrusion prevention systems (NGIPS) play key roles in the cloud. Cisco Secure firewalls and NGIPS provide much-needed inspection capabilities at multiple segments of the network and help propagate strong macro-segmentation practices. The capabilities also extend to cloud-native solutions such as SSE.

Direct Internet Access for Branches and Campuses with SSE

As covered previously in Chapter 4, "Always Start with 'Why,'" Internet backhauling is a common tactic performed by employees who want branches to maintain the same policy as headquarters. In a Zero Trust network architecture, the same connection performed over an MPLS setup can be directly provided over the Internet with TLS encryption. This greatly improves connection and still gives the option to maintain the same policy in the cloud.

SD-WAN for Branches

Software-defined wide area network (SD-WAN) is a technology that allows organizations to connect their branch offices and data centers to the cloud using software rather than proprietary hardware. SD-WAN uses a combination of multiple Internet connections, including broadband, LTE, and MPLS, to create a hybrid WAN infrastructure that provides high availability and optimal performance.

SD-WAN is an answer to the long and tedious question of how to securely connect branches to the headquarters. SD-WAN allows an enterprise to implement strong security communications between branches and headquarters. For example, in a Cisco SD-WAN, control communications are encrypted in TLS and data communications are protected by IPsec. In a Zero Trust environment, SD-WAN can help enforce security policies and ensure that only authorized users and devices are allowed to access corporate resources. By combining SD-WAN with Zero Trust access policies, organizations can create a secure and scalable network architecture that supports remote work and cloud adoption without compromising security.

As an extension, SD-WAN solutions integrate with Security Services Edge capabilities (providing Zero Trust policy creation and management) to achieve a larger Secure Access Service Edge (SASE) architecture.

Zero Trust Doesn't Need Decryption (End-to-End Encryption)

Another advantage in a Zero Trust environment is the encryption between the PEPs. Currently without PEPs, all traffic between the user and application within the network is essentially unencrypted. Now, with a PEP at both the user application and network level, all the traffic is encrypted, preventing unwanted disclosure. Essentially any enforcement point can deny access to the application if encryption is missing, if there is a need to reauthenticate with MFA or to even revalidate single sign-on (SSO) credentials via Security Assertion Markup Language (SAML).

The key factor here is striving to achieve an end-to-end encryption rather than stopping it at a VPN concentrator and then assuming trust between the concentrator and the server. These are common pitfalls that lead to lateral movement and data loss. Deploying HTTPS/TLS/DTLS tunnels also removes the need to have a specific VPN concentrator because the data itself is protected through the TLS tunnel.

Flow Visibility and Flow Stitching

A connection is a tangible network communication channel facilitated by physical ports, and a session is an information exchange channel and is entirely logical. Independently or as a combination, the term **context** is misaligned with connections and sessions. A context is assigned based on broader logical attributes and not specific to just details limited by the OSI layer. A **network flow** aims to alleviate this to an extent, but a network flow is still restricted to characteristics identified within the OSI layer and is specific to the current traffic that the device processes. Hence, for a security model to be able to process and evaluate context and intent, visibility in the network must be mature enough to stitch traffic from various parts of the network together.

Usually most enterprises have packet brokers, inline taps, or SPAN ports that send packets to a central processing solution like an advanced threat solution or a sandbox. The concern with this method is that the central solution does not get clear information relating to the scope of the flow, the changing identities, or other traffic-specific characteristics. Hence, a continuous monitoring solution is critical to monitor various strategical points in the network. Cisco Secure Analytics (previously Stealthwatch) is an example of one such product that provides this capability, whereas Splunk is another example for visibility solutions across the enterprise. These solutions serve as PIPs. They aggregate all flow information and send the attributes of a flow to the PDP so that the PDP can make an informed decision based on available information.

Visibility solutions are useful not only after Zero Trust implementation but also before a Zero Trust assessment. They are commonly used to understand all the relevant business flows in the scope of the assessment, which helps architects and business analysts to understand the key gaps in the security architecture and implementation. With better flow visibility, granular access control can be implemented based on the intent of the flow rather than the tuples of the packet. Flow visibility also supports and feeds the segmentation efforts in an enterprise.

Additionally, flow visibility helps threat hunting and incident response to a great extent. With more clarity in the overall flow path, it is easier to identify critical points where an attack or an infection can be quarantined. Most modern compliance frameworks require all assets in the enterprise to send logs to a centralized visibility solution. For example, PCI-DSS regulation requires that audit logs must be retained for a year and that 90 days of logs must be available for immediate analysis. Similarly, there are incident response and time to detect metrics that need to be in place to comply with specific industries. For these reasons, visibility is an important driver and critical mechanism in achieving not only compliance but also alignment to the Zero Trust vision.

Network Access and Control

From the perspective of Zero Trust, network access control is an umbrella capability that covers a wide range of enforcement points that receive their decisions from a common decision point. NAC helps enforce the principles of Zero Trust by providing visibility and control during network access.

NAC solutions can identify and profile devices seeking network access, including laptops, smartphones, IoT devices, and servers. This information enables organizations to establish a baseline of trust for each device and apply appropriate access policies. NAC solutions further integrate with IAM systems to authenticate users seeking network access. By leveraging strong authentication methods such as MFA, organizations can ensure that only authorized users gain access to network resources.

Consider NAC as one enforcement mechanism that utilizes all the identity and attributes that have been extracted from the user and device segment. NAC helps enforce access policies based on the Zero Trust principle of least privilege. Access policies can be granular and dynamically applied based on the device's security posture, user context, and network conditions. NAC ensures that devices and users are granted access only to the resources they need and under the appropriate conditions. Ranging from validating security posture of the endpoint workstations to policy management and access control, NAC has various functions that makes it relevant in controlling access to a dynamic subject profile.

Most enterprises usually have some level of user identification in the network; however, NAC is usually a second thought. Security posture is either an operational overhead or impacts the user experience with more wait times. However, most compliance and regulatory bodies mandate the validation of security posture. With NAC, a wide range of compliance checks (such as minimum OS level or presence of security patches) is validated before providing access to the network. As an extension of the visibility capability, NAC logs provide much needed insight into the subject profile, which enhances the context of access requested and subsequently provides more insights to the PDP to make a valid decision to allow or deny access. This applies not only to end users accessing the network via the enterprise infrastructure but also to headend devices and IoT devices that cannot produce a username. Printers and phones get differential treatment thanks to the NAC capability.

NAC solutions work with existing enforcement mechanisms and identity solutions to rapidly block a threat in the environment. If an endpoint has been deemed malicious and has been identified by a security solution, NAC can then integrate with this solution to get the metadata for the flow and block the entire endpoint from accessing the network. It is a versatile capability to build and maintain in the network to support the Zero Trust strategy.

Zero Trust Mechanisms for Data and Applications

Discussions around objects/workload or data storage focus on a more static set of entities:

- They do not use dynamic IP assignment and are usually assigned to a fixed IP.
- They do not have a user assigned to them. Their identities are usually machine identities.

- Their traffic patterns are fairly constant. The server runs an application, and the traffic will likely be consumers accessing that application.

- An application does not try to circumvent security controls.

Keeping these characteristics in mind, some common mechanisms in Zero Trust specific to adding more context are as follows:

- Micro-segmentation

- Data security

- Asset management and automation

Micro-Segmentation and Its Implication

Micro-segmentation is highly relevant in Zero Trust architectures because it helps to reduce the attack surface and limit lateral movement of attackers within a network. By combining micro-segmentation with Zero Trust principles, organizations can create a highly secure network that limits the access of users and devices to only the resources they need to do their job. This approach makes it more difficult for attackers to move laterally within a network, as they would need to gain access to multiple segments to do so. If we were to look back at some of the principles of Zero Trust, the key aspect micro-segmentation solves is reducing the blast radius of an attack and avoiding lateral movement.

Micro-segmentation allows organizations to apply more granular security policies to different segments of their network, based on the sensitivity of the data or the importance of the application hosted within each segment. This approach improves security and compliance by limiting the exposure of sensitive data and ensuring that security policies are consistently applied across the network.

Consider the server farm implicit trust use case. Most modern enterprises have huge implicit trust in their server farm simply because of assumed trust. Micro-segmentation allows you to maintain business flows without compromising on security. A typical micro-segmented server farm would look like Figure 9-9.

The diagram in Figure 9-9 should be familiar, but it highlights some interesting aspects. One is that the NGFW placed at the border of the server farm doesn't have to cater to all the east-west communication control. This is a common advantage that comes with distributed enforcement points coupled with a centralized policy management server and decision-maker. By effectively moving the policy decision-making capability to a centralized entity and decoupling the decision-making from the data path enforcement points, it becomes safe to remove east-west control from the NGFW and shift it towards the applications and their microsegments, which is exactly the advantage that micro-segmentation provides. With effective micro-segmentation, it is possible to enforce east-west flows in a more fine-grained mechanism. Applications still hold the final enforcement, but the fine-grained network control allows for greater scope of automation as well as scale, especially in enterprises with large application counts. What generally happens in these networks is that the coarse-grained north-south and east-west flows end up overloading the NGFW. With micro-segmentation and technology like TrustSec or VMware NSX, the control can be transitioned to a different point in the overall end to-end access control architecture. With a common PDP, this fine-grained enforcement can also be centralized, making the end-to-end flow truly Zero Trust.

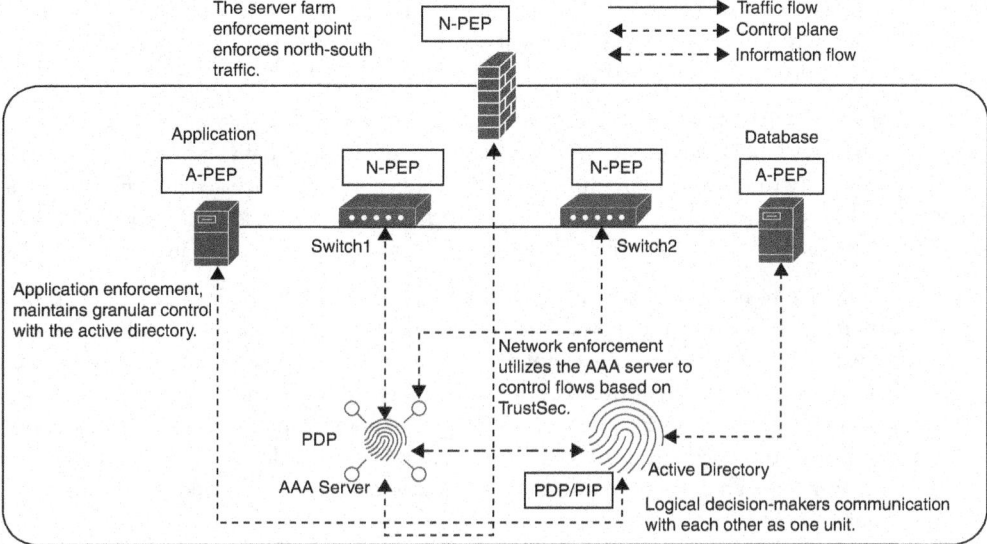

Figure 9-9 *A Simple Visualization of Micro-Segmented Security Control*

Data Security

Data security in a Zero Trust model involves implementing various security measures to protect sensitive data, including:

- **Data classification:** Identify the data, classify which types are sensitive, and tailor security controls accordingly.

- **Data encryption:** All sensitive data should be encrypted both in transit and at rest to prevent unauthorized access.

- **Data loss prevention:** Data loss prevention tools should be used to prevent data leaks, whether accidental or malicious, and to monitor and control the flow of data within an organization.

- **Access controls:** Access controls should be implemented to ensure that only authorized users can access sensitive data. This is done through the use of role-based access control (RBAC), MFA, and other techniques.

- **Monitoring and logging:** Continuous monitoring and logging of data access and usage will help identify potential security incidents and prevent data breaches.

Data security has much in common with how privileged management works in a Zero Trust environment. Similar to PAM, data security tools have the same bidirectional relationship with the PDP. The PDP is the decision-maker in the entire flow and therefore the key aspect of data security is the classification levels. When considering data security, it is imperative that the PDP makes decisions considering the data classification levels as well. For example, a privileged user is allowed to make changes to the data stored in a resource classified as confidential but not top secret or secret. Various data security models like Bell-LaPadula and Biba's model will come into play, and the PDP must consider these characteristics when making a decision.

Similarly, the PDP will become an information source for data security controls like DLP, which will use the PDP to get the verdict about certain types of data and then make a decision locally to allow or deny the flow of that specific data. Since security mechanisms like DLP and PAM cannot be avoided in a network, they will still maintain capabilities to take decisions but will consult the PDP as well as provide information to it.

It is worthwhile to note that even though these capabilities might be indispensable now, they still represent a fairly old methodology of protection. For example, a DLP would be needed when data exfiltration is attempted using SSH or other known methods of extraction. Similarly, PAM stores passwords, which overall is no longer the direction of the security community. Zero Trust changes the way we authenticate as well as the way data is protected. With strong encryption and more granular access, loss of data is possible only if a user set is compromised and the compromised user has the need to know access to the data.

In general, the simplest approach to data security with Zero Trust is to enhance the security controls of the other four segments: users, devices, networks, and applications. This automatically covers a large chunk of data security. DLP capability is an integral part of the cloud-native SSE solution.

Asset Management and Automation

This is a no-brainer, and we need to accurately be able to catalog the "what" aspect of the equation. Workload assets or applications as well as network assets like devices, network devices, and so on are key subjects and objects, and they need to be cataloged to make sure each attribute is captured. With these attributes, their access policies will be derived and then put into the Zero Trust policy needed to create granular access. Data cataloging is also part of the asset management and inventory, and the same must be part of asset inventory to begin with. This is an important step of either beginning or augmenting the asset management to support the attribute creation of the resources. Most Zero Trust implementations begin with asset discovery, as it is critical to know what subjects, network assets, and objects exist to create policies for them based on flows.

In a typical enterprise network, asset management is usually manual unless the software explicitly performs automated asset discovery or at least automated asset monitoring. Another aspect is that enterprises usually have different solutions for endpoints and applications. For example, an enterprise might use Endpoint Central to manage endpoint asset inventory but use Flexera to manage server and application inventory. The plan should be to unify the asset management to a centralized content and asset management database solution like ServiceNow.

Asset management need not be an isolated software solution. For example, a AAA server or an AD server manages the user assets. MDM manages mobile assets. A software-defined network controller manages your application and workload assets. A database management system manages your data assets. The aim of asset management as part of the Zero Trust vision is to provide visibility into all assets with a single pane of glass to manage and automate asset discovery.

Zero Trust Mechanisms for Visibility with Security Orchestration and Automation

Similar to user segmentation and overall macro-segmentation, visibility mechanisms are most likely already present in your network and are just not tuned enough to provide you the right orchestration and automation to match your Zero Trust strategy. Some of the key aspects of visibility, security automation, and orchestration will be explored in this section. It is important to note that when topics like segmentation or uniform policy enforcement are being considered, the desired outcome is automated enforcement at all PEPs based on a dynamic changing context and risk. Visibility and related topics are more informational and less enforcing. Due to this reason, many enterprises do not put in a lot of effort to develop these capabilities in the network. The reality is that these capabilities are not specific to a PEP or PDP but are more relevant to a PIP. Without knowing what to protect and how they change their access characteristics, how will PEPs keep up with enforcement? You cannot protect what you cannot see. At a high level, some key factors to consider when designing good visibility, orchestration, and automation mechanisms include the following:

- Security and threat intelligence

- Security operations center (SOC)

- Threat hunting

- Handling of malicious insiders

- Security analytics

- Security orchestration and automated response (SOAR)

Investing in Security Intelligence

Security intelligence is not just a list that serves as a checkpoint before more detailed inspection is done. It is the sum total of all observations around the enterprise boundary, which can feed into actions to at least identify a potential problem. It is basically identifying what information you need as an enterprise by considering all static and dynamic boundaries and then start gathering and processing information from various sources. Some common sources include the following:

- **Open source intelligence (OSINT):** This includes information from publicly available sources such as news articles, blogs, social media, and other online forums.

- **Closed source intelligence (CSINT):** This refers to intelligence that is not publicly available, such as data gathered by government agencies, law enforcement, and intelligence services.

- **Human intelligence (HUMINT):** This refers to intelligence gathered through human sources, such as informants, undercover agents, and espionage.

- **Technical intelligence (TECHINT):** This involves the collection and analysis of technical data, such as network traffic, system logs, and malware samples.

- **Signature-based intelligence:** This involves the identification and analysis of known malware signatures and patterns, such as those used by antivirus software.

- **Behavior-based intelligence:** This involves the analysis of the behavior of threats, such as malware, to identify new and emerging threats.

- **Indicator-based intelligence (IoCs):** This involves the collection and analysis of data related to specific indicators of compromise, such as IP addresses, domains, and file hashes.

- **Threat feeds:** These are services that provide continuous updates on new and emerging threats, based on various sources of intelligence.

- **Dark web intelligence:** This involves monitoring the dark web for information related to cyber threats, such as new malware strains, stolen data, and hacker forums.

- **Industry-specific intelligence:** This involves the analysis of threats specific to a particular industry, such as financial services, healthcare, or government.

Creating intelligence out of such a plethora of information is not easy and must go through an intelligence cycle. The intelligence cycle is a systematic process of collecting, analyzing, and disseminating intelligence information in order to support decision-making. The cycle typically involves five stages, as illustrated in Figure 9-10 and described in the list that follows.

1. **Planning and Direction:** In this stage, the intelligence agency or organization establishes goals, priorities, and specific information requirements. This stage also involves identifying potential sources of information and assessing their reliability and credibility.

2. **Collection:** In this stage, the intelligence agency gathers raw information from various sources, including human sources, technical collection methods, and open source information. This stage might involve covert or overt collection methods, depending on the nature of the information being sought.

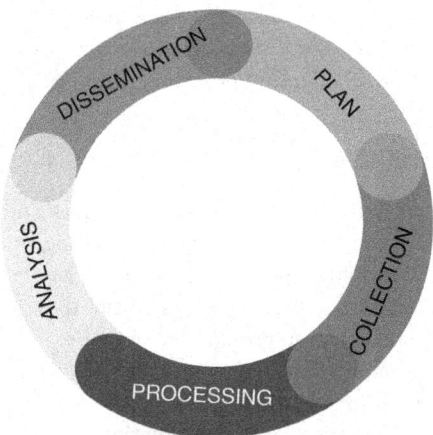

Figure 9-10 *Information Lifecycle*

3. **Processing:** In this stage, the raw information collected in the previous stage is transformed into a usable format, which may involve translation, decryption, and other forms of analysis. This stage might also involve filtering out irrelevant information and prioritizing information based on its importance.

4. **Analysis:** In this stage, the processed information is analyzed in order to produce meaningful intelligence products, which may include assessments, estimates, and other forms of intelligence analysis. This stage might also involve identifying patterns and trends in the data, as well as assessing the credibility and reliability of the information.

5. **Dissemination:** In this final stage, the intelligence products produced in the previous stage are disseminated to the appropriate decision-makers. This might involve presenting the information in a variety of formats, including written reports, briefings, and other forms of communication. The intelligence agency may also provide feedback to sources and collect feedback from recipients to improve future intelligence collection and analysis. Cisco TALOS as an organization follows and disseminates intelligence for all consumers of Cisco products.

All this information is collected, processed, analyzed, and converted into intelligence. Once this becomes intelligence, it helps provide much more relevant information about the threats and serves as a viable input into the overall decision-making for the PDP. It is an important and critical aspect of visibility, as it gives a view of many aspects of the enterprise and not just threats. It is key to prioritize your intelligence reports and also important to understand the different type of team members who can pull in the right people and understand various threats, which may not fully be dependent on the IT infrastructure. Some threats are targeted to reputation or maybe the destruction of a building. The focus should be to get the right information and identify the right threat. For example, at a CxO level, geopolitical issues such as a war are sources of information that could influence the way a CxO makes decisions. These are strategic in nature. There might be tactical information like day-to-day operations or even daily authentication access attempts that feed into overall threat hunting and control selection process.

Security Operations Center (SOC) Capabilities

An enterprise security operations center (SOC) solution is responsible for monitoring and analyzing security events and alerts generated by various security tools and technologies such as firewalls, intrusion detection systems, and endpoint protection solutions. The SOC team can quickly identify and respond to potential security incidents before they become major security breaches. In addition, SOC analysts can provide ongoing threat intelligence, conduct vulnerability assessments, and maintain security policies and procedures to ensure that the organization's security posture remains effective over time. As is with all tuning activities, it is not recommended to integrate every information source into a SOC platform, as this would lead to a large number of false positives. One must spend time and identify the actual sources and types of information required to run daily operations to make sure that all incidents/events viewed are logged with relevant context; however, this is not a simple automated task. Apart from looking at logs and figuring out if there is an attacker hiding in there, you need to identify which information sources are relevant to the investigation, as not all the sources would be useful for the security incident in question. Once this information is collected, it will get processed and fed back into the intelligence cycle for enhancement.

It's crucial to ensure that security operations teams have the necessary maturity to handle your organization's security challenges. There isn't a single solution that can solve all visibility issues. The efficacy of a SOC solution, just like any other security solution, depends on the expertise of the analysts using them. Therefore, before investing in any technology, it's important to make the necessary investments in people and processes. Advanced technologies are more beneficial to more mature security programs. This is a key feedback that is given to most enterprises, especially managing a SOC. A SOC capability is key to process multiple sources and provide key information to decision-makers such as a PDP or threat hunting activities.

Be mindful to choose a solution that not only meets current needs but that can evolve with your organization over time. Consider the following before choosing your solution:

- Does the solution accommodate both on-premises and cloud deployment models?

- Can the solution ingest raw format logs from a variety of sources, including less common or homegrown sources?

- Does the solution accommodate application-level use cases or is it limited to network and traditional security solution use cases?

- Does it address emerging use cases such as operational technology (OT), industrial control system (ICS), or Internet of Things (IoT) deployments within the enterprise?

- Does it use modern technology like artificial intelligence?

Deploying an enterprise SOC is not easy and can easily take more than a year to develop a fully functional SOC solution, especially if the organization is complex. Even if deployment is completed within a year, it will take much time and effort to learn and improve detection capabilities along with building skillset and processes to support the overall solution.

Zero Trust greatly enhances the data that a SIEM processes within a SOC. One of the important capabilities for any SOC platform is to corelate data. With technology like VPN and NAT, an end-to-end flow will need multiple extraction points to be stitched together to get the full context along with added identity. With Zero Trust, this is no longer a requirement because the access model is similar to the cloud where applications are accessed with their real IP. Hence, log processing is not only simplified, but also enhanced with the additional attributes and contexts that are collated by the SOC solution. Most compliance and regulations mandate logging to a centralized event management platform.

Threat Hunting

An essential activity in identifying security intelligence is threat hunting. Threat hunting is generally not a part of the overall security portfolio of the enterprise when it is just starting out. Threat hunting is essentially a subset of security intelligence in the sense that it still passes through an intelligence lifecycle, but the target here is to identify threats. Threat hunting closely aligns with Zero Trust in a way that it always assumes a breach and then begins hunting for threats. Lack of indicators of compromise does not imply the absence of a breach and definitely does not signify effective security controls. The assumption of breach is a key reason why threat hunting is an important tool in a Zero Trust architecture. By assuming a breach, you are recording a possible threat scenario that might not have been fathomed and then running an active assessment to validate the scenario and subsequently creating a security control to avoid it. Once a control has been identified, a quantitative analysis is done to make sure that the control implementation is justified. Quantitative analysis of security controls with asset value and exposure factor involves assessing the effectiveness of security controls in mitigating the potential impact of security incidents on assets. The process involves the following steps:

- **Asset valuation:** Determine the value of assets that are critical to the organization. This includes data, intellectual property, infrastructure, and other valuable assets.

- **Threat assessment:** Identify the potential threats that could compromise the security of the assets. This includes cyberattacks, natural disasters, and human errors.

- **Exposure factor:** Determine the percentage of asset value that is exposed to the identified threats. This is based on factors such as the likelihood of the threat occurring, the impact of the threat, and the existing security controls in place.

- **Risk assessment:** Calculate the level of risk associated with the identified threats and exposure factors. This is derived by multiplying the asset value by the exposure factor.

- **Security control analysis:** Evaluate the effectiveness of existing security controls in mitigating the identified risks. This includes controls such as access controls, encryption, backups, and disaster recovery plans.

- **Cost-benefit analysis:** Determine the cost of implementing additional security controls and compare it to the potential reduction in risk. This helps determine the most cost-effective security controls to implement.

Threat hunting processes multiple sources of traffic, including enterprise logs, security information and event (SIEM) logs, and even endpoint detection and response (EDR) logs. These static log analyses are coupled with an active penetration testing activity to validate the identified threat scenarios. This methodology confirms the identified threats and gives more context to an enterprise when selecting security controls. Keeping this in mind, one of the biggest threats to an enterprise is a malicious insider.

Threat hunting removes our dependency on signature-based technology because you go hunting for threats and try to bridge that gap between the known and unknown. A skilled attacker will try to make an attack as unique as possible, and threat hunting examines these anomalies with a target to identify a pattern that can be utilized to put specific controls in place.

Identifying Malicious Insiders

Malicious insiders have a special place in a Zero Trust conversation. They represent the epitome of everything Zero Trust is against. An insider is basically anyone who has access to information that is not otherwise accessible to external entities. Insider threats can come from various types of insiders, including the following:

- **Employees:** These are people who work for the organization, have access to sensitive information, and are authorized to use the company's resources.

- **Contractors and temporary workers:** These are individuals who are hired on a temporary or contractual basis to work for the organization. They might have limited access to sensitive information and systems, but their activities can still pose a risk.

- **Partners and vendors:** These are individuals or organizations that have a business relationship with the organization and might have access to its systems and data.

- **Board members:** Members of the board of directors or other governing bodies can also pose a risk if they have access to sensitive information and use it for personal gain or to benefit their own interests.

- **Privileged users:** These are individuals who have elevated privileges and access to critical systems and data, such as system administrators or IT personnel. They might be able to exploit their access for personal gain or malicious purposes.

- **Former employees:** Even after leaving the organization, former employees may still have access to systems or information and can potentially cause harm or share sensitive information with others.

- **Hackers:** In some cases, external actors can also become insiders by exploiting vulnerabilities or using social engineering techniques to gain access to the organization's systems and data.

If you see a close pattern here, that means some of the book's content has made sense to you. If not, this is the point that the context of Zero Trust will make more sense. Complacency regarding user segmentation is one of the key reasons insider threats are on the rise. Enterprises assume an implicit trust between administrators or other types of similar users and believe that they cannot be malicious insiders or disgruntled employees. Essentially, every user in the preceding list must be isolated at a user level from each other. Access should never be the same for any of these roles, especially board members who are most commonly targeted by spear phishing and whaling.

Zero Trust doesn't adopt the concept of trusted insiders because:

- Insiders propagate the concept of implicit trust.

- Access is almost always the entire scope of the enterprise assets.

- Risk of exposure is much higher.

Of course, insider threats are not always malicious. It is the responsibility of the individual to make sure they are educated with the social engineering attacks that external entities might post. A malicious actor may pose as a member of your enterprise IT team and extract your password and MFA token, essentially masquerading as you in the network. That doesn't mean you as a user are malicious. It just means you were careless and should still face consequences for your negligence. The impact of an insider threat materializing can vary based on what assets the company finds critical. As a security company, loss of data due to an insider threat could mean loss of intellectual property, reputation, money, and so on. Security control is not always about protecting customer information, though it is paramount to protect it. Security controls are always tailor-made to protect the assets in scope of the overall risk the enterprise faces.

With Zero Trust, you should essentially have identified all the subjects and their access restrictions to objects in the enterprise environment. If there is a malicious entity that has gained wrongful access via collusion or some other means of social engineering, the blast radius is still limited to a limited user, network, and workload scope, thus protecting you from a massive breach. Zero Trust has multiple tenets in play here:

- Identifying the right access for users and their objects

- Isolation of these objects and subjects

- Continuously monitoring the access to provide it just in time and just what is needed on a need-to-know basis

This is where designing a network with all the user segments, network macro-segments, and application micro-segments is an unavoidable task. Subject access requirements must be identified clearly, and multiple means of analysis such as syslog, SIEM, UEBA, heuristic IPS, and so on must be utilized to identify the context of a flow and to determine its anomalous nature. Two actions are key to remember here. An enterprise must understand who the insiders are and identify what each asset's business context is with respect to their insiders.

Insider threat is not just an IT issue; it is an organizational issue. There must be strong security awareness and training programs to mitigate some of these concerns.

Security Analytics

Security analytics (SA) is more than just looking at flows and identifying their business context. It is about additional monitoring of the same flows and comparing it to the original intended business contexts. It involves taking preventive measures when a specific behavioral pattern is breached, taking into consideration multiple identities and attributes. The Zero Trust strategy will always assume that a subject has already got access to the network. In the case of a malicious insider, they already had access and the assumption was accurate. The motive of behavior analysis is to understand the standard behavior of multiple users and identify anomalies. It is better to know how much access was gained and what abnormal behavior is being exhibited by a subject early during the attack phase rather than knowing it after data has been exfiltrated. The sooner you act, the better you can respond.

Secure analytics consolidates data from multiple segments and then provides visibility and context to the PDP. It is a key aspect of both a SOC solution and security orchestration and automation response (SOAR) capability. When it comes to SA, pulling together data from all over the enterprise and then combining that data is as important as preventing the attack itself. Examples of such data might include threat intelligence, telemetry from applications, user information, or information from data loss prevention products. The right analytics will give you contextual information on targets and assets that you have already identified in your network as part of asset discovery or threat hunting. The goal is to understand why the target is being attacked and how fast they must be protected based on their asset value, data classification, and other critical attributes. The second aspect is the ability to detect and respond to the threat itself because being an information point is useless if you cannot feed it into an enforcement point. It is important to make sure that the data being monitored is useful to the enterprise monitoring it. It is counterproductive to analyze or monitor something you cannot take an action on.

The final goal of an effective analytics deployment is to:

- **Avoid large-scale spread of an infection:** One of the first activities an infection does is spread. SA deployed in the right points in the network can give much needed visibility and also will help feed into incident response teams and controls to take timely action.

- **Cover more Zero Day threats:** The whole concept of a Zero Day threat is that there is no or limited information about the threat to be translated to a signature. With more focus on behavioral analysis, specific types of attacks or exfiltration attempts can be prevented.

- **Complement threat hunting and forensics:** SA feeds into threat hunting tools with much needed context and possible IoCs. These can be utilized by threat hunting teams to identify new threats or strengthen the controls against existing threats. During an incident response situation, SA can help identify where an attack is and what is the current behavior being exhibited by the entity. Added context, visibility, and threat intelligence give security analysts more information on which information source to act on.

- **Monitor activities inside the network:** SA systems collect and analyze data from various sources, including applications, endpoints, data loss prevention (DLP) systems, IAM systems, and network flow data. By utilizing features such as user and endpoint behavior (UEBA), these platforms can detect malicious users and compromised accounts through the analysis of user activity. Additionally, analyzing network traffic enables the identification of suspicious behavior, such as infected endpoints or compromised accounts.

- **Make SOC capability and SOAR capability improvements:** Many enterprises employ SIEM solutions at the heart of their SOC. To help increase analyst productivity, SA platforms typically include or integrate with SOC solutions and SOAR tools, making operations more efficient and moving toward automated incident responses.

- **Support compliance efforts:** Compliance support for standards and regulations (like the Payment Card Industry – Data Security Standard [PCI-DSS], ISO 27002) is a standard use case even in today's security architecture. SA platforms can provide much needed capabilities like archiving, incident tracking, audit logs, and so on, which support an overall compliance initiative in the enterprise.

Before deploying any analytics, make sure they fit the business needs of the enterprise and match its maturity. Specifically, consider the following:

- The specific **threat** or **attack type** that monitoring needs to detect.

- The **assets** or **capabilities** vulnerable to that threat. Develop and be able to articulate the impact and business value that monitoring provides. Align this to the needs and language of the business.

- The **data sources** required to support this use case. Your data sources will vary by use case but may broadly include network flow data, alerts, and telemetry from applications and security solutions. Efficacy usually increases when an SA platform also receives external threat intelligence, vulnerability information, and other data about the assets at risk.

- **Data dependencies.** Ensure that you can support ongoing monitoring for the specific use case as long as it remains relevant to the priorities of your risk portfolio.

Understand potential upgrade (or divestment) paths for your data sources in order to sustain the flow of appropriate information to the SA platform.

Just like security metrics, use cases encompass three distinct levels of concern: strategic, operational, and tactical. Your use cases must reflect the mandatory requirements of all three of these components, from the business need and threat landscape to the implementation of the analytics within your environment.

When you're initially adopting security analytics, it is recommended to choose a single most valuable use case that will provide immediate benefit. Implement this first and consider everything else as a "nice to have" capability that will augment your team's continuous improvement projects. Ultimately, your use cases must deliver value. Deploying analytics for analytics sake will seldom benefit anyone other than the vendor that provides your data storage solutions. Similarly, low-risk assets that require expensive operational changes to monitor are unlikely to drive the value you seek. Finally, as part of your management process, consider removing use cases that are unaffected by security incidents or that yield excessive false positives because those use cases may consume resources better dedicated elsewhere.

It is important to note that automation in the realm of cybersecurity heavily relies on security analytics. Without the insights and confidence provided by security analytics, it would be challenging to automate breach response. As the security landscape moves toward an automated future, it is crucial to start preparing early.

Security Orchestration and Automated Response

Security orchestration and automated response (SOAR) is an approach to cybersecurity that combines security orchestration, automation, and response to streamline and automate security operations. This involves using technologies to integrate and automate the processes and workflows associated with threat detection, investigation, and response. SOAR is a key component of enterprise security, not because it protects you from an attack, but it keeps you running at the same pace as threats. With the right security analytics and a mature SOC, SOAR can greatly support and enhance an enterprise's response to an attack. It is not common in growing enterprises but is a more mature and deliberate action when the overall security operations are performing above expectations. SOAR also helps automate some of the basic SOC analysis and incident responses to alleviate some of the work done by T1 analysts so that they can focus on more pressing and high priority alerts.

SOAR fundamentally relies on security processes and how those processes interact with the rest of the organization. It is important to acknowledge that SOAR in general is not a technical control. It doesn't involve an implementation of technology but is more focused on enabling automation for existing technology like SOC capability, IR capability, threat hunting, and so on. Hence, the capability of SOAR to support the enterprise will greatly depend on how mature its SOC and other operational and administrative activities, especially governance policy and process definition, are. With an effective SOAR, or at least a viable SOAR strategy, SOC response times to incidents are greatly be improved. Consider the following example to understand this concept.

Consider a user is accessing the Cisco website. When the user logs in, they will be authenticated using Duo MFA and will be provided access based on SSO. An external website flow is illustrated in Figure 9-11, and an internal website flow is illustrated in Figure 9-12.

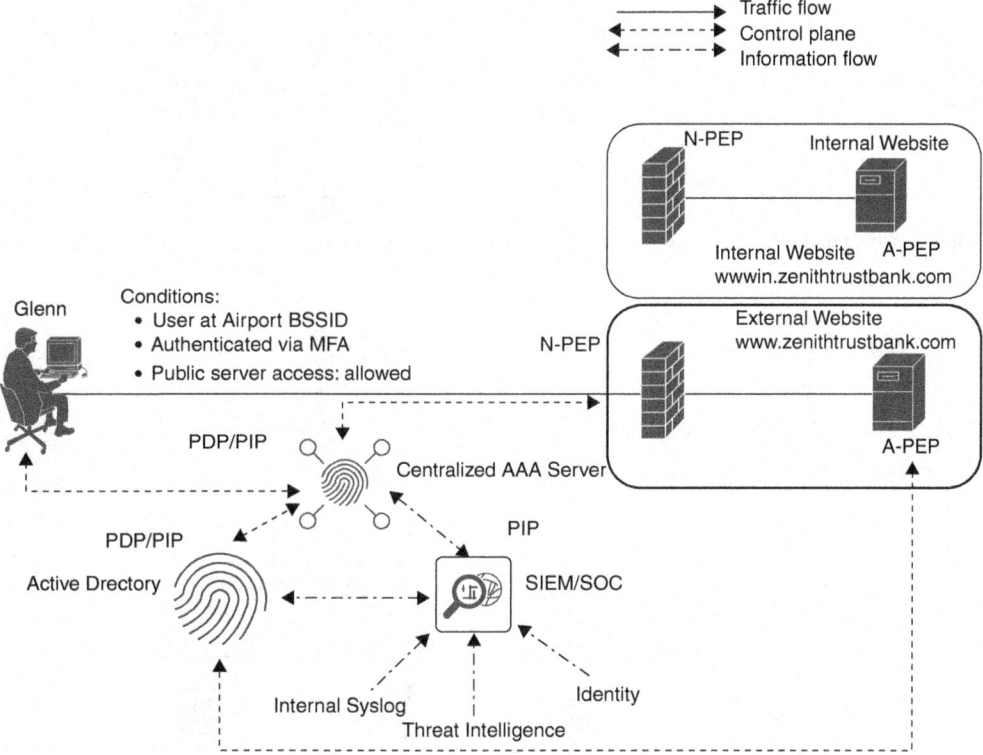

Figure 9-11   *External Server Access Flow on a Public Wi-Fi*

Now when the user moves to another SSID, maybe like an airport Wi Fi and so on, the BSSID changes. This will be used as a trigger to reevaluate the access conditions, and a SOC solution powered by a strong SIEM will be a great asset in this case. The enterprise SOC solution will collate all data from multiple silos and provide a context to the PDP. The PDP queries the SOC, gets the required context, and makes a decision. When the location changes, it triggers an automated playbook to validate context again. Figure 9-13 illustrates what the revalidation flow would look like.

It is also possible to trigger revalidation via any incident reported by the SOC solution. For example, change in network, reauthentication, session timeout, accessing a specific server at a specific classification, and so on are some triggers that can force the PDP to reevaluate access.

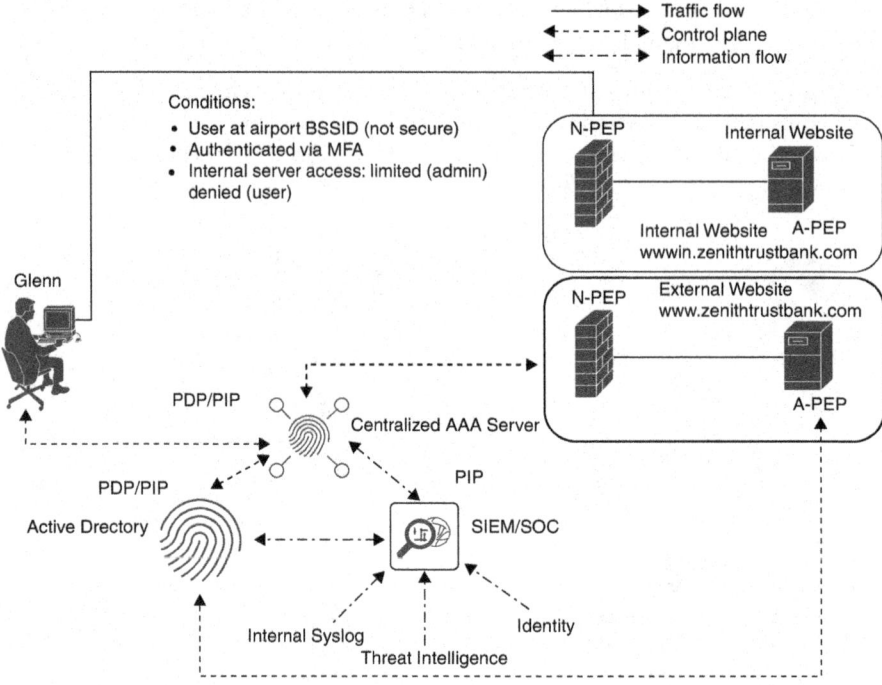

Figure 9-12   *Internal Server Access Flow on a Public Wi-Fi*

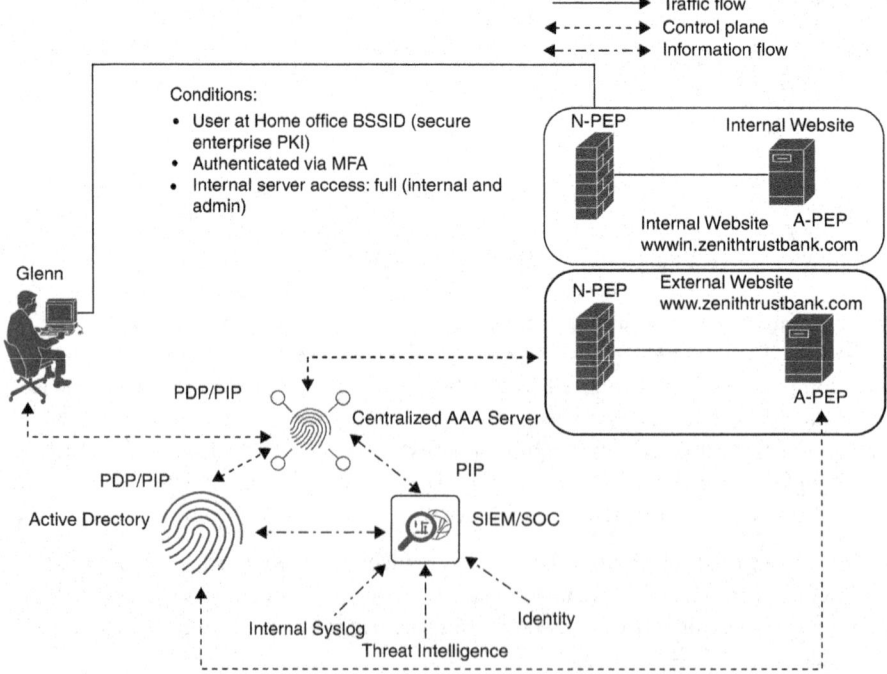

Figure 9-13   *Revalidation When Location Changes*

Another aspect to understand is that SOAR is not a single pane of glass for visibility. Most likely this "single pane of visibility" is almost always custom-made by enterprises. A SOAR interface is a common location for all automations and orchestration playbooks in an enterprise. It takes in information from multiple sources but itself is not an overarching common information point.

We are at a strong juncture to understand that automation and orchestration are not the same. The use case for automation is to be able to repeat simpler or even slightly complex tasks, depending on the maturity. As manual tasks make an activity more cumbersome and time-intensive, automation helps alleviate that excess time spent in doing tasks manually. Orchestration, on the other hand, is performing tasks in a specific order and then providing a tangible response or outcome to a specific trigger. For example, a simple orchestration playbook would be to process all intrusion events and identify whether they are false positives. With the advent of AISecOps, some of this processing and analysis, which is part of the playbook, can be done using machine learning. Automation, on the other hand, would be a simple task of deduplicating logs and storing them as part of the SOC platform architecture.

The SOAR should also trigger the PDP to enforce certain actions. For example, consider the extension of the website access use case, which is blocked based on an IoC elsewhere in the network. Figure 9-14 illustrates this scenario.

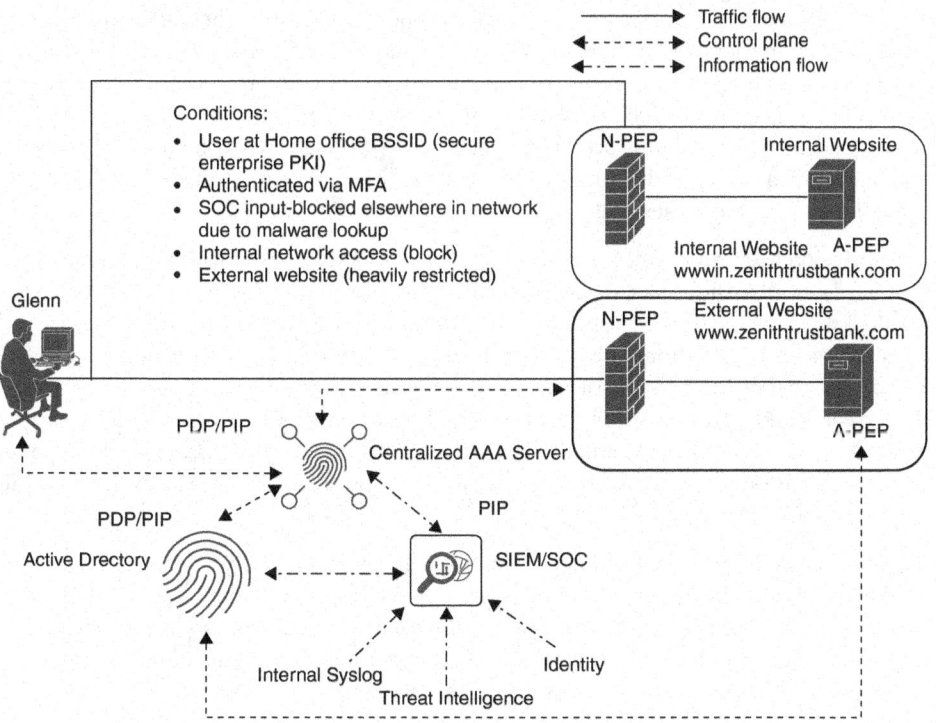

Figure 9-14 *User Blocked Access When Triggered by SOC*

When a user has been identified as compromised, there will be a high-priority incident registered in the incident management platform, and this will trigger a playbook, which will in turn let the PDP know that any access for this user should be blocked. Zero Trust thus adds additional versatility to decision-making. Of course, triggers to deny access will be more than just a compromised user. Lack of privilege will also lead to a reevaluation and potential denial of access.

When an enterprise is planning to enhance existing automation, it finds tasks that take time to perform and greatly rely on humans. These are usually repetitive and are likely a technical task like enabling a specific user or validating access and so on. When an enterprise wishes to embark on enhancing overall orchestration, it will look for a process to improve like incidence response and then create an automation playbook for it.

Finally, security can no longer function in a silo. Security must do what the business requires, and hence it will build policies to support it. The operations team will then translate, design, configure these policies, which will then be audited by the business to make sure all business requirements are incorporated into the operations. This is called a rule of engagement. A rule of engagement in security orchestration refers to a predefined set of guidelines or instructions that govern how security operations, network operations, and incident response teams should respond to security incidents. These rules are created to ensure that everyone involved in the security process understands their roles and responsibilities and follows a consistent approach in responding to security incidents. Primarily, network and security teams need to work together and cannot continue working in siloed environments. With more adoption of Zero Trust strategies and a keen understanding of the cooperation and teamwork required to successfully deploy and manage a Zero Trust architecture, it is clear that infrastructure and security teams must spend less time deciding scope and more time working toward the common vision of maintaining a secure and operable infrastructure. The concept is detailed in Chapter 7, "Zero Trust Avengers, Assemble!"

In security orchestration, rules of engagement can be defined using various techniques, such as playbooks, runbooks, or decision trees. They typically specify what actions should be taken in response to specific security events, who should be notified, what tools and technologies should be used, and what escalation procedures should be followed. The goal of having rules of engagement in security orchestration is to improve the effectiveness and efficiency of incident response, reduce the time it takes to detect and respond to security incidents, and minimize the impact of security threats on the organization. With SOAR in place, teams can automate response actions, and analysts can focus on more complex incidents.

Cisco Secure Analytics uses advance technology like machine learning and artificial intelligence to identify and correlate multiple aspects of a flow. Coupled with SIEM technology like ArcSight and Splunk, along with security orchestration like Cisco XDR, automation and orchestration is made easy for a larger number of customers looking to move their SCA and SOAR initiatives to the next level.

The Deep Dive

[Glenn completes his mechanism discussion and pauses to let the information sink in. It is a common strategy he uses when he has been explaining a lot of technical information. The entire virtual team, including the operations team, the deployment and security architecture teams, as well as specialists, are processing the required information mechanisms. He turns to Mr. Jonathan Smith and continues the discussion.]

Glenn: In summary, you have a lot of options and security mechanisms to implement, but, holistically, the security mechanisms that you will definitely need to consider based on the road map we provided are as follows:

- **Users and Devices**

    - Identity and access management enhancements

    - Profiling and posturing

    - MDM for mobiles

    - BYOD

    - Privileged access management

- **Network and Multicloud**

    - SD-WAN

    - Direct Internet access

    - Macro-segmentation

- **Applications and Data**

    - Asset management for devices and data

    - Micro-segmentation with data classification

    - Data security

- **Visibility, Analytics, and Orchestration**

    - Secure analytics

    - Invest in security intelligence and threat hunting

    - Security operations center

    - Automation initiatives with SOAR

    - Identify malicious insiders.

Mr. Smith: *[Turns to Mr. Chen.]* This looks like a very huge set of mechanisms and security controls to set up. *[He turns to Ms. Lee, Mr. Eaton, Mr. Jed Moon, the operations director, and Mrs. Arianne Straum, the technical director.]* Are you across all of these mechanisms? Have you built out a deployment road map for this and what we need to achieve this?

Glenn: Mr. Smith, actually, this step is more to identify the capabilities and subsequently the vendors you want to partner with to deploy your target architecture based on the gap analysis that we had done previously. You already have a fair set of security controls in place. It is easier than you would imagine. All you need to do is follow the road map that we have created and, over time, the entire network will be transformed to a Zero Trust network architecture that you envision.

Mr. Smith: Great! Give us some time and we will digest this information, identify the right vendors for each mechanism we need to deploy, and send out RFPs as needed.

Glenn: Thanks, Mr. Smith. As part of your strategic discussions, please consider Prolink Solutions as your strategic partner and trusted Zero Trust advisor. Most enterprises find it useful to stick to the consultants that have crafted their road maps. If you would like I can give you a rundown of the products portfolio that Prolink Solutions provides under each of the security mechanisms provided.

Mr. Smith: I think we will consider that as well when we discuss this internally. For now, I would like to really thank you for all the information and knowledge that you have shared as well as your insights into our network. We will reach out to you shortly once we have finalized how we will approach the next steps.

Endnote

1. "CISO Perspectives and Progress in Deploying Zero Trust," https://cloudsecurity-alliance.org/zt/resources/ciso-perspectives-and-progress-in-deploying-zero-trust/

Present the Zero Trust Strategy and Metrics

After rigorous architecture designing, metric evaluation, and team building, the next step is to present all the hard work to stakeholders. Some important approaches will be considered when presenting the strategy to the leadership. The following chapter is covered in this phase:

Chapter 10 Presenting the Zero Trust Strategy

Chapter 10

Presenting the Zero Trust Strategy

Until this point, the focus of embarking on the Zero Trust journey has been gathering the right presentation points to represent the overall Zero Trust strategy. The information prepared includes the following:

- The business driver for Zero Trust adoption

- Key performance metrics

- Key risk metrics

- Tactical roadmap

- Important executive stakeholders

- Critical Zero Trust mechanisms

- Vendors and products to fit the tactical road map

Armed with this information, it is time to prepare the final pitch. Optimization of the vendor budget should already be in place, but the conversation with leadership is not just going to be about budget. It is going to be a larger discussion about the paradigm shift and how the entire enterprise needs to come together to help achieve the overall vision. The point that needs to be driven home is that Zero Trust is much more than an access control model. It is an amalgamation of all aspects, including people, processes, and technology, and hence has its roots in all spheres of an enterprise security strategy. Adopting a Zero Trust strategy gives the enterprise a chance to look in retrospect at its security practices.

Unfortunately, leadership cannot just flip a switch and get a Zero Trust–enabled network. Setting up and converting the existing infrastructure to Zero Trust would take some time, but getting people to support and maintain the momentum of practicing Zero Trust will take longer because people are more difficult to convince than network devices. Zero Trust is a philosophical change rather than a technical change and therefore the budget will be more than the standard digital transformation. Yet, the outcome of the initial spend is measurable and the impact will last much longer.

This chapter covers what it takes to get the Zero Trust ball rolling, especially with presenting to leadership. Unlike adopting a new security technology, Zero Trust requires a perspective change, which has been covered extensively in Chapter 2, "The Zero Trust Kaleidoscope," and Chapter 4, "Always Start with 'Why.'" Presenting Zero Trust is not about convincing the enterprise that it can be done. It is more about explaining why the shift is needed and how the enterprise can benefit from it.

Presenting Zero Trust to the Enterprise

Presenting the Zero Trust strategy to leadership is the culmination of all concepts that have been discussed so far. It starts with conceiving the idea and then slowly building a set of leadership supporters for your cause. As you garner more support, the need for specialists arises, and over time, with the right guidelines, a strategy and architecture are formulated. Of course, this will need the strong support of a lot of teams that usually do not work together. Legal, infrastructure, security, and many more teams need to join forces to build a virtual team to plough forward toward the business vision. This is by far the most important prerequisite before presenting a strategy. A clear strategy, a viable architecture, a deployment road map, and a coalition of all the critical teams are imperative for the successful presentation of the Zero Trust initiative in an enterprise.

Once the basic prerequisite work has been done, it is time to create and present a viable Zero Trust proposal to leadership. To be able to successfully talk about Zero Trust to leadership by speaking their language is important. The presenter must become a translator, depending on who they are communicating to. Leadership cares about performance and risks and wants to understand how Zero Trust aligns with the business. Compliance segments concern themselves with how Zero Trust helps them maintain or augment compliance initiatives. Operation segments aim to reduce downtimes with the adoption of Zero Trust and want to understand how Zero Trust will impact overall tactical activities. When you present metrics to explain the advantages and risks of the Zero Trust concept, keep the target audience in mind.

Some key aspects and mandatory content that should be in your presentation are described in the sections that follow.

Target the Perception

Since the main aspect of the presentation is changing the mindset, time needs to be spent on making sure that all the stakeholders and target audience are on board with your strategy and are open to the idea of the large change that needs to be undertaken. The sections that follow cover some common topics to drive the mindset conversation.

Convey That Security Is a Business Risk, Not an IT Concern

After multiple security breaches and data loss, security is gradually being amalgamated with the overall business strategy. However, security is still almost always overlooked when it comes to data center transformation or campus initiation projects. To be clear, the projects and initiatives being considered in this example are end-to-end enterprise initiatives and not just product deployments. When campus networks, for example, are being newly designed and discussions around macro-segmentation are initiated, one of the first questions asked is, "Does the enterprise have an existing segmentation strategy?" The answer to that question is almost always, "Yes, a bit, but we do not manage it, and it is handled by the security team." An anomaly that you should have picked up on as a reader is that the consultant is speaking to the infrastructure teams for security feedback. The overall picture that infrastructure leads have is that security strategy is as simple as "protect the network" and that this can be achieved by strong security hardware and the right security controls. Subsequent strategies usually involve acquiring new technology and implementing more hardware to fix gaps in the network that are best addressed with more strategic controls involving people skills and processes.

This viewpoint is slowly changing as more lines are blurred between the infrastructure and security. Creating a security strategy is a very personal activity for the enterprise and is incepted and driven at a very senior level. Irrespective of what the business vertical is (and I do mean any vertical, ranging from wide-scale IT enterprises to a shopkeeper down the road), security is always a business driver. There are at the very least two entities you want to protect—your own data and your customer's data. Loss of this data is not just a technical concern but a business issue. It could lead to reputational loss, revenue loss, and much more bigger concerns. Hence, strategizing the right security capabilities to fit the business vision and mission is key.

From a Zero Trust perspective, this would be a good opportunity to involve the larger enterprise into the overall security strategy. Usually, most teams just outsource security strategy to security teams. This mindset needs to change, and as leaders, there needs to be an inclusion of all the business owners to clearly articulate their business requirements and help practitioners and architects construct strategies around business requirements rather than just create asset-focused policies.

As part of the overall conversation about business risk, identify the key use cases that greatly enhance the current status quo or that alleviate current pain points in the network. This will help drive home the idea that embarking on a Zero Trust journey is not just a security control but a need for effective realization of the business.

Know Your Target Audiences and Push the Right Buttons

Remember that you might not always be successful in pushing a concept to leadership's door. A great deal depends on you pushing the right buttons. For example, if the enterprise has had a recent breach, then a lot of eyes will be on you to explain how a Zero Trust architecture could have reduced the damage. However, if the enterprise specializes in medical equipment, the perspective is not to enhance security but rather to enhance operations. Whatever you present about Zero Trust easing incident response will not have the desired impact. Understand your target audience and make sure the message is received in alignment with the right business driver that will benefit from the Zero Trust vision.

More Devices Does Not Mean More Secure

Focus on building a strategy rather than a large technology base. This topic was covered in Chapter 2, "The Zero Trust Kaleidoscope." When you're proposing a Zero Trust strategy, the vendor or technology matters only to optimize the budget and derive the right features based on business drivers. What is important is to identify a technology partner who can guide you throughout your journey and then utilize their products that closely align with both your security and business strategies. Focus on understanding "why" and "what." The "how" will follow automatically. The business strategy is important to focus on before technology drivers are selected, as discussed in Chapter 4, "Always Start with 'Why.'"

Showcase the Security Landscape and Relevance of the Strategy

This might essentially look like a scare tactic, but it is more of a broader visibility enhancement approach into the enterprise and its ecosystem. The introduction must cover the broader security landscape to showcase the current direction of the security community. An enterprise might have multiple business drivers and might (or might not) be aware of the complex attacks being launched against other enterprises. The landscape must be clear to the enterprise leads, and the message we need to drive home is that an attack is not a matter of *if* but *when*.

Some other topics to cover include modern threats and the market for specific types of attacks and controls. It would be a plus if some of these threats can be aligned with the existing threat hunting in the enterprise. This section must end with explaining the overall mission of the proposal, the mission statement, and the guidelines to implement them.

Showcase Alignment to Overarching Business Strategy

This might seem to be a repetitive theme, but it's only so because of how important it is to the overall business strategy. Identify the business drivers and strategy and then map the Zero Trust initiatives to the strategy. This will give leadership an idea of how the initiative will add value to the existing drivers and also how the strategy can augment it. If the enterprise wants to do a rapid expansion to the cloud, then maintaining the security of its employees and customers will be key. If the enterprise wants to change the business model, the Zero Trust strategy will adapt if needed. The presentation must showcase how the initiative aligns with overall business drivers.

Explain why you are embarking on each of the listed tactical initiatives. This provides a perspective of where each initiative stands in the overall big picture for the enterprise. In the end, leadership wishes to keep the board and stakeholders happy. This leads them to create certain business drivers, which are the core fundamental policies that become the base for most security and end-user policies. When you're proposing Zero Trust, especially to senior leadership, the alignment to business strategies must be clear. For example, if the business strategy is to provide fast and seamless access for employees to enterprise resources and Zero Trust breaks that strategy, you will find it hard to push the idea and get approval.

As you identify the key use cases to support the business, you need to showcase how Zero Trust supports, enhances, and aligns itself with the overall business strategy. Qualitative or quantitative analysis can be used, but it is always useful to give a dollar value to the value-add that Zero Trust can bring in. Alignment with senior leadership is an important piece of the puzzle because if an entire enterprise has to change the way they perceive security, it has to be deployed top-down.

As an additional step, set your vision, mission, and goals early on in the discussions. This is important so that down the line when you provide the proposal and begin the implementation, all your stakeholders must be able to track the goals you had set earlier. These must be tangible goals that can be measured so that the progress boosts more initiatives in the overall Zero Trust strategy. Present the vision, mission, and finally approach to present the Zero Trust strategy with each business driver aligned with a tactical initiative.

After having a meaningful conversation with the customer and finally understanding what they want to do and why, most consultants will be eager to jump in and implement immediately and start planning vendors; however, this is not always a good idea. One must build a strategy where security controls, vendors, and many other aspects are deeply analyzed. By now, each control or strategic mission must be aligned to a business driver and therefore consultation work with the appropriate teams must have begun and the strategy should be begin following the change cycle. The cycle typically involves four key steps:

Step 1. **Assess the current situation:** This involves taking a critical look at the organization's current strategy, as well as its strengths, weaknesses, opportunities, and threats. The gap analysis that was performed during metrics creation in Chapter 5, "Measuring Zero Trust Success," is one such step.

Step 2. **Identify new strategic options:** Based on the assessment, the organization can identify new strategic options that may be more effective in achieving its goals. Metric creation will lead to key initiatives being initiated. This will suggest a strategy shift if needed.

Step 3. **Evaluate and select new strategies:** The organization then evaluates the potential benefits and risks of each new strategy and selects some that are more likely to succeed. The strategy building and identification of critical mechanisms that you performed in Chapter 8, "Building a Zero Trust Architecture," and Chapter 9, "Critical Security Mechanisms for Zero Trust Architectures," cover these topics.

Step 4. **Implement and monitor the new strategy:** The final step involves putting the new strategy into action and monitoring its progress over time. This might involve making further adjustments and adaptations to ensure the strategy remains effective and aligned with the organization's goals.

This will need deep foresight to implement, and you must always remember that you cannot deploy Zero Trust in a couple of months and then forget it. It is a journey, and not a small one. It is not **revolution** but more like **evolution** to a better, secure access model. Hence, the sponsor's overall predilection for the concept, and the capability of the organization to adopt the concept, will be key in the success factor of Zero Trust. Zero Trust strategies must at a bare minimum be adaptive and dynamic. Your strategy may vary if business needs change, and that should be the way the initiative functions. When business direction changes, so must the business strategy, but the security strategy's principles must stay aligned.

Measure the Existing Maturity

Generally, when the maturity of an enterprise is measured, it is mapped against how many attacks or how much budget was spent on the security. This is no longer relevant to the current security landscape. The CISA has mapped out a maturity matrix that has been covered extensively in Chapter 5, "Measuring Zero Trust Success." These metrics are extremely useful to set goals to achieve as part of the road map. A gap analysis is a common entry point into a Zero Trust conversation to explain what is missing and what strategy needs to be adopted to achieve the vision.

As mentioned throughout the book, Zero Trust is not just technology-focused; the enterprise's people and processes need to be considered as well. If there is a value-add in implementing security and additionally improving the existing processes, that will be a stronger business driver. As long as it can be measured, the metric must be presented. The more you can measure, the more effective the conversation with leadership will be.

Showcase the Best Zero Trust Team

Know the people who make a difference. Being an introvert, I cannot stress enough that Zero Trust is a people (technically identity-driven) model. You cannot sit in a bar and talk to a security implementation engineer about Zero Trust and get their approval. Engineers might agree with you over a drink that Zero Trust makes sense, but they will not have a say in the overall direction of a company. It is important to understand the right people in your enterprise who will be on the side of this huge Zero Trust paradigm shift. This is because they will have the power to approve or reject the idea, and if you can convince them, they will help you convince others as well. Make sure these enablers in your enterprise clearly understand the science and reasoning behind what you are trying to do. Remember that before you can implement Zero Trust, you need to sell it.

It is important to showcase all the stakeholders that are involved in the overall success plan for Zero Trust. You should have already identified the key specialists and leadership representatives that will support the enterprise to achieve the vision and sponsor the Zero Trust initiative. The main motive of showcasing the stakeholders is to show that it is not just the security team that holds the responsibility of successful implementation of Zero Trust, but the support of other groups like compliance, governance, business owners, audit, and so on is needed as well. The accountability, of course, stays with the security business unit. Each stakeholder is aware of the risk when Zero Trust is not implemented, and the presentation will show how risk-based decisions are made to support the initiative.

Sometimes it might not always be under security to secure approval or budgets. If you are able to position Zero Trust as a business enabler rather than a security enabler, it becomes easy to convince non-security business units as well. For example, application and business owners constantly complain about security being restrictive with access and that each time they need access to other applications or the Internet they need to ask for more ports to be open or more access control. What if Zero Trust could give them full freedom as long as identity and flow-based controls are in place and the business has a clear understanding of what the application has to achieve and not what ports need to be open? When business owners understand the advantage of getting this tangible freedom when deploying applications, you will get more backing from multiple sources, and then moving around budgets with existing projects will not be an issue. Collective momentum helps push larger initiatives when the concept becomes appealing to a lot of stakeholders. This is especially useful to influence people who are not entirely sure about whether to adopt Zero Trust.

Alternatively, use external factors like increase in attacks, a vendor discount, or something that can emphasize to a sponsor the dire need to move to Zero Trust. Overall, the right folks need to back you to be able to sell the idea. Remember that there are people, processes, and technologies involved in the overall Zero Trust proposal. Showcase how much time and effort you have spent to gather the right team to help propagate the Zero Trust vision.

Present the Metrics

Once the alignment has been established and the team has been introduced, the next step is to showcase how to plan to measure success in the overall Zero Trust initiative. At this stage, you present the performance and risk metrics that you have crafted. The goal here is to show that the business requirements are clear and there is a defined method to measure the success of the overall initiative.

The presentation will explain the current gaps that have been identified in the enterprise regarding people, processes, and technologies. There are usually questions around how the gap was measured and how it can be mitigated. Once details of the gap analysis have been shared, the measurable metrics are presented to show the different levels of metrics (strategic, tactical, and operational) along with how they will be measured (qualitative and quantitative). Leaders share feedback on the efficacy of the metrics presented, and then the presentation will move on to the strategy and road map. Additional details on metrics are covered in Chapter 5, "Measuring Zero Trust Success."

Present the Strategy and Road Map

Once the metrics have been presented, the target audience should have mapped out the key metric that suits them and helps them with augmenting their overall business plan. At this stage, you will present the overall strategy that has been decided based on multiple analysis channels. Once the strategic road map presentation is complete, you must include the vendors you believe will be the right fit for the enterprise to realize their vision.

Showcase the Augmentation of Overall Security Posture

With a clear analysis of the gaps in the network, the final step before the budget is to show the overall augmentation that the adoption of Zero Trust would bring to the enterprise. Irrespective of the specific business goals that an enterprise has, increasing the security posture will always be one of the most important enhancements when adopting a new security strategy.

The primary presentation strategy must be to merge the gap analysis, the potential improvements, their metrics, and how it all comes together in the strategy and subsequently the budget. The board will be interested to know how much improvements are expected to be achieved in the overall enterprise security strategy with the adoption of Zero Trust. The presentation must concisely explain how various aspects of the Zero Trust strategy have helped improve the overall posture of the enterprise. These must fit in with the pain points that were initially highlighted by the enterprise. VPN overload being a problem must map to a more scalable solution to access internal services without compromise in security. The identity of each actor must be enhanced with more detailed attributes for the device from which they are connecting. The transport of communication must add in more context to the overall network flow. Services must showcase their capability to protect themselves from unauthorized access, and the network must show resilience to a malware infection. Improvements must be visible in overall process automation and orchestration to support a more proactive response to incidents. Augmentations must align with business and security strategy.

There are other use cases discussed in Chapter 4, "Always Start with 'Why,'" where access to a SaaS application could be a pain point. The Zero Trust strategy must showcase how it has enabled the users to migrate to the cloud seamlessly with the same or better security posture. These are but a few examples of how you should present the impact of the Zero Trust strategy to the enterprise.

Present the Budget

A security budget is an indication that your metrics have been crafted in an optimized manner and you have identified the key mechanisms and the products to support you on your Zero Trust journey. When you consider senior leadership, their agenda is simple: perpetuate the business and increase their customer base. Therefore, all their strategies will be around how to increase revenue. Good leaders make sure their decisions do not hinder the business, and better leaders make sure their decisions augment the business. Simply put, at the end of the day, when you present your strategy and metrics and how you plan to achieve them, there will be some expense attached to it. Based on the priority of how you want to showcase your metrics, the budget must be presented. Expect questions that challenge the need for a specific strategy. No one likes to spend early capital, and you must be able to justify each of your products or strategic mechanisms. Every justification you provide for the metrics should indirectly answer, "How will this strategy help the business to do better?" When you propose a strategy or when you bring this up with leadership, the relevance of all the recommended initiatives should be the key top-of-mind for you as a consultant to justify to your customer or leader.

When you are building a budget or proposing a budget, make sure you consider all the existing projects, their pending costs, and their current status. It is possible as part of your discovery that you find an initiative that's bleeding money, or perhaps there is a project that is completely rendered redundant by implementation of the Zero Trust strategy. Identifying these projects and diverting money then becomes critical and an important task you need to perform so that the leadership knows how the current strategy adds value. Leadership must know why you selected specific security mechanisms as well as why you selected certain vendors and how this helps optimize the budget.

An important aspect of budgeting to remember is that the budgeting exercise being performed is not a sure-shot reservation of funds. It is a way of forecasting possible expenditures in subsequent years. This is the reason why road maps for Zero Trust are rarely for a year but are mapped out for multiple years. The usual path for most enterprises it to plan a pilot in the first year and secure funding for the same. Once the pilot is successful and all prerequisites for the large-scale adoption are complete (mass software distribution, end-user communications, and so on), the additional budget may be requested. The motive of the budgeting exercise is to make sure there is an awareness of how much budget is expected each year for the successful adoption and implementation of the strategy.

Showcase an Attack Kill Chain and Zero Trust Relevance

The best way to drive home a point in security is to show how effective it is against an attack. Showcase how modern attacks can be devastating to enterprises and how the new initiative helps to mitigate some of that risk. With good strategy and achievable metrics, Zero Trust can be the deadliest weapon against advanced persistent threats. Make the kill chain presentation personal by selecting one of the many flows discovered in their network. Showcase to leadership how easy it could be to infiltrate their network and then drive the need to move to Zero Trust.

As part of showcasing the kill chain, the important security constructs like risk, vulnerabilities, and threats become real and measurable for most enterprises. This helps leaders to understand how managing vulnerability and patches for all assets as part of Zero Trust initiatives helps propagate a broader compliance alignment.

Highlight the Importance of User Education

Emphasize that a Zero Trust strategy requires the participation and cooperation of all employees. Explain how user education and training will be essential to the success of the strategy. Present how you will craft more employee-engaging training sessions and create a safe environment without the usual blame-game tactics when someone lapses in maintaining security. Showcase how you will engage every team at the grassroots level with rotating security champions and collective responsibility.

Future Readiness

As an extension of a previous point, explain the current threat landscape and threats specific to the type of asset that's critical for the business. Leadership needs to know that the initiative is good for the enterprise and that the technology and strategy being proposed will last for the enterprise and support the business for a long time. The proposal must also showcase how it can adapt to new technological challenges in the network. For example, if there is a new compliance rule regarding the storage of PII, explaining how the new Zero Trust principles and strategy streamlines the process for the enterprise would be a key talking point.

The Presentation

After the follow-up session with Mr. Jonathan Smith, Mr. David Chen, Mr. Christopher Eaton, and Ms. Samantha Lee, Glenn spent time preparing his presentation that had all the important information needed to persuade Zenith Trust Bank to take the Zero Trust journey. His motive was to make sure that Zenith Trust Bank understands the value of Zero Trust and was able to request for the right funding to support its vision.]

[Figure 10-1 through Figure 10-11 show the presentation with some key talking points. Note that the flow is minimal to showcase only the important aspects of the presentation. Use this as a reference only.]

Zero Trust Perception

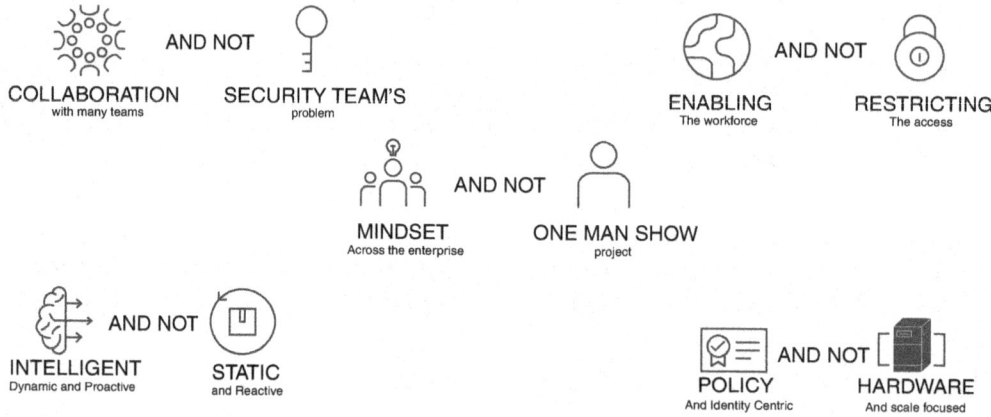

Figure 10-1   *Zero Trust Paradigm Shift*

[Glenn has spent quite some time convincing the Zenith Trust team about how important Zero Trust is. He summarizes this in his introductory slide (see Figure 10-1), where he explains what Zero Trust is and isn't. The goal is to encourage his audience to keep an open mind when he broaches topics like secure private access. The motive here is to make sure the target audience understands what Zero Trust really is at its core.]

Demands from Security Systems

Figure 10-2   *Demands on Modern Security Systems*

[Glenn then moves on to explain what the common drivers are to move toward a Zero Trust Model, as highlighted in Figure 10-2. Some references to business and technical drivers, along with delays in adoption, have been extracted from Chapter 4, "Always Start with 'Why,'" and Chapter 2, "The Zero Trust Kaleidoscope," respectively.]

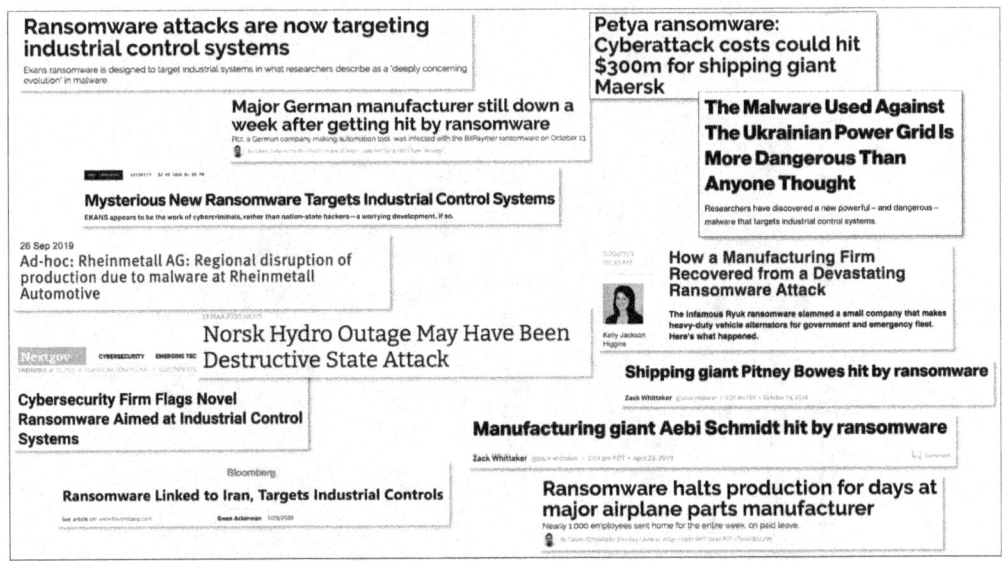

Figure 10-3   *Current Threat Landscape*

[Glenn pivots toward showcasing the modern threat landscape, as visualized in Figure 10-3. As an extension of the first slide, he targets the general complacency enterprises have with their existing security architecture and drives the point home that irrespective of the technology in place, a breach is a matter of when and not if. Modern threats need modern approaches to security.]

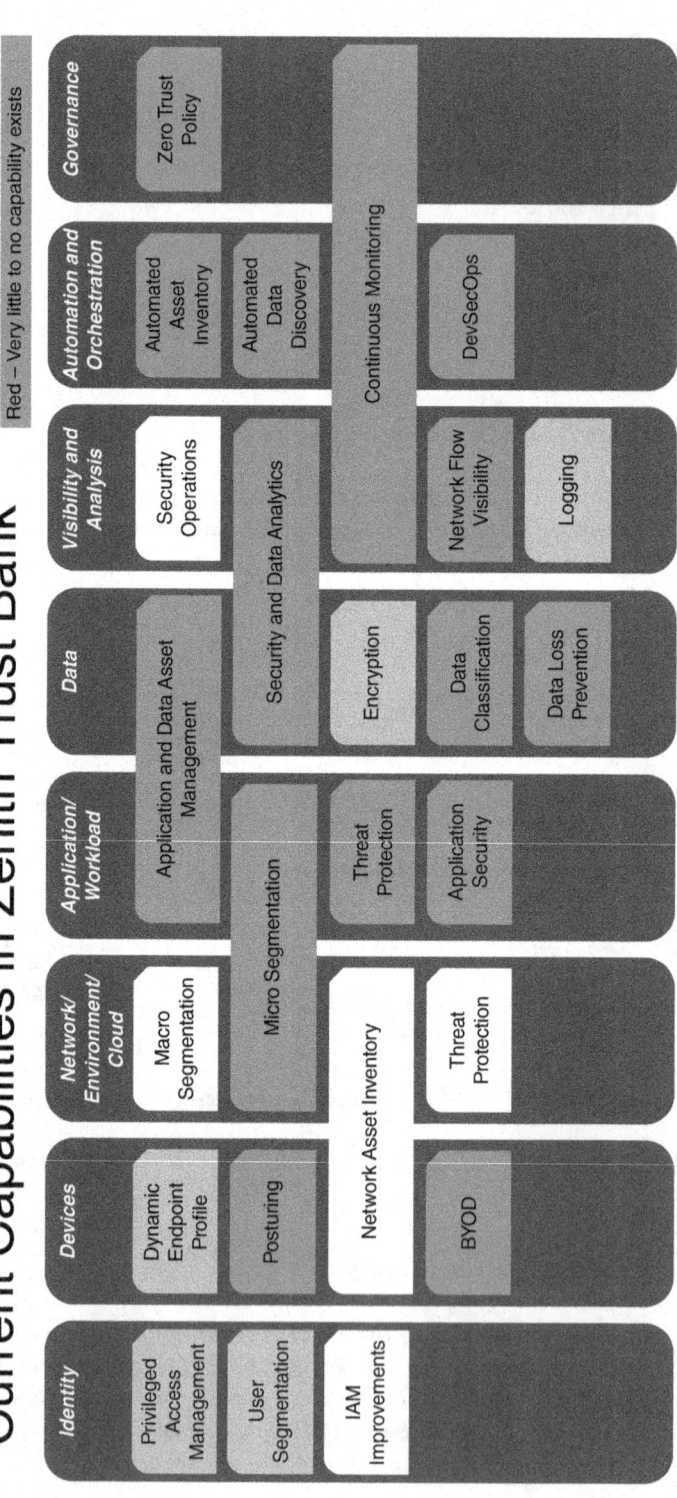

Figure 10-4 *Current Security Capabilities in Zenith Trust Bank*

[Glenn targets the gaps relevant to Zenith Trust Bank. By citing the gap analysis performed with Version 2 of the CISA Zero Trust maturity model, as illustrated in Figure 10-4, Glenn explains that although the enterprise has some strengths, it has a long way to go to achieve the broader Zero Trust vision and mission. Usually, more time is spent on this slide to explain to leadership the rationale behind some of the observations and recommendations.]

Primary Threat Scenarios

 Scenario 1: An insider gets access to customer data and exfiltrates it to a malicious actor.

 Scenario 2: An external entity gains access to a non-secure DMZ server and then moves laterally to get access to a server farm resource. Subsequently it can move freely in the flat data center subnet and get access to customer data and exfiltrate it to a malicious actor.

Figure 10-5   *Primary Threat Scenarios*

[With gaps identified, Glenn uses the threat scenario approach depicted in Figure 10-5 to showcase a viable threat to the enterprise and how it can materialize into something more sinister if the right strategy is not in place. The motive of the slide is not to scare but rather to inform and get the enterprise to accept that a breach is inevitable and that it is the enterprise's responsibility to prepare for the worst-case scenario with the right controls supported by compliance alignment, vulnerability management, and threat hunting.]

What it takes to get Zero Trust right

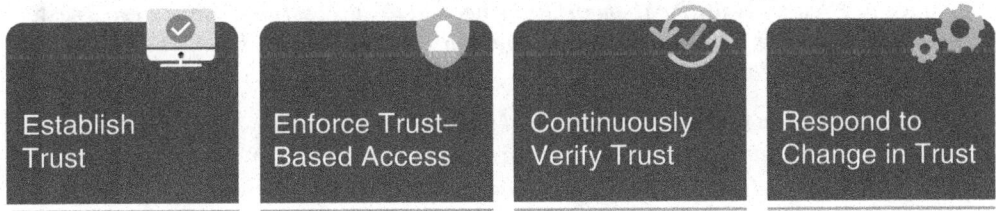

Figure 10-6   *High Level Strategy Steps*

[With a clear understanding of threats and gaps, Glenn moves on to explain a high-level strategy on what needs to be performed in the enterprise to be able to achieve the Zero Trust vision. Glenn stresses in Figure 10-6 that Prolink Solutions can help craft a vision and mission statement that aligns with the industry standard for Zero Trust architectures and implementation strategies, which are aligned to NIST 800-207 and CISA maturity model v2.]

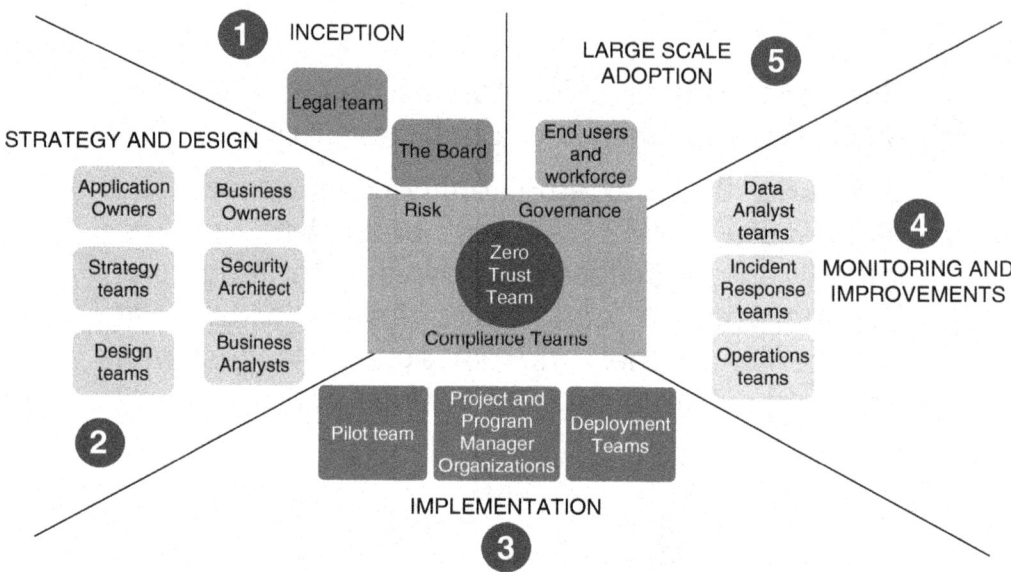

Figure 10-7   *The Zero Trust Team*

[After explaining what the Zero Trust journey would look like and how specific initiatives can help, Glenn moves on to explain the work done to set up and align a strong Zero Trust virtual team, as illustrated in Figure 10-7. The presentations shows various teams involved at each phase of the Zero Trust journey. This is to make sure that Zenith Trust Bank is aware of who needs to be mobilized at which stage to effectively utilize the workforce needed to successfully adopt Zero Trust in the enterprise. Substantial time is spent at this junction to clarify the extent of involvement for each team.]

[Glenn moves on to present the derived vision, mission, and Zero Trust tactical enablers for Zenith Trust Bank, as illustrated in Figure 10-8. With the support of the technology office, Glenn explains the details of what the enterprise wants to achieve and how Zero Trust, as a concept, architecture model, and implementation approach aligns with the overall vision. Each tactical enabling pillar is an initiative that will spawn into various projects that will be tracked individually but aligned globally.]

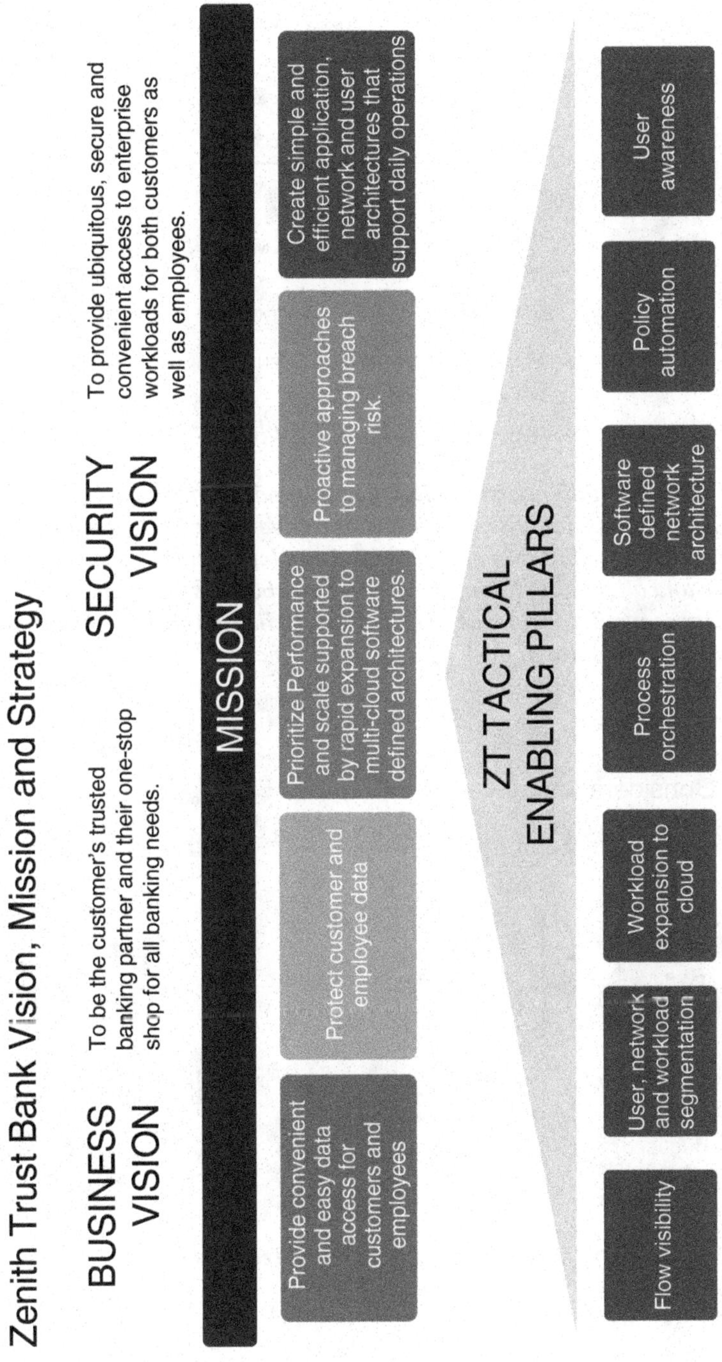

Zenith Trust Bank Vision, Mission and Strategy

Figure 10-8   *The Vision, Missing, and Strategy*

Metrics to Support the Zero Trust Initiative

	STRATEGIC	TACTICAL	OPERATIONAL
QUALITATIVE	Ease of access of user data Critical data control strength	Success of tactical initiatives that include • SDA/SD-WAN • Policy automation • Multicloud adoption and migration • Flow visibility	User awareness and skillset Personnel work satisfaction
QUANTITATIVE	Organizational agility to adapt Total cost of ownership	Reduce incident response time Endpoint posture and visibility Segmentation initiative to reduce the blast radius	Reduce downtime Reduce fraudulent activities Reduce mean time to detection Reduce troubleshooting time
	ADAPTABILITY AND DATA CENTRIC	INITIATIVE CENTRIC	WORKFORCE CENTRIC

Figure 10-9 *Metrics to Measure the Strategy and Implementation*

[After ensuring his audience has a clear understanding of the gaps, threats, and the overall vision and mission, Glenn elaborates on the identified metrics. The goal is to showcase to leadership that no stone has been left unturned and all metrics are measurable and align with all the necessary stakeholders in the enterprise. Clearly defined metrics instill confidence in the target audience that the initiatives can be measured and monitored over time to build efficiency. This is illustrated in Figure 10-9.]

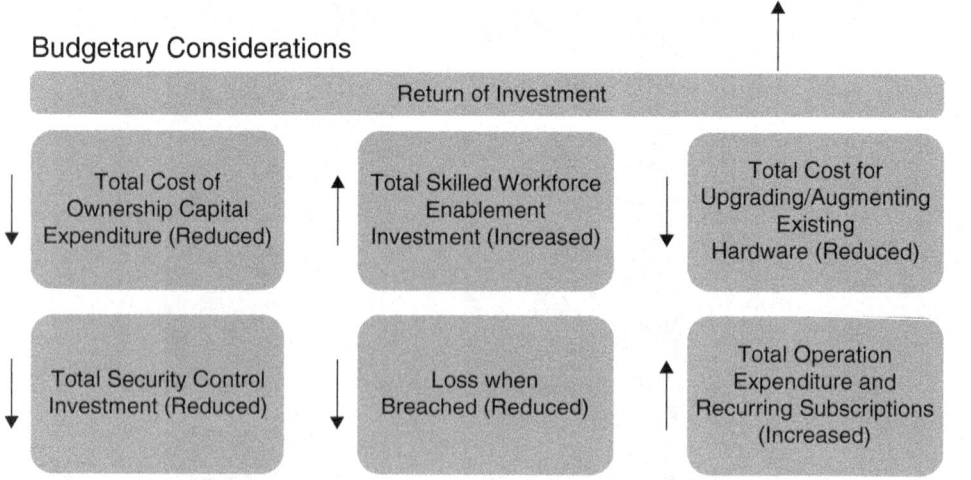

Budgetary Considerations

Return of Investment

Total Cost of Ownership Capital Expenditure (Reduced)

Total Skilled Workforce Enablement Investment (Increased)

Total Cost for Upgrading/Augmenting Existing Hardware (Reduced)

Total Security Control Investment (Reduced)

Loss when Breached (Reduced)

Total Operation Expenditure and Recurring Subscriptions (Increased)

Figure 10-10 *Budgetary Considerations for Zenith Trust Bank*

[Glenn proceeds to justify the overall monetary impact of Zero Trust for the enterprise, as illustrated in Figure 10-10. This is not just a product-level comparison but a complete capital and operational expense impact analysis. Based on experience with deploying Zero Trust architectures for multiple customers, Glenn showcases how Zero Trust can provide more return of investment in the long run for Zenith Trust Bank.]

Closing thoughts

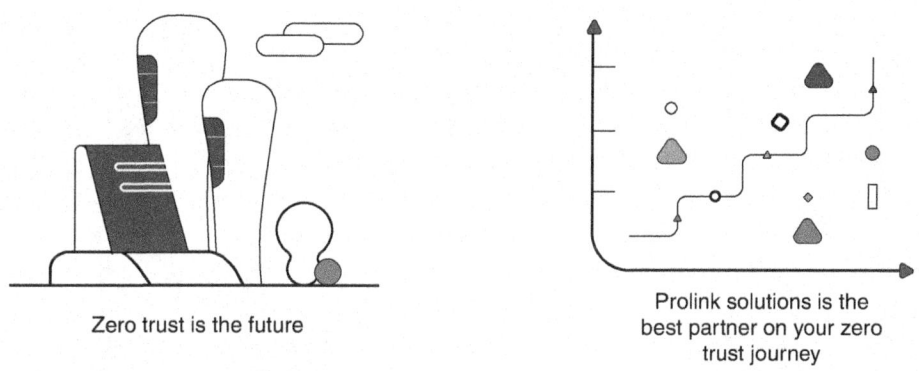

Zero trust is the future

Prolink solutions is the best partner on your zero trust journey

Figure 10-11   *The Road Ahead*

[Finally, Glenn ends on a positive note, citing that Zero Trust is the future and that to be able to wade through the sea of misconceptions and inconsistencies, a strong partnership with a technical advisor would be ideal. Over the course of a month, Zenith Trust Bank was able to secure the required funding for its Zero Trust initiative, supported by Prolink Solutions as its trusted Zero Trust advisor.]

Implement, Monitor, Feedback, Repeat

This is the final phase of your Zero Trust journey. It does not end here; instead, it is just the beginning. The following chapters are covered in this phase:

Chapter 11

Implementation and Continuous Monitoring

When the enterprise reaches the implementation phase, essentially the bulk of the work (drivers, design, metrics, and architecture) in the entire Zero Trust lifecycle has been completed. Implementation is a daunting task, but it always follows a predefined plan or guideline that engineers refer to. This is different from designing a concept, a network, or even an access model from scratch. Designing involves a lot of theories and assumptions. The implementation phase is where those assumptions and theories are validated. Overall, the following are some important prerequisite steps before entering this phase:

1. Clearly understand the reason to move to Zero Trust access models (business drivers).

2. Identify the key pain points and craft performance metrics.

3. Identify gaps in the people, processes, and technologies and craft risk metrics in response to specific threat scenarios.

4. Identify the key stakeholders and build the Zero Trust strategy. Involve all stakeholders based on their interest and stake in the strategy.

5. Select the partners, products, and vendors based on a Zero Trust architecture and implementation road map.

6. Communicate the change in organizational direction to the workforce and customers.

7. Identify tactical initiatives to drive Zero Trust implementation.

8. Present the strategy and metrics and acquire sufficient funding.

The next logical step is to stick your hands in the mud and start implementing. Implementation is a unique experience and depending on various factors such as available skillset, vendor maturity, and budget, the time taken to implement might vary. Irrespective of the time, there are some aspects to remember and consider when it comes to implementation strategies, as detailed next.

Do Not Ignore Your Current Gaps

Zero Trust is not a magic wand that will fix all the gaps in your network. Although reflecting on Zero Trust may reveal multiple gaps in the network, there are also gaps it may not cover. For example, Zero Trust is at its core a philosophy that influences the access of subjects to objects, making it one of the most versatile information security access models. Capacity generally does not get considered as a Zero Trust performance metric; however, capacity is a general enterprise performance metric that impacts the organization. It might not impact Zero Trust strategies, but it definitely will impact performance. Zero Trust is a larger initiative covering multiple domains, whereas capacity is device and vendor specific. Do *not* ignore any metrics not covered by the Zero Trust strategy.

Frameworks Are Only a Reference

Think outside of frameworks and regulations. Zero Trust is a product-agnostic information security model, which means, as a viable framework itself, it can provide security control for any other regulation or framework as well. Operationalize your frameworks wisely and within the context of your business drivers. Look at each control that a framework specifies and see if it is relevant to you. For example, if you do not have a MySQL database, then database access monitoring doesn't make sense. Technology might give you a lot of information, but you need to validate whether it fits the gap in your environment. As part of the Zero Trust journey, the main business drivers and their alignment with compliance should already be complete. The implementation should take reference from existing frameworks like NIST or CISA, and a baseline implementation plan should be tailor made for the enterprise. The core tenets of Zero Trust are immutable and should be the base for any policy created.

Adopt Agile for Initiatives and SAFe for the Strategy Delivery

Changing the mindset of the people around you is not easy and doesn't happen in a day. If you need to make a full enterprise follow in your footsteps, there must be a large base of followers of the concept itself. Because you are the authority on the subject, your enterprise will look up to you to provide guidance on how the Zero Trust concept itself will change everyone's perception, and you must make sure that the influence in the enterprise is top-down. To do this, you need to have a team that shares the same ideals as you and strives for the betterment of the enterprise, and not just pushes Zero Trust as an agenda. Build a team that likes to grow and is delivery focused.

When it comes to delivery of the implementation, one of the best-suited strategies to implement the Zero Trust vision is to use the Agile method. Scaled Agile Framework (SAFe) is the most suited to scale Agile across multiple initiatives. Overlaying some of the

Agile principles with the Zero Trust journey gives us more insight into why Zero Trust is more efficiently deployed with Agile:

- **Customer satisfaction through early and continuous delivery:** As part of Zero Trust, a pilot application or initiative is always the first checkpoint. Zero Trust helps run multiple initiatives, which allows you to showcase results almost immediately. Once the pilot is verified and validated, budgets can be released with each successful initiative being implemented.

- **Welcome changing requirements, even in late development:** Zero Trust is all about being flexible in implementation and strategy. Because the final goal is clear, keeping deployment options open facilitates faster Zero Trust adoption. This includes situations where the technology might not be fully able to support the vision.

- **Deliver working solutions frequently, with a preference for shorter timescales:** Since there are usually multiple initiatives involved in a Zero Trust project, successful completion of an initiative or part of the initiative provides a huge boost to leadership, and the overall success factor of the Zero Trust initiative increases.

- **Collaboration between business stakeholders, developers, and infrastructure owners throughout the project:** DevSecOps as a concept helps drive larger security and Zero Trust initiatives within the application domain with feedback from business owners and senior stakeholders.

- **Successful initiatives is the primary measure of progress:** This principle emphasizes the importance of measuring progress based on the delivery and completion of specific initiatives, rather than relying on other metrics such as documentation or design. Zero Trust is best measured with visible outcomes rather than paper concepts.

- **Agile processes promote sustainable development:** This principle emphasizes the importance of maintaining a sustainable pace of development that allows the team to maintain a high level of productivity and quality over the long term. Zero Trust initiatives have a large timeframe to adopt, and this expectation is set early on with stakeholders. Hence, the implementation resources do not need to focus on the larger timeline but just need to complete the immediate initiative within its timeline.

- **Simplicity is essential:** This principle emphasizes the importance of keeping things simple and avoiding unnecessary complexity, which can lead to increased costs and reduced quality. Zero Trust at its core is a simple access model that controls traffic from subject to object based on context.

- **Self-organizing teams make the best decisions:** This principle emphasizes the importance of building self-organizing teams that are empowered to make decisions and take ownership of their work. A Zero Trust team is enabled to not only make decisions but to provide feedback as to how the strategy can be improved. Due to the larger scope of teams involved, Agile is the best approach.

- **Regular reflections and adaptations to improve effectiveness:** This principle emphasizes the importance of continuously monitoring the deployment process and making adaptations to improve the effectiveness of strategy and metrics and meet changing customer needs.

Agile can be extended across multiple initiatives in the enterprise (segmentation, AD improvements, and so on) using SAFe. Some SAFe principles that align with Zero Trust are as follows.

- **Take an economic view:** SAFe Lean Agile emphasizes the importance of considering the economic impact of decisions made during the development process. This principle overlays Zero Trust by taking into account the cost benefit of implementing Zero Trust security measures and how it aligns with the overall business drivers. For example, if the general direction of the enterprise is to cut costs and move to the cloud, Zero Trust initiatives can be the best alternative to security control in multi-cloud architectures.

- **Apply systems thinking:** SAFe Lean Agile encourages teams to think about the larger system in which they operate, rather than just the individual components. This principle aligns with Zero Trust, which also takes a holistic view of the people, processes, and technologies, and its components, rather than just focusing on implementing larger security devices.

- **Assume variability; preserve options:** SAFe Lean Agile recognizes that requirements can change over time, and it is important to preserve options for adapting to change. Zero Trust also acknowledges that threats can change, and it is important to have flexible security measures that can adapt to evolving threats as well as changing business drivers.

- **Implement incrementally with fast, integrated learning cycles:** SAFe Lean Agile promotes a cycle of continuous learning, where feedback is incorporated into the implementation process in real time. This principle aligns with Zero Trust by ensuring that security measures are continually evaluated and updated based on feedback from all relevant teams. Similarly, if an implementation project becomes a roadblock for the overall release trains, then a new strategy can be considered. Release train in the context of SAFe refers to a dedicated cross-functional team that works towards the common goal (Zero Trust vision) for a long time till the vision has been achieved. This virtual Zero Trust team was discussed in detail in Chapter 7. "Zero Trust Avengers, Assemble!"

- **Base milestones on objective evaluation of deployed initiatives:** SAFe Lean Agile emphasizes the importance of measuring progress based on the objective evaluation of deployed initiatives. This principle aligns with Zero Trust, which also relies on objective evaluation and monitoring of strategy success and security control effectiveness. This, along with metric measurement, will provide the much-needed visibility into how the implementation is progressing and how aligned it is with the expected road map. The expected milestones can be modified accordingly.

- **Visualize and limit work in progress (WIP), reduce batch sizes, and manage queue lengths:** SAFe Lean Agile recognizes the importance of limiting work in progress (WIP) and reducing batch sizes to improve efficiency. This principle aligns with Zero Trust by separating the overall strategy into multiple tactical initiatives such as segmentation and software-defined access, and then further breaking them into separate projects such as network and security. Some of these projects can then run in parallel to provide a minimum viable solution to showcase to leadership. The road map created provides a bird's-eye view of which projects can run simultaneously and which projects and initiatives have dependencies.

- **Apply cadence, synchronize with cross-domain planning:** SAFe Lean Agile promotes a cadence of regular planning and coordination between teams. A Zero Trust team is almost always a virtual team, where team members are part of various aspects of the enterprise group. Creating a cadence to sync up and align on vision and implementation strategy is an important task. This principle aligns with Zero Trust by ensuring that security measures are coordinated across different domains and teams.

- **Unlock the intrinsic motivation of knowledge workers, decentralize decision-making:** SAFe Lean Agile recognizes that knowledge workers are motivated by autonomy, mastery, and purpose. This principle aligns with Zero Trust by empowering users to take responsibility for their own security and providing them with the tools and resources they need to do so. Security awareness and the general organizational culture propagate the Zero Trust mentality. By empowering users to make security decisions based on their knowledge and expertise, SAFe principles help implement the overall Zero Trust initiatives by providing autonomy to the Zero Trust team as well as the workforce and the customers. True Zero Trust is when the users and employees are empowered to access their resources with granular and apt security controls.

- **Organize around value:** SAFe Lean Agile emphasizes the importance of organizing teams around delivering value to the customer. This principle aligns with Zero Trust by ensuring that security measures are aligned with the business drivers that have been identified with the help of the customer, and that this is not seen as a hindrance to delivering value.

Implement the Vision

Expect a lot of changes technologically in the enterprise. You might start seeing failed projects or extended feature support timelines from selected vendors, but what is key here is to keep an eye on the goal. As long as the base tenets of Zero Trust are maintained, keep moving forward. Make sure that the vision statement is clear to all members of the team, and each time there is a roadblock, showcase the goals and pivot the strategy to satisfy the common Zero Trust vision. Make sure the strategy is fluid enough to meander around showstoppers.

Each Zero Trust journey is unique, and hence the implementation approach is usually dependent on the road map and strategy. The logical steps, however, will be similar, and this section will highlight typical approaches that most enterprises adopt when implementing Zero Trust strategies.

Pillar-Based Approach

A pillar-based approach is usually undertaken by enterprises that have performed a discovery or a gap analysis and know existing gaps in each pillar of the CISA maturity model v2. The approach would then be to bridge gaps per pillar based on priority and comfort level of the enterprise. The following subsections depict a typical implementation flow for a pillar-based approach of Zero Trust adoption.

Discovery

Most Zero Trust implementations begin with a discovery. Based on a gap analysis, certain security controls have already been recommended, and implementation begins with what the current state is. For example, what is my current user structure in Active Directory? Does it need additional user segmentation? The enterprise must look to identify its critical applications as well as finalize data classification. Usually, asset inventory is manual for most enterprises when beginning the Zero Trust journey, so implementing some automation around assets and data goes a long way to support the overall implementation. It is important to note, however, that automation is not mandatory. An enterprise may choose to perform asset inventory when it is planning to implement a workload-specific initiative such as micro-segmentation. This is just to showcase a common starting point. You need to know who can access what in the enterprise and derive your scope of protection.

Users and Devices

The next logical step is to strengthen what most enterprises already have—its user segment. Enterprises are most certainly on the path to Zero Trust even before they realize it, and when a gap analysis is performed, it is a simpler task to augment an initiative that is already established in the organization. Users are segmented and authenticated with SSO and MFA, and if the capabilities are not in place, initiatives and projects are spawned to provide the capabilities.

Devices, however, might not be as well maintained. Most enterprises will begin seeing a need to design and implement endpoint security and posture policies. Some automation initiatives like automated patch management and vulnerability management may be put off until more maturity is achieved. The primary goal is to make sure the device is adding to the context of the user and hence initiatives such as posture deployment will be prevalent as the next step to support the identity initiatives. As maturity around devices grows, initiatives like baselining and automation will take precedence.

Applications

The next step might be different for various enterprises and based on their customized road map. Most enterprises move on to identify critical applications and their business and network flows. A common understanding here is that Zero Trust is about granting need to-know access to an object, which means that visibility into applications is a logical next step. It is worth noting that any form of application dependency mapping is usually a large initiative for enterprises, so application-specific verticals are often also performed after network-based initiatives. This is, however, a typical approach to implementation, which involves providing access to applications without a VPN or NAT. Usually, a pilot system or application set is identified and then, based on the subject's access, the implementation for the pilot is initiated. This is the phase where most application security is revaluated. Local authentication is removed, and more process-relevant controls are applied.

Mature activities like policy centralization and deployments are moved to subsequent years. Depending on the maturity of the enterprise, DevSecOps and secure coding might be considered as key tactical initiatives for applications early in the implementation lifecycle. With APIs being the new language for communication, secure API access is also an initiative that most enterprises focus on.

Flow Visibility

Once subjects and objects are identified, it is important to model the flows between these constructs. The flows usually feed into the Zero Trust governance initiatives that are needed to create access policy. Security operations center (SOC) solutions and security information and event management (SIEM) solutions are prioritized to make sure there is a central log collector in the enterprise and there are resources to react to incidents. Secure analytics is another typical initiative that is implemented to support the overall visibility vertical.

Because most of the traffic is encrypted, initiatives to detect and inspect encrypted traffic are usually deployed in this phase, along with asset health monitoring.

Automate and Orchestrate

Once visibility into flows is achieved, most enterprises move on to identifying key processes in the overall security architecture. Activities like critical data classification and false positive identification for incidents are included under the automation and orchestration umbrella.

Full Zero Trust Network Architecture

With the visibility of flows, clear subject segmentation, and critical data and application assets being identified, the final step is to provide secure access for subjects to the respective objects. This phase usually covers the micro-segmentation and macro-segmentation initiatives. The enterprise is already beginning to assume breach and hence network analytics and subsequent proactive monitoring and reaction to events is a common initiative that is run in parallel with segmentation. The network will be more enabling than restrictive, so initiatives to secure any access to applications will be considered at this phase. Initiatives like DNSSEC and MFA coupled with NAC for servers are common implementation priorities with increasing maturity.

It is important to highlight once more that this is a typical implementation plan, which means most customers follow the same or similar implementation aspects; however, some of these might run in parallel. A customer might implement posture for all end users along with a macro-segmentation improvement initiative. Hence, the implementation order and priority will depend on the identified road map, which in turn is fed by current projects and initiative priority.

Use Case–Based Approach (with SASE)

Secure Access Service Edge (SASE) utilizes Zero Trust Network architecture (Security Services Edge) at its core. Zero Trust identifies as an information security model, and SASE identifies as a network architecture. SASE combines network security functions with wide area networking (WAN) capabilities to provide secure access to enterprise applications and services from anywhere, at any time, and from any device. Note this is a key aspect of Zero Trust as well and closely follows the cloud model of application deployment. SASE aims to address the challenges posed by the increasing use of cloud services, mobile devices, and remote workforces by delivering a unified and consistent security framework across all locations and devices, including data centers, branch offices, cloud services, and mobile users.

In a SASE architecture, Zero Trust network architecture via SSE provides granular control and visibility over network traffic, allowing security teams to apply security policies based on the user's identity, device type, location, and context. Zero Trust network architecture also promotes micro-segmentation, which divides the network into smaller, isolated segments to contain breaches and prevent lateral movement by attackers. All of these security mechanisms have been covered in detail in Chapter 8, "Building a Zero Trust Architecture," and Chapter 9, "Critical Security Mechanisms for Zero Trust Architectures," but the essence of the conversation for you to take away is that multicloud architectures are the future of networking, and next-generation software-based WAN is connecting all these multicloud architectures.

In light of larger adoption of SASE architectures, a use case–based approach to implementation is when the enterprise has already selected a product and intends to implement Zero Trust via a specific use case, usually after selecting a platform that provides Zero Trust network architecture capabilities like SSE. Even though this approach is uniquely different from the pillar-based approach, it is important to note that specific use cases are the technical drivers to Zero Trust adoption, and the prerequisites in each Zero Trust pillar for that specific use case must still be achieved. The following subsections describe typical use cases.

Secure Outbound Internet Access

One of the first use cases that customers target to achieve for quicker Zero Trust adoption is to move the Internet access from the on-premises security controls to cloud-native capabilities like SSE. This greatly reduces the total cost of ownership and drives the sustainability discussion for most enterprises. However, it might involve initial migration of policies to the SSE and will need enhancements to all aspects of Zero Trust, especially identity, network, and application capabilities. The expectation would be that the enterprise has identified all of its users and onboarded them to an identity provider so that identity-centric unified access control can be achieved for all segments utilizing SSE for outbound Internet access.

VPN as a Service (VPNaaS)

VPN as a Service is a common use case for customers that want to adopt a Zero Trust platform like SSE but do not want to move away from traditional VPN. The advantage with this hybrid approach is to be able to transition to Zero Trust network access seamlessly from VPN. The approach also involves some level of migration but does not need additional activities like application onboarding. Identity controls, however, are mandatory, and the expectation would be that the enterprise has identified all of its users, onboarded them to an identity provider, and created granular segmented roles or groups that refer to the Active Directory user segmentation or reflect the existing segmentation structure for VPN access.

Zero Trust Network Access (ZTNA) for Applications

A typical use case of how enterprises approach Zero Trust implementation is to move toward access of private applications from anywhere with Zero Trust Network Access (ZTNA). This requires the most prerequisites, where all applications, networks, and user identities must be robust and well-designed before the overall ZTNA can be implemented. Users must be segmented and onboarded to the identity provider. Network security capabilities like DLP, IPS, malware inspection, and secure web gateways must be crafted for each user segment or individual user. Users must be able to view only applications that they have need-to-know access to, which infers that the Zero Trust policy as well as the application onboarding must be complete. The implementation of this use case follows the software device perimeter (SDP) model of implementation for Zero Trust access. SDP enables not only network access design but influences application access design as well, which makes the secure private access of applications via ZTNA a very versatile use case for enterprises to embark on.

Figure 11-1 illustrates a typical SASE network and its relevant use cases.

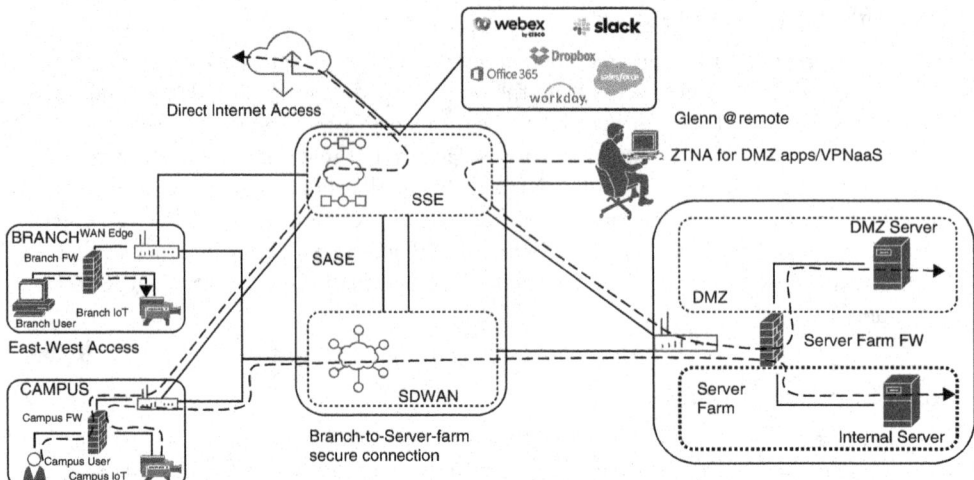

Figure 11-1 *Typical SASE Adoption Use Cases Driving Zero Trust Network Architecture*

Monitor and Enhance

After spending time implementing the overall strategy, you should not just sit back and relax because the strategy is implemented successfully. Implementation of the Zero Trust strategy is a precursor to a long journey of self-improvement. Detailed monitoring systems must be in place to measure metrics, performance, and overall success of the strategy. The journey truly begins at this junction.

Monitor Your Metrics

This is the final stage of the metrics lifecycle discussed in Chapter 5, "Measuring Zero Trust Success." The strategy has been translated to technology. Now you need to monitor all your crafted metrics and make sure they are measuring accurately what they were expected to measure. Monitor the metrics along with the success of the strategy.

Challenge the Strategy

Once the implementation is complete, as part of the continuous monitoring and enhancement scheme, the strategy must be challenged at all aspects. A core tenet of Zero Trust is to assume a breach has occurred. Perform penetration testing activities to identify backdoors or weaker security controls. Run simulation and tabletop exercises for critical flows and make sure security is preserved at each junction. Some of these flows should already have been found as part of the business flow identification and gap analysis efforts. Diamonds are created only when coal is put under pressure. Test the limits of your implementation and identify what can be done to improve the implementation further.

Continue Education and Awareness

Sending out a communication for the adoption of Zero Trust is the beginning of a constant communication channel between the stakeholders and the adopters. Not everyone is security savvy, and new additions to the workforce might not be as keen to share your vision of a more secure enterprise. Through the raising of awareness, users can better understand their roles in maintaining a secure environment and become more vigilant about potential threats. The workforce needs to be enabled enough to understand that they are responsible for their security, and not just the enterprise security team.

It is important to continuously monitor the workforce landscape and keep improving the trainings to make it engaging and relevant. End users are likely to be interested in how the strategy will help them with their daily work, not on strategic gaps that have been identified. Similarly, operations engineers will relate to how the Zero Trust implementation improves the business continuity. Know your target audience and pivot accordingly.

Maintain Momentum

It's easy to reach the top, but it's difficult to stay there. Once you have a pilot or a minimal viable solution (MVS) to showcase to leadership, make sure the lifecycle remains iterative. Perform the same actions, and keep your leadership informed of all activities being performed.

Make sure a successful initiative completion doesn't make you complacent. For example, if you are able to fully achieve user segmentation, you should stop to celebrate, but keep the momentum going. Consistent and constant monitoring and improvement of the enterprise adds incremental value, and the performance multiplies exponentially. This includes performing continuous risk assessments as well as setting a baseline to showcase improvement over the years. Achieve your quick wins, celebrate, check for the next task, and move on. Attackers will not wait, and you shouldn't either.

The Serendipitous Meeting

[Over the course of five years, Zenith Trust Bank deployed its Zero Trust strategy, and it was greatly welcome in the banking community. As the pioneers of the Zero Trust implementation concept, Zero Trust Bank was asked to present its Zero Trust journey at various events. At one such conference was none other than Glenn the architect from Prolink Solutions, who ignited the spark that Zenith Trust Bank had regarding Zero Trust.

[The CIO of Zenith Trust Bank, Mr. Jonathan Smith, was nearing the end of his presentation.]

Mr. Smith: It wasn't an easy journey, but for anyone planning to follow in our footsteps, the following are the key tactical initiatives we prioritized:

  1. We identified our business drivers.

2. We identified our performance and risk metrics.

3. We built an awesome team and proved to our leadership it can be done.

4. We built a road map and got the necessary funding.

5. We vigorously implemented and deployed over these five years with continuous monitoring and feedback.

6. We set up a Zero Trust improvement team to validate all metrics to make sure we were continuously improving.

We used the Agile method to make sure that all stakeholders were happy and satisfied at frequent intervals. SAFe helped us scale across the enterprise. Our Zero Trust strategy is to continuously monitor what we designed and adapt to any changes accordingly without changing the base tenets.

Of course, none of this would have been possible if it hadn't been for Prolink Solutions. Their technical expertise has been instrumental in our adoption success.

[When Glenn heard about Prolink Solutions, he was taken aback. After the conference he walked up to Mr. Jonathan Smith.]

Glenn: Mr. Smith. I heard your talk and, honestly, I am very happy for Zenith Trust Bank for having achieved your Zero Trust vision. I thank you for letting Prolink Solutions be part of your journey. I was merely a medium, but thank you for the kind words.

Mr. Smith: Hey, Glenn, I meant every word I said. Zero Trust can be daunting for someone who doesn't know where to start or what to expect. I am glad you were with us throughout the journey. We have seen the benefits that Zero Trust brought to our enterprise, and I am sure you can convince many other enterprises to take up this journey to achieve their own Zero Trust vision. Best of luck and let's keep in touch. Zenith Trust Bank and Prolink Solutions' partnership is stronger than ever, and we have many more milestones to achieve together.

[They shook hands, and Glenn walked off while contemplating that this experience sounded like a good idea for a book.]

Chapter 12

The Road Ahead

Time has flown, and we are now at the tipping point of our Zero Trust journey. The enterprise has deployed its Zero Trust architecture and the metrics are measuring its efficacy. The enterprise's entire security outlook has been transformed and attackers need to spend more time and efforts to get in. The goal of this entire book was to take you, the reader, on a multiyear journey from the inception of a Zero Trust idea to the successful implementation and monitoring of an entire Zero Trust strategy. Before you leave to apply what you have learned, I would like to introduce some more concise imagery of what you have learned so far. This is called the Zero Trust lifecycle framework. It was intentionally not introduced at the beginning of our journey so that your learning process was not biased with jargon and terminologies. The illustration in Figure 12-1 is basically a summary of the entire book in a reusable framework. It is cyclic in nature, because all Zero Trust initiatives must finally incorporate feedback from the field about what has been deployed. The activities listed in Figure 12-1 cover the entire Zero Trust inception to implementation lifecycle.

You will notice that all of these phases have been covered throughout the book, but you explored each phase as a point-in-time activity. The reason this was placed at the end of the book is to showcase how the culmination of multiple years is covered with the Zero Trust lifecycle framework. All parts of each phase are important and must be met before you can move on to the next phase. For example, you cannot move to identifying key leadership stakeholders if you cannot showcase to your supporters that you have the right metrics to measure success.

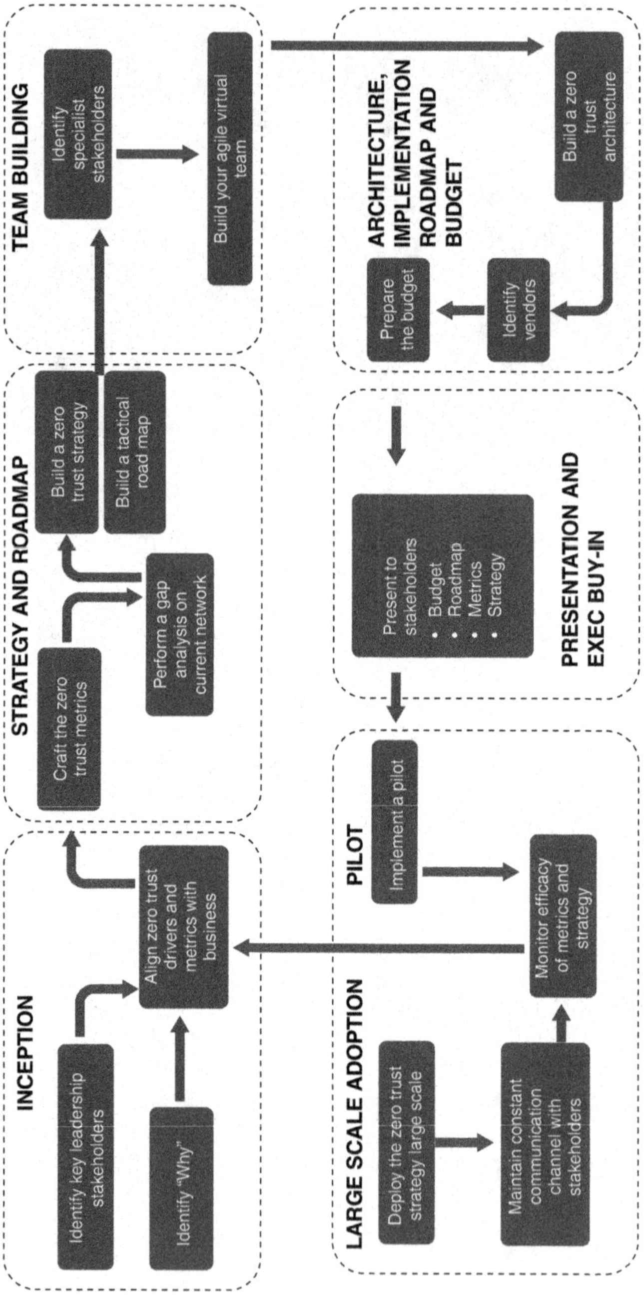

TEAM BUILDING

- Identify specialist stakeholders
- Build your agile virtual team

ARCHITECTURE, IMPLEMENTATION ROADMAP AND BUDGET

- Build a zero trust architecture
- Prepare the budget
- Identify vendors

STRATEGY AND ROADMAP

- Build a zero trust strategy
- Build a tactical road map
- Perform a gap analysis on current network
- Craft the zero trust metrics

PRESENTATION AND EXEC BUY-IN

- Present to stakeholders
  - Budget
  - Roadmap
  - Metrics
  - Strategy

INCEPTION

- Identify key leadership stakeholders
- Identify "Why"
- Align zero trust drivers and metrics with business

PILOT

- Implement a pilot
- Monitor efficacy of metrics and strategy

LARGE SCALE ADOPTION

- Deploy the zero trust strategy large scale
- Maintain constant communication channel with stakeholders

Figure 12-1 *The Zero Trust Lifecycle Framework*

In summary, the Zero Trust lifecycle framework has six phases:

1. **Inception:** Focus on understanding the business driver for Zero Trust transformation and align the Zero Trust metrics and strategic requirements to the overall business. You will begin strategic discussions with leaders and start building strong leadership support. You will also try to win over initial skeptics and manage the detractors. Once these tasks have been completed, you will then start building a strategy and road map. Without the right leadership support, the project cannot start.

2. **Strategy and road map:** Focus on crafting metrics based on initial conversations and requirements from your primary leadership stakeholders. This will be your success currency to use to convince other stakeholders who are still sitting on the fence. Also perform a gap analysis to be able to identify a tactical road map to showcase the seriousness of the current state of security and the final state the enterprise needs to be in (performance and risk).

3. **Team building:** Reel in your stakeholders. You have charmed your leadership already. Now rope in specialists and build an Agile team that is ready to strategize, design, and implement the collective vision in an Agile manner.

4. **Architecture, implementation road map, and budget:** Build a Zero Trust strategy. Questions that must be answered include what security mechanisms are needed, where to focus, and so on. Use the gap analysis as a reference. Decide which products to use and build a budget based on the products, control requirements, skills, and process improvements.

5. **Presentation and executive buy-in:** Present the Zero Trust strategy, budget, road map, and metrics. Sell the Zero Trust vision to the board to get funding.

6. **Pilot and large-scale adoption:** Once the strategy and road map are accepted and the budgeting is approved, the enterprise will proceed to implement a pilot application that fits the desired business use case. This will, in turn, provide insights that will provide feedback to the overall strategy. There will also be some more alignment exercises based on the implementation results. You will understand whether there are people, process, or technology issues, which you can feed back into the Zero Trust lifecycle framework. In this phase, communication is usually with the pilot application and user group.

Implement on a larger scale and monitor for more information. With the pilot providing a detailed understanding of possible challenges, the subsequent phase is a replication of the pilot phase but at a larger scale. Monitor for challenges with the large-scale implementation and deployment and use any feedback to improve the overall Zero Trust lifecycle framework. This is the phase where most of the end users will need to be directly involved in the implementation. A constant communication channel should be maintained between all stakeholders and the Zero Trust virtual team deploying the solution.

With these six stages, the Zero Trust initiative might take multiple years to complete, but the results will last much longer. Note that the Zero Trust lifecycle framework is a constant cycle where dynamic monitoring will lead to new use cases being discovered, which eventually will lead to a realignment exercise. This realignment will need additional specialist stakeholders, possibly an amendment in the strategy, or even alignment to a new leadership stakeholder. Constant optimization and improvement are key to the overall success of the Zero Trust initiative.

A Trusted Zero Trust Partner

Cisco has been a pioneer in security technology and offers a wide range of products that can be used to fit into each of the critical security mechanisms discussed in Chapter 9, "Critical Security Mechanism for Zero Trust Architectures." Figure 12-2 illustrates the overall Cisco portfolio.

I do not want to sound like a YouTuber citing a sponsor, nor do I want to sound like a sales representative pitching the utilization of Cisco products. The point being driven home here is that no vendor can provide Zero Trust as a product itself. It is an amalgamation of a large set of security mechanisms, and generally it is helpful to select a single transformational partner to help you along the journey with the right feedback and guidance. With Cisco's innovative security product portfolio, you can rest assured that your security transformation is in good hands.

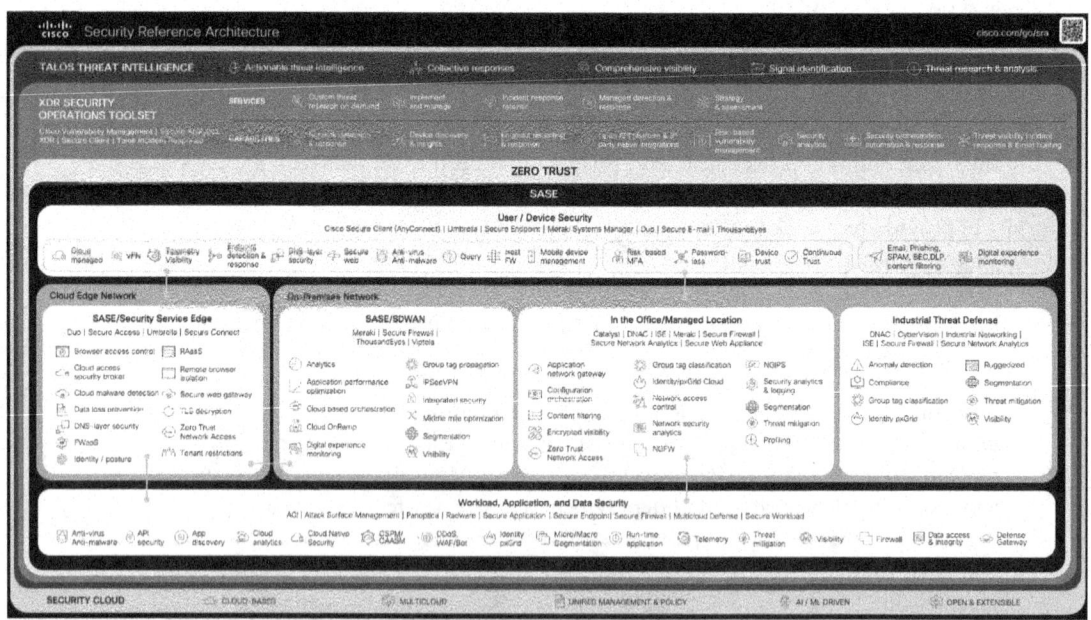

Figure 12-2   *The Cisco Critical Security Mechanisms Portfolio[1]*

Aim Higher, Together

As a closing note, I would like you to pat yourself on the back for taking the first step in embarking on your own Zero Trust journey and embracing this emerging paradigm shift. I hope this book helped you understand and achieve the Zero Trust architecture as part of your vision. The one constant in this entire strategy is always to aim higher. As we implement various mechanisms in the network, we are constantly challenged to add in more security and deploy more solutions. Aim higher to enable and empower your employees and customers so that they can enable you to be a better version of yourself.

When I talk to customers, the first point I like to bring up, irrespective of the project or technology, is that there is no need to take any step alone. Most enterprises have an underlying fear that they are taking a leap of faith that may fail or that may adversely impact their primary business. The purpose of this book is to assure you that the nature of Zero Trust is to bring more people together to work toward a common goal. Threats will keep getting more complex, and as professionals, we are all at the cusp of a transformational philosophy that aims to liberate and secure. We only need to find the courage to take that first step forward.

Endnote

1. "Cisco Security Reference Architecture," https://www.cisco.com/c/en/us/products/security/cisco-security-reference-architecture.html

Index

A

E

S

Register Your Product at informit.com/register

Access additional benefits and save up to 65%* on your next purchase

- Automatically receive a coupon for 35% off books, eBooks, and web editions and 65% off video courses, valid for 30 days. Look for your code in your InformIT cart or the Manage Codes section of your account page.
- Download available product updates.
- Access bonus material if available.**
- Check the box to hear from us and receive exclusive offers on new editions and related products.

InformIT—The Trusted Technology Learning Source

InformIT is the online home of information technology brands at Pearson, the world's leading learning company. At informit.com, you can

- Shop our books, eBooks, and video training. Most eBooks are DRM-Free and include PDF and EPUB files.
- Take advantage of our special offers and promotions (informit.com/promotions).
- Sign up for special offers and content newsletter (informit.com/newsletters).
- Access thousands of free chapters and video lessons.
- Enjoy free ground shipping on U.S. orders.*

* Offers subject to change.

** Registration benefits vary by product. Benefits will be listed on your account page under Registered Products.

Connect with InformIT—Visit informit.com/community

 **Pearson**   **informIT**